A to Z
of
Scientists in Space and Astronomy

NOTABLE SCIENTISTS

A TO Z
OF
SCIENTISTS IN SPACE AND ASTRONOMY

DEBORAH TODD AND JOSEPH A. ANGELO, JR.

☑®
Facts On File, Inc.

A TO Z OF SCIENTISTS IN SPACE AND ASTRONOMY

Notable Scientists

Copyright © 2005 by Deborah Todd and Joseph A. Angelo, Jr.

Facts On File, Inc.
132 West 31st Street
New York NY 10001

Library of Congress Cataloging-in-Publication Data

Todd, Deborah.
    A to Z of scientists in space and astronomy / Deborah Todd and Joseph A. Angelo, Jr.
        p. cm.—(Notable scientists)
    Includes bibliographical references and index.
    ISBN 0-8160-4639-5 (acid-free paper)
    1. Astronomers—Biography.   2. Space sciences—Biography.   I. Angelo, Joseph A.
II. Title.   III. Series.
  QB35.T63 2005
  520'.92'2—dc22

                                                    2004006611

Facts On File books are available at special discounts when purchased in bulk quantities for businesses, associations, institutions, or sales promotions. Please call our Special Sales Department in New York at (212) 967-8800 or (800) 322-8755.

You can find Facts On File on the World Wide Web at http://www.factsonfile.com

Text design by Joan M. Toro
Cover design by Cathy Rincon
Chronology by Sholto Ainslie

Printed in the United States of America

VB TECHBOOKS 10 9 8 7 6 5 4 3 2 1

This book is printed on acid-free paper.

"I know nothing with any certainty, but the sight of stars makes me dream."

—*Vincent Van Gogh*

"All of us are truly and literally a little bit of stardust."

—*William Fowler*

To the loves on this very tiny planet who have sprinkled their stardust into my life and believed in my dreams.

To Jason, my original and brightest star.

To A from T.

To Rob, as always, with love.

To Jenny.

To Gregg, thank you for everything.

To Rusty, our beloved star.

To Mom and Dad.

To my Sister.

Out of stardust comes poetry . . .

To Jeb, for the music and art.

To Omar, my favorite poet of all time.

—*Deborah Todd*

To my special friend, Michael Joseph Hoffman, his brother, Daniel, and all the other wonderful children of planet Earth who will inherit the stars. At just three years old, with the same unquenchable curiosity that inspired so many of the interesting people in this book, Michael asked me: "Where does the Sun go at night?" This book is written to help him discover an incredible universe that is now both a destination and a destiny through space technology.

—*Joseph A. Angelo, Jr.*

CONTENTS

LIST OF ENTRIES

ACKNOWLEDGMENTS

The creation of the universe took a matter of seconds. The creation of this book took somewhat longer. These acknowledgments are to thank those wonderful people who helped develop it from its original flicker of light into the shining example of the completed collaboration that it is today. First and foremost: to Chris Van Buren, who is now living the life of his dreams in a land far, far away, thank you for working us through the spark phase and into the realm of reality; to Margo Maley Hutchison for picking things up in the middle of it all with kindness and grace; to Kimberly Valenti for taking it the final mile—the hardest part of the journey—you are deeply appreciated; and to Bill Gladstone for insisting that I meet your "first writer," because we both live in the same county—thank you for introducing me to someone who has surely become a lifelong friend, and now part-time collaborator.

To Ed Addeo, a fabulous friend, and an equally wonderful writer. Your help was above and beyond the call of duty and/or friendship. You kept me afloat and made this book possible. And to think we did it all without being in the same room, except of course for those occasional lunches that kept us going and kept me sane. Thank you, thank you, thank you.

To Rob Swigart for the use of your office, your house, your kitchen, your desk, your other desk in the library and that wonderful chair, and your apartment in Paris. I'm still not sure about the change you made in my computer situation, but I guess honestly I am liking it more and more every day, and my computer and I have both been virus free since the switch! I know—you told me so. Thank you.

To Roger Eggleston for his great research, at a moment's notice, with a smile. To Elizabeth Eggleston for always smiling and always believing in this project and me. Thank you to both of you for your always-present support. Who's in charge? I know!

Special thanks to Kasey Arnold-Ince for her research skills, Heather Lindsay for her photo expertise on both books, Gregg Kellogg for again discussing biography candidates with me and the multitude of offers to help, and Curtis Wong for always asking how my book was coming along and the many varied discussions we had on this topic and others.

Thank you to John Newby, who continues the mantra "Keep moving forward." I am glad you were passing through. You are a collection of very special stardust.

Thank you to Matt Beucler. You know why, coach.

A special thank you to the editorial staff at Facts On File, especially Frank Darmstadt, for pulling the project through and helping bring it to completion with patience and understanding through cancer, physical disabilities, and the passing away of loved ones, and Laura Magzis, for an excellent copyediting job. To Joe Angelo for his expertise, dedication, and years of experience in this book-writing process that proved invaluable in getting it completed.

To my sister, Drena Large, I simply say "Thanks, Sis" for being the great listener and problem solver that you are. You didn't know it, but you helped tremendously.

Thank you to Jennifer Omholt. You prove to me daily that laughter is the best medicine. Thanks for listening and for the perspective.

My fondest thank you to Jeb Brady, for being an amazing collection of stardust and for continuing to shine brilliantly. Thank you for introducing me to the concept of "Life on Planet Ziploc." You make me laugh. I love you.

Finally, thank you to my son, Jason Todd, the star that forever shines brightest in my universe.

—*Deborah Todd*

This book could not have been prepared without the generous support and assistance of Ms. Heather Lindsay, curator of the Emilio Segrè Visual Archives (ESVA) of the American Institute of Physics (AIP). Special thanks also go to the staff at the Evans Library of Florida Tech, whose assistance in obtaining obscure reference materials proved essential in the success of this project. Finally, a special thanks is extended to the editorial team at Facts On File, particularly executive editor Frank K. Darmstadt, who kept a steady hand at the wheel while the draft manuscript traveled through uncharted, often turbulent, waters.

—*Joseph A. Angelo, Jr.*

INTRODUCTION

The best thing about writing a book on the *A to Z of Scientists in Space and Astronomy* is discovering that many of the most amazing contributions were made by ordinary people who loved gazing at the stars. Musicians, philosophers, priests, physicians, people who came from impoverished backgrounds, and those with untold wealth, have all contributed to our understanding of what we see when we look into the night sky, and where we fit into the scheme of things. Astronomy, unlike any other field, can be and has been significantly affected by amateurs. The stars are for everyone.

Throughout time, the stars have guided people in all walks of life in a variety of ways—to navigate ships, determine when wars should be waged, decide when marriages should take place (and to whom one should be married), where buildings should be erected, and when fields should be planted and harvested. Astronomy and astrology were once a single discipline, and important aspects in medical care. Even today, the medical profession admits that emergency rooms get a little crazy during full moons. And of course there is the term *lunatic*, which comes from the Latin *luna*, "the Moon."

Astronomy has allowed us to standardize our clocks and our calendars, and has provided us with some sense of order in our very busy lives. And as our bodies and internal clocks work in time with solar and lunar and stellar events, we cannot deny that we are all inexplicably connected to this tiny planet and to the place it occupies in a vast space that we are still striving to understand.

Our unquenchable thirst for knowledge in the galaxy and the universe has propelled us into space, traveling in search of answers. We have sent new kinds of ships into a new unexplored territory, gathering samples and pictures from places that are too far away for any human to travel, yet. And just in case, hoping that we might not be alone, we beckon a response from any other travelers who might be out there also exploring.

This book was written to introduce you to some of the inspiring people who have dedicated themselves to this field and have made it possible for us to learn more. Many have given their lives to their studies, and far too many have lost their lives for their beliefs. Some pursued finding answers in the stars as a hobby over a lifetime. Others made great contributions during a short period of time, and then left the field to pursue other interests. Yet their commonality is one that we all share—the wonder of looking up at the night sky. At least once during your lifetime, look through a telescope at the Moon, or a star, or a comet, or a planet. And when you do, remember that even Galileo could not have seen the wonders that you can see. It would have been beyond his wildest dreams.

ENTRIES

The entries in this book run the gamut from relatively unknown amateurs to the most highly regarded scientists in astronomy, cosmology, and physics. If there were more time and more space, there would be many, many more worthy people included in this book. Perhaps future editions will have further additions.

Each entry here is arranged in alphabetical order by surname or, in the case of the ancients, by the name by which they are most commonly known (for example, Ptolemy). In the instances of those whose names have had many different spellings, a variety of spellings have also been included. Following the name, you will find birth and death dates, nationality, and field of specialization.

The biographies are typically 750 to 1,200 words, usually containing some information about the subjects' childhood and family background, and exploring influences, progression into the field, educational background, and, most important, work and contributions to the study of space and/or astronomy. Where there is a name listed in SMALL CAPS within the body of the biography, that name will be listed as its own biography within the book.

CROSS-REFERENCES AND BIBLIOGRAPHY

Following the alphabetical entries are a number of cross-reference indexes. The Chronology lists entrants by the periods in which they lived. They are also grouped by birth dates in the Year of Birth section, plus by Country of Birth, Scientific Field, and Country of Scientific Activity.

An extensive bibliography lists resources used in the research of this book, including both printed material and Web sites. As you whet your appetite reading about these extraordinary people, you are invited to use these resources as a starting place to help you learn more about scientists in space and astronomy.

—*Deborah Todd*

Adams, John Couch
(1819–1892)
British
Astronomer, Mathematician

John Couch Adams, a brilliant scientist often described as shy, reserved, and even self-effacing, is best known as a central figure in a scientific fiasco that sparked an international debate, pitted nations against one another, and brought the world of astronomy out of academia and into the life of the average citizen. Adams's involvement in the missed discovery, and subsequent codiscovery, of the planet Neptune became one of the most important events in the history of astronomy, from both a scientific and a political point of view.

Born in 1819, the son of a farmer, Adams reportedly showed great mathematical ability from the time he was young, and at the age of 16 calculated the time of an upcoming solar eclipse. Four years later, in October 1839, Adams became a student at St. John's College in Cambridge, a town that Adams stayed tied to for the rest of his life.

Shortly after the discovery of Uranus by SIR WILLIAM HERSCHEL, astronomers discovered a baffling problem with that planet's orbit. Adams, in his second year of college, suspected that the problem was another planet, writing in 1841

that "the irregularities of the motion of Uranus" had him questioning "whether they may be attributed to the action of an undiscovered planet beyond it." His studies, and later his work, kept Adams busy, but he planned on "investigating, as soon as possible after taking my degree" (another two years away), the possible explanation for the planet's orbit, believing that he would then "if possible . . . determine the elements of its orbit, etc. approximately, which would probably lead to its discovery."

Unfortunately, life got in the way, protocol got in the way, and ineffective tools and people got in the way. The chain of events that took place between Adams's documented idea in July 1841, in which he suspected the existence of another planet, to his computations of the planet's location, to the actual discovery of the planet in September 1846, was a slow and painful progression. The time span of more than 5 years allowed plenty of time for outside forces and people to complicate matters, and the result was a full-blown scandal. The idea that "he who publishes first gets the credit," which was the protocol of the time, did not help Adams either. The professional positions and personal obligations of the people involved further complicated the events that followed.

In 1841 Adams was a second-year student convinced that there was another planet beyond

The British astronomer John Couch Adams, who is cocredited with the mathematical discovery of the planet Neptune in the 19th century. From 1843 to 1845, he investigated irregularities in the orbit of Uranus and used mathematical physics to predict the existence of a planet beyond. However, his work was ignored in the United Kingdom until Urbain J. J. Leverrier made similar calculations that allowed Johann Galle to discover Neptune on September 23, 1846. *(Courtesy of AIP Emilio Segrè Visual Archives)*

Uranus. He was overworked with his studies, but still reportedly spent hours by candlelight, working through the night on mathematical computations to pinpoint the planet's location. While this was unknown to the rest of the world at the time, Adams was the first person to use Newton's theory of gravitation as a basis for calculating the position of a planet.

Two years later, in 1843, Adams graduated with extraordinarily high marks from Cambridge, receiving the award of First Smith's Prizeman and a fellowship at Pembroke College. He soon became a curator of the Cambridge Observatory, and his career was underway. In October he

completed computations for the location of the planet, and shared his theory with the Astronomer Royal, Sir George Airy. But Airy disagreed, and thought that the problems with the orbit were related to the inverse square law of gravitation, in which gravitation begins to break down over large distances. Airy's dismissal of the computations, compounded by new teaching responsibilities, kept Adams from pursuing his theory.

Despite the setbacks, Adams had a brilliant mind, and although he was not what anyone would ever call aggressive, he was persistent about calculating the planet's position. Adams requested additional data on Uranus from Greenwich. Through the director of the Cambridge Observatory, Professor James Challis, Adams requested additional data from Airy in Greenwich, and in February 1844, the information was sent.

DOMINIQUE-FRANÇOIS-JEAN ARAGO, the director of the Paris Observatory, was also interested in Uranus's orbital problems, and he also thought another planet might be involved. In June 1845 he requested that one of his astronomers, URBAIN-JEAN-JOSEPH LE VERRIER, work on the solution to find this planet. Arago did not know about Adams's work, and Airy did not know that Arago was looking for a planet.

By September Adams's calculations were complete, and included the location of the planet, its mass, and its orbit. Adams personally took his findings to show Airy, but Airy was away in France, so Adams left a letter of introduction written on his behalf by Challis. Despite the fact that it was common for people from all walks of life to write to and drop in on the Astronomer Royal, Adams's behavior of showing up without an appointment is often highly criticized and cited as a contributing factor to the fiasco.

The following month, Airy wrote to Challis saying that he would see Adams, and Adams traveled once again without an appointment to see the astronomer. When he arrived, Mrs. Airy told Adams that her husband was out, so he left his card saying that he would return later in the day. She forgot to give the card to Airy, and when Adams returned as promised in the afternoon, he was told that the family was having dinner, which they did every afternoon at 3:30, and he was turned away again. Adams left his papers for Airy to review, and returned to Cambridge.

Airy still questioned Adams's theory, even after looking at his calculations and explanations, and later wrote to him asking specifically whether the new planet could explain some of the discrepancies of Uranus's orbit. Many think that Adams was bothered by the question because he thought it was missing the point—he had identified a planet and gave complete and thorough calculations. For whatever reason, Adams did not write back.

In November 1845 LeVerrier published a paper stating that Jupiter and Saturn definitely did not cause Uranus's orbit, insisting that another factor had to be involved. On June 1, 1846, he published another paper, this time stating with certainty that another planet beyond Uranus was the only possible explanation for Uranus's orbit. Airy got a copy of LeVerrier's paper on June 23, and wrote to him asking the same question he had asked Adams approximately six months earlier. This put Airy in the possession of information making him the only one who knew that Adams and LeVerrier were working on the same prediction, and that they in fact came to the same conclusion. Airy told neither of them about the other's work. On June 29 Airy received LeVerrier's reply. With this information, he met with Challis and another astronomer, Sir John Frederick William Herschel (son of Sir William Herschel, discoverer of Uranus) to discuss the "extreme probability of now discovering a new planet in a very short time . . ."

"It was so novel a thing to undertake observations in reliance upon merely theoretical deductions; and that while much labour was

certain, success appeared very doubtful," wrote Challis, but on July 29, he started the search by recording stars in the area, which he would have to observe and reobserve over several different evenings, to compare locations and see if one had moved. Since Challis did not have a current star map at the observatory, he could not make quick comparisons. He made his second observation on July 30, his third on August 4, and his final observation on August 12, then started comparing his data. Unfortunately, he only looked at the first 39 stars he recorded on August 12 when he made the comparisons to the July 29 data. It was the 49th "star" that later proved to be the planet, but Challis stopped his comparisons before finding this result.

In late August 1846 Hershel, who believed in the existence of the planet, suggested to a friend with a telescope that they could probably see the planet just by looking for it in the suggested location. His friend, William Dawes, decided that the telescope he owned was not adequate enough, and searching for it would be a waste of time.

As it turned out, LeVerrier had the same idea. On August 31 he published a paper with the planet's orbit, mass, and angular diameter, and suggested that using the data, they could look at the stars in this location, find the one that was a disc rather than a point of light (planets are disc-shaped in a telescope), and quickly find the planet.

On September 2 Adams sent Airy new computations with a more precise location of the planet. Meanwhile, Dawes wrote to a friend, William Lassell, asking if he would use his larger telescope to help look for it. But Lassell had a badly sprained ankle and could not get around, so he gave the letter to his maid for safekeeping while he recovered. She promptly lost the letter, so Lassell never looked.

On September 10 Herschel gave a speech at a British Association event, in which he discussed his belief in the new planet, and at which Adams was scheduled to present a paper on his findings. As if things could not become a more dreadful comedy of errors, it turns out that there was some confusion as to the scheduled date of the presentation, and Adams arrived at the session the day after it closed.

LeVerrier's experience in getting the French government to search for the planet soon began to resemble Adams's experience with the British government: Interest was lacking. So, on September 18 he wrote to the astronomer JOHANN GOTTFRIED GALLE at the Berlin Academy, giving him the calculations and asking him to search with his telescope. Galle and his assistant, Heinrich d'Arrest, received LeVerrier's letter on September 23, and since they had a telescope and new star maps success seemed likely. Within the hour, the planet was found. Galle immediately wrote to LeVerrier, "Monsieur, the planet of which you indicated the position really exists."

Back in England, Challis finally received a copy of LeVerrier's August 31 paper, and he did some more observations. This time he actually saw that one of the stars was a disc, but he preferred to confirm that it was really a planet by observing and comparing movement, which would take time, so he did nothing.

To the surprise of everyone involved, on October 1 London's prestigious newspaper *The Times* announced Galle's sighting of "LeVerrier's planet." Challis immediately looked through his telescope again and confirmed that the disc he saw on July 29 had moved. By October 3 the scandal had begun.

Herschel informed the public that Adams had made the same prediction earlier than LeVerrier. The humble Adams was called upon to confirm the time and nature of his computations, and Airy and Challis had to explain themselves to the press and the public in light of Adams's virtually ignored, yet significant, work.

Naturally, the British government wanted the credit for the finding, and so it claimed that

since they were the first to know about the planet, it was a British discovery. The French government, absent in helping LeVerrier look for the planet, suddenly decided it was time to take action, so it claimed ownership of the discovery. And the Germans, who were the first to actually *see* the planet, received no credit at all.

Additional investigations revealed that "LeVerrier's planet" had been observed and catalogued by other, earlier astronomers, including Galileo Galilei on December 28, 1612, and January 27 and 28, 1613, who observed that it had traveled but did not pursue it as a planetary discovery. The French astronomer Joseph-Jérôme Le Français de Lalande recorded it as a star on May 8 and 10, 1795, as did Herschel on July 14, 1830, and the Scottish-German astronomer Johann von Lamont on October 25, 1845, and September 7 and 11, 1846.

To add to the debate, American scientists attacked both Adams and LeVerrier, saying that their calculations of the planet's orbit were so far off (which they were) that its discovery was purely a coincidence. None of this seemed to really affect the mild-mannered Adams, who simply did not get embroiled in any of it.

In 1847 John Herschel invited both Adams and LeVerrier to meet with him at his home in Kent. Despite the insistence by the British government that Adams had discovered the planet, Adams himself apparently never made that claim, and it is considered common knowledge that he was never bitter about any of it. He was, instead, a great admirer of LeVerrier's, and upon meeting at Herschel's, the two apparently became good friends.

Adams survived the great fiasco, and both he and LeVerrier became recognized as the codiscoverers of the new planet, named Neptune. Adams went on to participate in other significant astronomical work, including computations on the acceleration of the Moon, and the relationship of the Leonids (meteors that originate from a region in the "head" of the constellation Leo) with a comet.

In 1848 Adams received the coveted Copley Medal from the Royal Society of London, and in 1859 he became the Lowndean Professor of Astronomy and Geometry at Cambridge University. He stayed in this position for the next 32 years. In 1861 he succeeded Challis as the director of the Cambridge Observatory, and later served two terms as president of the Royal Astronomical Society. Adams passed up two very deserved opportunities for fame. He declined Airy's job as Astronomer Royal when Airy retired, and he refused a knighthood offered by Queen Victoria because he apparently could not afford to keep up the standard of living of a knight.

Adams married Eliza Bruce (descendant of Scottish king Robert Bruce) in 1863. He retired from Cambridge in 1891 and died in 1892. In 1895 a memorial was placed in Westminster Abbey near the memorial to Newton, and many believe that Adams was the greatest English astronomer and mathematician since SIR ISAAC NEWTON. Adam's memoir, which included the story of his predictions, was published at some point during his life, under the title *An explanation of the observed irregularities in the motion of Uranus, on the hypothesis of disturbances caused by a more distant planet; with a determination of the mass, orbit, and position of the disturbing body.* A crater is named in honor of Adams on the surface of the Moon.

Astronomers Walter Sydney Adams (left), James Hopwood Jeans (center), and Edwin Powell Hubble (right) relax for a moment outdoors in spring 1931 at the Mount Wilson Observatory, in the San Gabriel Mountains about 30 kilometers northwest of Los Angeles, California. Adams specialized in stellar spectroscopy and served as the director of Mount Wilson from 1923 to 1946. *(Courtesy AIP Emilio Segrè Visual Archives)*

⊠ Adams, Walter Sydney
(1876–1956)
American
Astronomer

Walter Sydney Adams specialized in stellar spectroscopic studies and codeveloped the important technique called spectroscopic parallax

for determining stellar distances. In 1915 his spectral studies of Sirius B led to the discovery of the first white dwarf star. From 1923 to 1946 he served as the director of the Mount Wilson Observatory in California.

Walter Adams was born on December 20, 1876, in the village of Kessab, near Antioch in northern Syria, to Lucien Harper Adams and Dora Francis Adams, American missionaries from New Hampshire. This area of Syria, where Adams spent the first nine years of his childhood, was then under Turkish rule as part of the former Ottoman Empire. By 1885 his parents had completed their missionary work and returned home to the United States.

As a student at Dartmouth College, Adams earned a reputation as a brilliant undergraduate. He selected a career in astronomy as a result of his courses with Professor Edwin B. Frost (1866–1935). Adams graduated from Dartmouth in 1898 and then followed Frost to the University of Chicago's Yerkes Observatory. While learning spectroscopic methods at the observatory, Adams also continued his formal studies in astronomy at the University of Chicago, where, in 1900, he obtained a graduate degree. The famous American astronomer GEORGE ELLERY HALE founded this observatory in 1897, and its 40-inch refractor telescope was the world's largest. As a young graduate, Adams had the unique opportunity to work with Hale as he established a new department devoted to stellar spectroscopy. This experience cultivated Adams's lifelong interest in stellar spectroscopy.

From 1900 to 1901 Adams studied in Munich, Germany, under several eminent German astronomers, including KARL SCHWARZSCHILD. He then returned to the United States and worked at the Yerkes Observatory on a program that measured the radial velocities of early-type stars. In 1904 he moved with Hale to the newly established Mount Wilson Observatory, in the San Gabriel Mountains about 19 miles (30 km) northwest of Los Angeles, California. Adams became assistant director of this observatory in 1913, and served in that capacity until 1923, when he succeeded Hale as director.

Adams married his first wife, Lillian Wickham, in 1910. After her death 10 years later, he married Adeline L. Miller in 1922, and the couple had two sons, Edmund M. and John F. Adams. From 1923 until his retirement in 1946, Adams served as the director of the Mount Wilson Observatory. Following his retirement on January 1, 1946, he continued his astronomical activities at the Hale Solar Laboratory in Pasadena, California.

At Mount Wilson, Adams was closely involved in the design, construction, and operation of the observatory's initial 60-inch (1.5-m) reflecting telescope, and then the newer 100-inch (2.5-m) reflector that came on line in 1917. Starting in 1914 he collaborated with the German astronomer Arnold Kohlschütter (1883–1969) in developing a method of establishing the surface temperature, luminosity (the amount of radiation a star emits), and distance of stars from their spectral data.

In particular, Adams showed how it was possible for astronomers to distinguish between a dwarf star and a giant star simply from their spectral data. As defined by astronomers, a dwarf star is any main sequence star, while a giant star is a highly luminous one that has departed the main sequence toward the end of its life and swollen significantly in size. Giant stars typically have diameters from 5 to 25 times the diameter of the Sun and luminosities that range from tens to hundreds of times the luminosity of the Sun. Adams showed that it was possible to determine the luminosity of a star from its spectrum. This allowed him to introduce the important method of "spectroscopic parallax" whereby the luminosity deduced from a star's spectrum is then used to estimate its distance. Astronomers have estimated the distance of many thousands of stars with this important method.

Adams is perhaps best known for his work involving Sirius B, the massive but small companion to Sirius, the Dog Star—the brightest star in the sky after the Sun.

The German mathematician and astronomer FRIEDRICH WILHELM BESSEL first showed in 1844 that Sirius must have a companion and he even estimated its mass to be about the same as that of the Sun. In 1862 the American optician ALVAN GRAHAM CLARK made the first telescopic observation of Sirius B, sometimes called the Pup. Their preliminary work set the stage for Adams to make his great discovery.

In 1915 Adams obtained the spectrum of Sirius B—a very difficult task due to the brightness of its stellar companion, Sirius. The spectral data indicated that the small star was considerably hotter than the Sun. A skilled astronomer, he immediately realized that such a hot celestial object, just eight light-years distant, could remain invisible to the naked eye only if it was very much smaller than the Sun. Sirius B is actually slightly smaller than Earth. Assuming all of his observations and reasoning were true, Adams reached this important conclusion: Sirius B must have an extremely high density, possibly approaching 1 million times the density of water.

So in 1915 Adams made astronomical history by identifying the first "white dwarf"—a small, dense object that is the end product of stellar evolution for all but the most massive stars. A golf ball–sized chunk taken from the central region of a white dwarf star would have a mass of about 35,000 kilograms—that is, 35 metric tonnes, or some 15 fully-equipped sport utility vehicles.

Almost a decade later, Adams parlayed his identification of this very dense white dwarf star. He assumed a compact object like Sirius B should possess a very strong gravitational field. He further reasoned that according to ALBERT EINSTEIN'S general relativity theory, Sirius B's strong gravitational field should redshift any light it emitted. Between 1924 and 1925 he successfully performed difficult spectroscopic measurements that detected the anticipated slight redshift of the star's light. This work provided other scientists with independent observational evidence that Einstein's general relativity theory was valid.

Adams also conducted spectroscopic investigations of the Venusian atmosphere in 1932, showing that it was rich in carbon dioxide (CO_2). He retired as director of the Mount Wilson Observatory in 1946, but continued his research at the Hale Laboratory. He died peacefully at home in Pasadena, California, on May 11, 1956. His numerous honors and awards for his contributions to astronomy include the Henry Draper Medal of the National Academy of Sciences (1918), the Gold Medal of the Royal Astronomical Society (1917), and the Bruce Gold Medal from the Astronomical Society of the Pacific (1928). He became an associate of the Royal Astronomical Society in 1914 and a member of the National Academy of Sciences in 1917.

⊠ **Alfonso X (Alfonso el Sabio,
Alfonso the Learned, Alfonso the Wise)**
(1221–1284)
Spanish
Astronomer, Historian

Alfonso X, dedicated to creating a Spanish-language encyclopedia filled with the knowledge of all of mankind, and particularly committed to the study of astronomy, became an important figure in this scientific field during his reign (1252–84) as king of León and Castile. Armed with the power and money to attract the most brilliant minds in the known world, Alfonso ordered and oversaw the creation of astronomy-based texts and tables that became the definitive work on the subject for three centuries.

Astronomy and astrology were disciplines that were so closely intertwined during this era that they were not considered two separate fields of study. The Moon and stars were consulted for a variety of practical purposes, including religion, timekeeping, medicine, and navigation. Accurate and complete texts, charts, and tables were essential to govern these important matters, and the key to creating the definitive source was to amass the largest quantity of information and then turn it into the most accurate translations.

These translations required experts in both the languages and the subject matter. Whether the document was in Arabic, Greek, or Latin, each was to be translated into Spanish. Aside from simply transferring the information from one language to another, translators were required to correct and update the mathematical computations. Experts in astronomy and geometry were essential to this process. It is said Alfonso X brought together more than 50 of the greatest minds from around the world to work on just one of these translations.

The most significant work to come out of this regime was the Alfonsine Tables. These tables were calculations that, using Alfonso's year of coronation as the base year, gave exact rules to compute the position of any of the planets at any time. Derived from the Arabian *Toledo Tables*, created in the middle of the 11th century by Abu Ishaq Ibrahim Ibm Yahya al-Zarqali (better known by his European name of Arzachel), the Alfonsine Tables resolved some discrepancies of the earlier tables and were the "holy grail" of astronomical tables for the next 300 years.

It is estimated that the Alfonsine Tables were completed sometime between 1252 and 1280, and it is known for certain that they reached Paris in 1292, staying in use well into the mid-1500s. NICOLAS COPERNICUS learned about the Alfonsine Tables while studying at the university in Krakow, and he used them in his computations as he developed his theories of a heliocentric universe. With Copernicus's calculations, a new set of tables, the Prutenic Tables, were developed in 1551 by Erasmus Reinhold, and these became the new standard in astronomy.

In 1280 another huge undertaking was completed, the publishing of *Libros del Saber de Astronomia*, in which all of the known stars were identified and listed with their coordinates. Astronomical instruments were described, a variety of clocks were identified and detailed, and

PTOLEMY'S celestial spheres were explained in great detail in this text. Due to the complexity of the Ptolemaic system, in which the planets, Sun, and Moon each existed in their own circle, each circled the Earth, and most required additional circles to explain their motion, Alfonso X was so perplexed by Ptolemy's explanation that he is quoted as saying, "Had I been present at the creation, I would have given some useful hints for the better ordering of the Universe."

Because of his dedication to astronomy and the decree to amass the definitive collection of texts, Alfonso X is responsible for the survival of information from the ancient texts, which allowed their introduction to European society and helped fuel the scientific advancements that began during the Renaissance and continue to this day.

⊠ Alfvén, Hannes Olof Gösta
(1908–1995)
Swedish
Physicist, Space Scientist, Cosmologist

Hannes Alfvén developed the theory of magnetohydrodynamics (MHD), the branch of physics that helps astrophysicists understand sunspot formation and the magnetic field-plasma interactions (now called Alfvén waves in his honor) taking place in the outer regions of the Sun and other stars. For this pioneering work and its applications to many areas of plasma physics, he shared the 1970 Nobel Prize in physics.

Hannes Olof Gösta Alfvén was born in Norrköping, Sweden, on May 30, 1908, to Johannes Alfvén and Anna-Clara Romanus, both practicing physicians. He began his higher education at the University of Uppsala in 1926, and obtained his Ph.D. in philosophy in 1934. That same year, he received an appointment to lecture in physics at the University of Uppsala.

The Swedish physicist and Nobel laureate Hannes Alfvén developed the theory of magnetohydrodynamics (MHD). This important branch of physics allows astrophysicists to study sunspot formation and the magnetic field–plasma interactions that take place in the outer regions of the Sun and other stars. He appears here at a press conference after being announced as a co-winner of the 1970 Nobel Prize in physics. *(AIP Emilio Segrè Visual Archives)*

He married Kerstin Maria Erikson in 1935, and the couple had five children: one son and four daughters.

In 1937 Alfvén joined the Nobel Institute for Physics in Stockholm as a research physicist. Starting in 1940, he held a number of positions at the Royal Institute of Technology in Stockholm. He served as an appointed professor in the theory of electricity from 1940 to 1945, professor of electronics from 1945 to 1963, and professor of plasma physics from 1963 to 1967. In 1967 he came to the United States

and joined the University of California at San Diego as a visiting professor of physics, a position he held until 1991, when he retired. In 1970 he shared the Nobel Prize in physics with the French physicist Louis Néel (1904–2000). Alfvén received his half of the prize for his fundamental work and discoveries in MHD and its numerous applications in various areas of plasma physics.

Alfvén is best known for his pioneering plasma physics research that led to the creation of the field of magnetohydrodynamics—the branch of physics concerned with the interactions between plasma and a magnetic field. Plasma is an electrically neutral gaseous mixture of positive and negative ions. Physicists often refer to plasma as the fourth state of matter, because the plasma behaves quite differently from solids, liquids, or gases.

In the 1930s Alfvén went against conventional beliefs in classical electromagnetic theory. He boldly proposed a theory of sunspot formation centered on his hypothesis that under certain physical conditions a magnetic field can be locked or "frozen" in a plasma. He built upon this hypothesis and further suggested in 1942 that under certain conditions—as found, for example, in the Sun's atmosphere—waves can propagate through plasma influenced by magnetic fields. Today solar physicists believe that at least a portion of the energy heating the solar corona is due to action of these waves, Alfvén waves, propagating from the outer layers of the Sun.

Much of Alfvén's early work on MHD and plasma physics is collected in the books *Cosmical Electrodynamics* (1948) and *Cosmical Electrodynamics: Fundamental Principles*, coauthored in 1963 with the Swedish physicist Carl-Gunne Fälthammar (born 1931).

In the 1950s Alfvén again proved to be a scientific rebel when he disagreed with both Steady State cosmology and Big Bang cosmology. Instead, Alfvén suggested an alternate cosmology based on the principles of plasma physics. He used his 1956 book, *On the Origin of the Solar System,* to introduce his hypothesis that the larger bodies in the solar system were formed by the condensation of small particles from primordial magnetic plasma. He further pursued such electrodynamic cosmology concepts in his 1975 book, *On the Evolution of the Solar System,* which he coauthored with Gustaf Arrhenius.

Alfvén also developed an interesting model of the early universe in which he assumed the presence of a huge spherical cloud containing equal amounts of matter and antimatter. This theoretical model, sometimes referred to as Alfvén's antimatter cosmology model, is not currently supported by observational evidence of large quantities of annihilation radiation—the anticipated by-product when galaxies and antigalaxies collide. So while contemporary astrophysicists and astronomers generally disagree with his theories in cosmology, Alfvén's work in magnetohydrodynamics and plasma physics remains of prime importance to scientists engaged in controlled thermonuclear fusion research.

After retiring in 1991 from his academic positions at the University of California at San Diego and the Royal Institute of Technology in Stockholm, Alfvén enjoyed his remaining years by commuting semiannually between homes in the San Diego area and Stockholm. He died in Stockholm on April 2, 1995, and is best remembered as the Nobel laureate theoretical physicist who founded the important field of magnetohydrodynamics.

⊠ **Alpher, Ralph Asher**
(1921–   )
American
Physicist, Cosmologist

In 1989 the National Aeronautics and Space Administration (NASA) launched a $150 million satellite to investigate cosmic background radiation in space, the existence of which was suggested some 41 years earlier by the young Ph.D. student Ralph Alpher. Known for the concept that would eventually be dubbed the big bang theory, Alpher was the first person to devise the equations that explained the origin of the universe.

Ralph Alpher was born in 1921 to Samuel and Rose Alpher, and by the time he was 16, in 1937, the inquisitive young man was ready to leave his family's home in Washington, D.C., to study science at college. But it was the Great Depression, and Alpher's dream to study at the Massachusetts Institute of Technology (MIT) on a full scholarship was vanquished when the scholarship was inexplicably rescinded. So, the perseverant Alpher did what he thought would best accomplish his goals. He enrolled in night school at George Washington University.

Working during the day as a secretary to earn enough money for school, books, and food, Alpher spent the next six years as an undergraduate, first studying chemistry, then switching to physics. Science had taken hold of Alpher and would not let go. Acknowledging that there was much more he wanted to learn and accomplish, Alpher spent the next two years working toward his master's degree, and then, in 1945, he jumped into the doctorate program.

Alpher's thesis adviser was the renowned ex-Soviet scientist GEORGE GAMOW, who had the perfect project for Alpher. Gamow had given a lot of thought to what might have happened at the beginning of time to create the elements, but he had dedicated little time to work on the problem. Under Gamow's supervision, Alpher was to figure out the origin of the elements for his thesis. To narrow it down, Gamow suggested that Alpher not worry about the exact instant the universe was created, but rather look at what was happening when things started to stabilize and cool down to the point of containing radiation and *ylem*, the Greek term Gamow assigned to the concept of the primordial soup of life.

Alpher worked on the timing and the mechanics, considering first how neutrons had existed and then radioactively decayed into protons, electrons, and neutrinos; expanding, cooling, combining, creating the universe and all of the elements in it, all of which happened within the first five minutes of creation. These were concepts that no previous scientist had even considered explaining.

Gamow was interested in making certain that Alpher's work would get as much attention as possible. Since *alpha, beta,* and *gamma* are the first three letters of the Greek alphabet, Gamow made a play on words to give Alpher's work a memorable title. By adding another scientist's name to the paper, HANS ALBRECHT BETHE, he could call it the Alpher, Bethe, Gamow theory. Bethe refused to take credit for work he had not done, but Gamow included his name on the paper anyway, and the publicity idea worked. Gamow sent the paper to *The Physical Review* for publication, and the article was printed on April 1, 1948. This became the hottest topic in the history of astronomy.

With Alpher's thesis public, the idea that the universe was created by a superhot explosion caused a scientific stir of great proportions. Alpher's next task was to defend the work in his oral exam, which under normal circumstances would take place in front of a handful of faculty members. In this case, however, Alpher showed up to an audience of about 300 people, including the press and some of the most important physicists of the time. When Alpher was asked how long the big bang took, and he responded about 300 seconds, he became instant fodder for headlines and cartoons, and the recipient of all kinds of mail offering help for his soul because, as he said, "I had dared to trample on their concept of Genesis."

This kind of notoriety required more work immediately. Alpher teamed up with a colleague, Robert Herman, and the two wrote and published 18 more papers, the first of which was published in *Nature,* a British scientific journal. This paper suggested that the best way to prove the theory was to look into space for the radiation that still existed from the event that occurred 15 billion years ago. The radiation, they insisted, would be within a blackbody spectrum that would show up at about 5 kelvins (K), approximately 450 degrees below zero Fahrenheit. The simple math for the proof boiled down to this: The amount of radiation divided by the amount of matter that existed immediately after the big bang is equal to the amount of radiation divided by the amount of matter that exists now. This equation took Herman and Alpher an entire summer to devise, calculate, and substantiate.

Alpher's biggest problems were twofold. First, a highly respected group of notable scientists had an entirely different view on how the universe worked. These "steady-state" physicists included Fred Hoyle, who, ridiculing Alpher, came up with the term "big bang." The second problem involved the technology, or lack of it, available at the time. Alpher and Herman simply could not get their results verified because background radiation in space, they were told, could not be measured. Despite their exhaustive efforts to find a way, they came up with no help.

In 1955 Alpher and Herman left academia to earn a living in the corporate sector. Herman worked for General Motors, and Alpher went to General Electric (GE). Alpher wrote hundreds of papers for GE, but he and Herman wrote only four more research papers together on their theory. In time, their revolutionary ideas were forgotten.

Technology changed dramatically over the next nine years. By 1964, two scientists, A. G. Doroschkevich and Igor Novikov, took up Alpher and Herman's cause in a paper they published, suggesting a search for blackbody radiation. But no one seemed interested in the topic.

At the same time, two other scientists, both radio astronomers working for Bell Laboratories, were testing a horn-shaped antenna hoping to measure radio waves and echoes from satellites, when they were plagued with a background microwave noise that they could not get rid of no matter what adjustments they tried. What actually happened was that they had stumbled upon the blackbody radiation Alpher and Herman predicted in their early work. Instead of getting rid of the noise, ARNO ALLEN PENZIAS and ROBERT WOODROW WILSON ended up discovering cosmic microwave background radiation—at 3.5 K, instead of the 5 K that Alpher and Herman had suggested.

Penzias and Wilson's discovery was published in 1965, and almost immediately ended the debate between the steady-state and the Big Bang theorists. This proved that the Big Bang theory was correct and sparked the birth of cosmology as a dedicated science. But everyone seemed to have forgotten the fact that Alpher, Herman, and Gamow had predicted and calculated this radiation 17 years earlier, when radio astronomy was in its infancy and this kind of proof was impossible.

The three original scientists began a writing campaign to reinform the scientific community of the hot debate that evolved around the original predictions in 1948, but no one was interested and they received little response. Although Alpher and Herman did eventually receive the Magellanic Premium Award from the American Philosophical Society in 1975 for their discoveries, they were not included in the most important recognition a scientist can achieve. In 1978 Penzias and Wilson won the Nobel Prize in physics for their discovery of cosmic microwave background radiation.

Alpher's Big Bang theory is considered one of the Top Ten Astronomical Triumphs of the Millennium, according to the *American Physical Society News*. He has been called by his colleagues "arguably one of the most important scientists of the century."

In 1989 Alpher and Herman were honored guests at the launching of COBE, NASA's Cosmic Background Explorer. The satellite found exactly what it was looking for—cosmic background radiation—precisely at 2.7° K, radiation emanating from everywhere.

Today, Alpher is a distinguished research professor at Union College in New York. He is the administrator of the university's Dudley Observatory, which has a mission to "support research in astronomy, astrophysics, and the history of astronomy." Ralph Alpher is part of its living history.

⊠ Alvarez, Luis Walter
(1911–1988)
American
Physicist

Luis Walter Alvarez, a Nobel Prize–winning physicist, collaborated with his son, Walter, a geologist, to rock the scientific community in 1980 by proposing an extraterrestrial catastrophe theory. Called the "Alvarez hypothesis," this popular theory suggests that a large asteroid struck Earth some 65 million years ago, causing a mass extinction of life, including the dinosaurs. Alvarez supported the hypothesis by gathering interesting geologic evidence—namely, the discovery of a worldwide enrichment of iridium in the thin sediment layer between the Cretaceous and Tertiary periods. The unusually high elemental abundance of iridium has been attributed to the impact of an extraterrestrial mineral source. The Alvarez hypothesis has raised a great deal of scientific interest in cosmic impacts, both as a way to possibly explain the disappearance of the dinosaurs and as a future threat to planet Earth that might be avoidable through innovations in space technology.

Luis Alvarez was born in San Francisco, California, on June 13, 1911, to Walter C. Alvarez, a prominent physician, and Harriet Smyth Alvarez. In 1925 the family moved from San Francisco to Rochester, Minnesota, so his father

The Nobel laureate physicist Luis Walter Alvarez collaborated with his geologist son Walter to rock the scientific community in 1980 when they proposed an extraterrestrial catastrophe theory. Also called the "Alvarez hypothesis," their popular version of an ancient catastrophe suggests that a large asteroid struck Earth some 65 million years ago, causing the mass extinction of most living creatures, including the dinosaurs. Luis Walter Alvarez received the 1968 Nobel Prize in physics for his outstanding experimental work in nuclear particle physics. (*Ernest Orlando Lawrence Berkeley National Laboratory, courtesy AIP Emilio Segrè Visual Archives*)

enthusiasm and passion that remained for a lifetime. In rapid succession, he earned his bachelor's degree (1932), master's degree (1934), and Ph.D. in physics (1936) at the University of Chicago. He married Geraldine Smithwick in 1936, and they had two children: a son (Walter) and a daughter (Jean). Almost two decades later, in 1958, Luis Alvarez married his second wife, Janet L. Landis, with whom he had two other children: a son (Donald) and a daughter (Helen).

After obtaining his doctorate in physics from the University of Chicago in 1936, Alvarez began his long professional association with the University of California at Berkeley. Only World War II disrupted this affiliation. Alvarez conducted special wartime radar research at the Massachusetts Institute of Technology in Boston from 1940 to 1943. He then joined the atomic bomb team at Los Alamos National Laboratory in New Mexico (1944–45). He played a key role in the development of the first plutonium implosion weapon. During the Manhattan Project, Alvarez had the difficult task of developing a reliable, high-speed way to multipoint-detonate the chemical high explosive used to symmetrically squeeze the bomb's plutonium core into a supercritical mass. During the world's first atomic bomb test, called Trinity, on July 16, 1945, near Alamogordo, New Mexico, Alvarez flew overhead as a scientific observer. He was the only witness to precisely sketch the first atomic debris cloud as part of his report. He also served as a scientific observer onboard one of the escort aircraft that accompanied the *Enola Gay*—the B-29 bomber that dropped the first atomic weapon on Hiroshima, Japan, on August 6, 1945.

After World War II Alvarez returned to the University of California at Berkeley and served as a professor of physics from 1945 until his retirement in 1978. His fascinating career as an experimental physicist involved many interesting discoveries that go well beyond the scope of this

could join the staff at the Mayo Clinic. During his high school years, Luis spent the summers developing his experimental skills by working as an apprentice in the Mayo Clinic's instrument shop. He initially enrolled in chemistry at the University of Chicago, but quickly embraced physics, especially experimental physics, with an book. Before discussing his major contribution to modern astronomy, however, it is appropriate to briefly describe the brilliant experimental work that earned him the Nobel Prize in physics in 1968. Simply stated, Alvarez helped start the great elementary particle stampede that began in the early 1960s. He did this by developing the liquid hydrogen bubble chamber into a large, enormously powerful research instrument of modern high-energy physics. His innovative work allowed teams of researchers at Berkeley and elsewhere to detect and identify many new species of very-short-lived subnuclear particles—opening the way for the development of the quark model in modern nuclear physics.

When an elementary particle passes through the chamber's liquid hydrogen (kept at a temperature of −250 degrees Celsius), the cryogenic fluid is warmed to the boiling point along the track that the particle leaves. The tiny telltale trail of bubbles is photographed and carefully computer-analyzed by Alvarez's device. Nuclear physicists then examine these data to extract new information about whichever member of the "nuclear particle zoo" they have just captured. Alvarez's large liquid-hydrogen bubble chamber came into operation in March 1959, and almost immediately led to the discovery of many interesting new elementary particles. In a quite appropriate analogy, Alvarez's hydrogen bubble chamber did for elementary particle physics what the telescope did for observational astronomy.

Just before his retirement from the University of California, Luis Alvarez became intrigued by a very unusual geologic phenomenon. In 1977 his son Walter (born 1940) showed him an interesting rock specimen that he had collected from a site near Gubbio, a medieval Italian town in the Apennine Mountains. The rock sample was about 65 million years old and consisted of two layers of limestone: one from the Cretaceous period (symbol K, after the German word *Kreide* for Cretaceous) and the other from the Tertiary period (symbol T). A thin (approximately 1 centimeter) clay strip separated the two limestone layers.

According to geologic history, as this layered rock specimen formed eons ago, the dinosaurs flourished and then mysteriously passed into extinction. Perhaps the thin clay strip contained information that might answer the question of why the great dinosaurs suddenly disappeared. Alvarez and Walter carefully examined the rock and were puzzled by the presence of a very high concentration of iridium in the peculiar sedimentary clay. Here on Earth, iridium is quite rare and typically no more than about 0.03 parts per billion are normally found in the planet's crust. Soon geologists discovered this same iridium enhancement (sometimes called the "iridium anomaly") in other places around the world in the same thin sedimentary layer (called the KT boundary) that was laid down about 65 million years ago—the suspected time of a great mass extinction.

Since iridium is so rare in Earth's crust and is more abundant in other solar system bodies, the Alvarez team postulated that a large asteroid—about 6 miles (10 km) or more in diameter—had struck prehistoric Earth. This cosmic collision would have caused an environmental catastrophe throughout the planet. Alvarez reasoned that such a giant asteroid would largely vaporize while passing through Earth's atmosphere, spreading a dense cloud of dust particles including large quantities of extraterrestrial iridium atoms uniformly around the globe. The Alvarez team further speculated that after the impact of this killer asteroid, a dense cloud of ejected dirt, dust, and debris would encircle the globe for many years, blocking photosynthesis and destroying the food chains upon which many ancient animals depended.

When Alvarez published this hypothesis in 1980, it created quite a stir in the scientific community. In fact, the Nobel laureate spent much of the last decade of his life explaining and defending his extraterrestrial catastrophe theory. Despite the geophysical evidence of a global iridium anomaly in the thin sedimentary clay layer at the KT boundary, many geologists and paleontologists still preferred other explanations concerning the mass extinction. While still controversial, the Alvarez hypothesis emerged from the 1980s as the most popular explanation of the dinosaurs disappearance.

Shortly after Alvarez's death, two very interesting scientific events took place that gave additional support to the extraterrestrial catastrophe theory. First, in the early 1990s, a ring structure 112 miles (180 km) in diameter, called Chicxulub, was identified from geophysical data collected in the Yucatán region of Mexico. The Chicxulub crater has been age-dated at 65 million years. The impact of an asteroid 10 kilometers in diameter would have created this very large crater, as well as caused enormous tidal waves. Second, a wayward comet called Shoemaker-Levy 9 slammed into Jupiter in July 1994. Scientists using a variety of space-based and Earth-based observatories studied how the giant planet's atmosphere convulsed after getting hit by a cosmic "train" of chunks, about 20 kilometers in diameter, of this fragmented comet that plowed into the southern hemisphere of Jupiter. The comet's fragments deposited the energy equivalent of about 40 million megatons of trinitrotoluene (TNT), and their staccato impact sent huge plumes of hot material shooting 1,000 kilometers above the visible Jovian cloud tops.

About a year before his death in Berkeley, California, on August 31, 1988, Alvarez published a colorful autobiography entitled *Alvarez, Adventures of a Physicist*. In additional to his 1968 Nobel Prize in physics, he received numerous other awards, including the Collier Trophy in Aviation (1946) and the National Medal of Science, personally presented to him in 1964 by President Lyndon Johnson.

⊠ **Ambartsumian, Viktor Amazaspovich**
(1908–1996)
Armenian
Astrophysicist

Viktor Ambartsumian founded the Byurakan Astrophysical Observatory in 1946 on Mount Aragatz, near Yerevan, Armenia. This facility served as one of the major astronomical observatories in the former Soviet Union. Considered the father of Russian theoretical astrophysics, he introduced the idea of stellar association into astronomy in 1947. His major theoretical contributions to astrophysics involved the origin and evolution of stars, the nature of interstellar matter, and phenomena associated with active galactic nuclei.

Ambartsumian was born on September 18, 1908, in Tbilisi, Georgia (then part of the czarist Russian Empire). His father was a distinguished Armenian philologist—a teacher of classical Greco-Roman literature—so young Viktor was exposed early in life to the value of intense intellectual activity. When he was just 11 years old, he wrote the first two of his many astronomical papers: "The New Sixteen-Year Period for Sunspots" and "Description of Nebulae in Connection with the Hypothesis on the Origin of the Universe." His father quickly recognized his son's mathematical talents and aptitude for physics and encouraged him to pursue higher education in St. Petersburg (then called Leningrad), Russia. In 1925 Ambartsumian enrolled at the University of Leningrad. Before receiving his degree in physics in 1928, he published 10 papers on astrophysics. From 1928 to 1931, he did his graduate work in astrophysics at the Pulkovo Astronomical Observatory. This historic observatory near St. Petersburg was founded in the 1830s by the German astronomer Friedrich Georg Wilhelm von Struve (1793–1864) and served as a major observatory for the Russian (Soviet) Academy of Sciences until its destruction during World War II.

During the short period from 1928 to 1930, Ambartsumian published 22 astrophysics papers in various journals, while still a graduate student at Pulkovo Observatory. The papers signified the emergence of his broad theoretical research interests, and his scientific activities began to attract official attention and recognition from the Soviet government. At age 26, Viktor Ambartsumian was made a professor at the University of Leningrad, where he soon organized and headed the first department of astrophysics in the Soviet Union. His pioneering academic activities and numerous scientific publications in the field eventually earned him the title of "father of Russian (Soviet) theoretical astrophysics."

Ambartsumian served as a member of the faculty at the University of Leningrad from 1931 to 1943, and was elected to the Soviet Academy of Sciences in 1939. During World War II, drawn by his heritage he returned to Armenia (then a republic within the Soviet Union). In 1943 he held a teaching position at Yerevan State University, in the capital city of the Republic of Armenia. That year he also became a member of the Armenian Academy of Sciences. In 1946 he organized the development and construction of the Byurakan Astrophysical Observatory. Located on Mount Aragatz, this facility served as one of the major astronomical observatories in the Soviet Union. Ambartsumian served as the director of the Byurakan Astrophysical Observatory and resumed his activities investigating the evolution and nature of star systems.

In 1947 he introduced the important new concept of stellar association—the very loose groupings of young stars of similar spectral type that commonly occur in the gas- and dust-rich regions of the Milky Way Galaxy's spiral arms. He was the first to suggest the notion that interstellar matter occurs in the form of clouds. Some of his most important work took place in 1955, when he proposed that certain galactic radio sources indicated the occurrence of violent explosions at the centers of these particular galaxies. The English translation of his influential textbook, *Theoretical Astrophysics*, appeared in 1958 and became standard reading for astronomers-in-training around the globe.

His fellow scientists publicly acknowledged Ambartsumian's technical contributions and leadership by electing him president of the International Astronomical Union (1961–64) and president of the International Council of Scientific Unions (1970–74). His awards included the Janssen Medal from the French Academy of Sciences (1956), the Gold Medal from the Royal Astronomical Society in London (1960), and the Bruce Gold Medal from the Astronomical Society of the Pacific (1960).

On August 12, 1996, this eminent Armenian astrophysicist died at the Byurakan Observatory. He is best remembered for his empirical approach to complex astrophysical problems dealing with the origin and evolution of stars and galaxies. He summarized his most important work in the paper "On Some Trends in the Development of Astrophysics," published in the *Annual Review of Astronomy and Astrophysics*, volume 18, 1980.

⊠ **Anderson, Carl David**
(1905–1991)
American
Physicist

Carl David Anderson began his studies in science in 1923 at the California Institute of Technology as an electrical engineering student. But he soon discovered physics and changed disciplines, and eventually changed the world with a scientific discovery that has been called one of the most momentous of the century.

Born in New York City on September 3, 1905, Anderson was the only child of Carl David Anderson and Emma Adolfinja Ajaxson, Swedish immigrants. The family moved to Los

Angeles, California, when Anderson was seven, and he lived with his mother following his parents divorce soon after the move. Following an education in public schools, Anderson enrolled at a new college in Pasadena, the California Institute of Technology (Caltech), in 1923, and soon became a student of Robert Millikan, who in December 1923 was honored for his work on the elementary charge of electricity and on the photoelectric effect, as a Nobel laureate in physics.

It was Millikan who kept Anderson at Caltech after his graduation in 1927. Despite the fact that Anderson had applied and been accepted for a fellowship in doctoral studies at the University of Chicago under Millikan's suggestion to get experience someplace other than Caltech, Millikan decided that he wanted the bright student to stay and work with him on cosmic rays. Since this subject was the one Anderson most wanted to pursue, he obliged, and worked on research and experiments under Millikan's direction at Caltech.

From 1927 to 1930, Anderson was a teaching fellow at the university while working on his doctoral dissertation. In 1930 he received his Ph.D. magna cum laude, again in physics engineering, with a thesis on the space distribution of photoelectrons ejected from various gases by X-rays. For the next three years, he worked with Millikan as a research fellow, excited about the possibility to delve into the world of gamma rays, of which little was known. It was during this time, in 1932, that Anderson made a discovery that would soon earn him the coveted Nobel Prize.

Cosmic radiation was relatively new ground for Millikan and his researchers. A young astronomer named PAUL ADRIEN MAURICE DIRAC had suggested that positive and negative states should exist for all matter, but not much was understood about it, and Milliken was extremely curious. He assembled a team of researchers, including William Pickering, Victor Neher, and Anderson, to oversee a group of experiments that

would lead to some insights and discoveries in this new frontier.

Anderson's responsibility was to head up the research with the Wilson chamber, a cylindrical glass "cloud chamber" filled with water vapor-saturated gas. According to Pickering, "the pressure is dropped suddenly so that the gas expands and cools to a supersaturated state." When an ionized particle passes through the cylinder, a visible path of water droplets appears along its path, which is photographed and analyzed by the scientific team.

Anderson needed to improve the conditions of the Wilson chamber, for which Wilson won a Nobel Prize in 1927, and decided to build a newer version that had the effect of dropping the pressure much faster with the help of a piston and a vacuum, changing the vapor to include a mixture of alcohol and water, and boosting the electromagnetic power to bend the path of ionized particles. This was a key both to getting better photographs and to finding the polarity of the particle's charge. When the new experiments with the modified chamber also included Pickering's Geiger counter readings of when a particle was present, Anderson ended up with more data, but also with new results that caused great concern. It appeared that the particles were negatively charged and positively charged. Electrons, of course, could only be negative.

Anderson added one more modification to his Wilson chamber, a lead plate that would slow down the rate of speed as a particle passed through it, and therefore determine the direction the particle traveled. He then photographed a particle that was absolutely a positively charged electron traveling upward through the cylinder. This was no longer cause for concern; it was instead the discovery of a lifetime. Reported in the *Proceedings of the Royal Society* in 1932, the newly discovered positron became one of the first new fundamental particles. Anderson had discovered antimatter.

Anderson still wanted more. Despite the fact that his research was taking place during the Great

Depression, when funding was virtually nonexistent, he wanted to see what else he could find at different altitudes and latitudes. Anderson enlisted the help of graduate student Seth Neddermey to perform what became known as the Pike's Peak experiment, in which they, with great difficulty and many setbacks, took their Wilson chamber to an elevation of 14,000 feet and spent six weeks photographing particles. Nearly 10,000 photographs later, Anderson had made his next great discovery, a new particle called the muon (originally called the mesotron, then the mu meson, then shortened to simply muon). For his work during his early years in physics, Anderson received his first award in 1935, the Gold Medal of the American Institute of the City of New York.

In 1936, at the age of 31, this assistant professor from Caltech borrowed $500 from Millikan to go to Sweden to accept his next award, the Nobel Prize in physics, following in the footsteps of his mentor. Anderson was corecipient of the award that year with VICTOR FRANCIS HESS (1883–1964) who discovered cosmic radiation. In presenting the award, the Nobel Committee for Physics applauded Anderson for "finding one of the building stones of the universe, the positive electron." Anderson used half of his nearly $20,000 prize money to pay for his ill mother's medical bills, and invested the rest in real estate.

In 1937 Anderson became an associate professor at Caltech, and the same year received the Elliott Cresson Medal of the Franklin Institute, as well as an honorary degree from Colgate University. The following year, he was elected to the National Academy of Sciences. Anderson became a full professor of physics at Caltech in 1940, and in 1945 won the Presidential Certificate of Merit.

During World War II Anderson was involved in several projects to benefit the U.S. government. He turned down the position of director of the atomic bomb project, but later helped the U.S. Navy, through Caltech, to develop a stable and reliable solid rocket propellant. From 1941 though 1945, he participated in research for the Office of Scientific Research and Development, the National Defense Research Committee, and even traveled to Normandy, France, to observe rockets in use.

Anderson added a family to his life in 1946 when, at age 41, he married Lorraine Bergman and adopted her three-year-old-son Marshall David. From 1947 to 1948 he was president of the American Academy of Arts and Sciences. In 1949 two more honors were bestowed upon him: an honorary degree from Temple University and a new son, David Anderson.

In his remaining years at Caltech, Anderson continued teaching, conducting individual research, and receiving awards for the work that was the foundation of cosmic physics. He earned the John Ericsson Medal of the American Society of Swedish Engineers in 1960, and joined President Kennedy at a special White House dinner held in 1962 in honor of the Nobel laureates. The following year brought an honorary degree from Gustavus Adolphus College, and his membership in the National Academy of Sciences became a chairmanship of the Physics section in 1963, a position he held for three years.

Anderson retired from Caltech in 1970, leaving behind his positions as chairman of the freshman admissions committee and chairman of the Physics, Mathematics, and Astronomy divisions, which he had occupied for the prior eight years. In 1976 Anderson was named professor of physics emeritus of the Caltech Board of Trustees. He died on January 11, 1991, at the age of 85 following a brief illness.

⊠   **Ångström, Anders Jonas**
(1814–1874)
Swedish
Physicist, Astronomer

This famous 19th-century Swedish physicist and solar astronomer performed pioneering spectral studies of the Sun. In 1862 Ångström discovered that hydrogen was present in the solar atmosphere and went on to publish a detailed map of the Sun's spectrum, covering other elements present as well. A special unit of wavelength, the angstrom (symbol Å), now honors his accomplishments in spectroscopy and astronomy.

Anders Jonas Ångström was born in Lögdö, Sweden, on August 13, 1814. His father was a

Portrait of Anders Ångström as a young scientist. This famous 19th-century Swedish physicist and solar astronomer performed pioneering spectral studies of the Sun. In 1862 he discovered that hydrogen was present in the Sun's atmosphere and published a detailed map of the solar spectrum, covering other elements present as well. *(AIP Emilio Segrè Visual Archives)*

chaplain for the timber industry. Anders studied physics and astronomy at the University of Uppsala—founded in 1477, the oldest of the Scandinavian universities—and graduated in 1839 with his doctorate. Upon graduation, Ångström joined the university faculty as a lecturer in physics and astronomy. For more than three decades, he remained at this institution, serving it in a variety of academic and research positions. In 1843 he became an astronomical observer at the famous Uppsala Observatory—the observatory founded in 1741 by Anders Celsius (1701–44). In 1858 Ångström became chairperson of the physics department and remained a professor in that department for the remainder of his life.

Ångström performed important research in heat transfer, spectroscopy, and solar astronomy. With respect to his contributions in heat transfer phenomena, he developed a method to measure thermal conductivity by showing that it was proportional to electrical conductivity. He was also one of the 19th-century pioneers of spectroscopy. Ångström observed that an electrical spark produces two superimposed spectra. One spectrum is associated with the metal of the electrode generating the spark, while the spectrum is from the gas through which the spark passes.

He applied LEONHARD EULER's resonance theorem to his experimentally derived atomic spectra data and discovered an important principle of spectral analysis. In his paper "Optiska Undersökningar" ("Optical investigations") presented to the Swedish Academy in 1853, Ångström reported that an incandescent (hot) gas emits light at precisely the same wavelength as it absorbs light when it is cooled. This finding represents Ångström's finest research work in spectroscopy, and his results anticipated the spectroscopic discoveries of GUSTAV ROBERT KIRCHHOFF (1824–87) that led to the subsequent formulation of Kirchhoff's laws of radiation.

Ångström was also able to demonstrate the composite nature of the visible spectra of various metal alloys. His laboratory activities at the University of Uppsala gave Ångström the hands-on experience in the emerging field of spectroscopy necessary to accomplish his pioneering observational work in solar astronomy.

By 1862 Ångström's initial spectroscopic investigations of the solar spectrum enabled him to announce his discovery that the Sun's atmosphere contained hydrogen. In 1868 he published *Recherches sur le spectre solaire* (Researches on the solar spectrum), his famous atlas of the solar spectrum, containing his careful measurements of approximately 1,000 Fraunhofer lines. Unlike other pioneering spectroscopists—for example, ROBERT WILHELM BUNSEN (1811–99) and Gustav Kirchhoff—who used an arbitrary measure, Ångström precisely measured the corresponding wavelengths in units equal to one ten-millionth of a meter. Ångström's map of the solar spectrum served as a standard of reference for astronomers for nearly two decades. In 1905 the international scientific community honored his contributions by naming the unit of wavelength he used the "angstrom"—where one angstrom (symbol Å) corresponds to a length of 10^{-10} meter (one ten-millionth of a millimeter).

Physicists, spectroscopists, and microscopists use the angstrom when they discuss the visible light portion of the electromagnetic spectrum. The human eye is sensitive to electromagnetic radiation with wavelengths between 4 and 7 × 10^{-7} meter. Since these numbers are very small, scientists often find it convenient to use the angstrom in their communications. For example, the range of human vision can now be expressed as ranging between 4,000 and 7,000 angstroms (Å).

In 1867 Ångström became the first scientist to examine the spectrum of the aurora borealis, commonly known as the Northern Lights. Because of this pioneering work, his name is sometimes associated with the aurora's characteristic bright yellow-green light. He was a member of the Royal Swedish Academy (Stockholm) and the Royal Academy of Sciences of Uppsala. In 1870 Ångström was elected a fellow of the Royal Society in London, and in 1872 received its prestigious Rumford Medal. Ångström died on June 21, 1874, in Uppsala.

⊠ Arago, Dominique-François-Jean
(1786–1853)
French
Mathematician, Astronomer, Physicist, Politician

Dominique-François-Jean Arago, born in 1786 just before the French Revolution, became one of the most renowned scientists ever to come out of Napoleon's famous Parisian school, École Polytechnique.

Arago was a mathematician first in his career, but a lover of science foremost. In 1809, at the age of 23, he became a full mathematics professor at the École Polytechnique with an appointment as chair of analytical geometry. This same year, he was also elected into the Académie des Sciences. His career was firmly set in place at this early age, and within two years, he was entrenched in scientific discovery.

Light and optics theories were fascinating to Arago's brilliant mind, and in 1811 he began to contemplate the theories of polarization, how light rays have different characteristics when they travel in different directions when reflected. Using quartz crystals, he conducted experiments on light that resulted in the discovery of chromatic polarization. Arago was contacted by another scientist, Thomas Young, who, using Arago's work, proposed the first transverse wave light theory. Another contemporary, Augustine Fresnel, also wrote Arago about wave optics, and by 1815 Fresnel presented his theory, *La diffraction de la lumière*, to the Académie. Despite opposition, Arago completely backed Fresnel's theory, which was in direct conflict with the better-known corpuscular theory of light, in which light is made up of particles, as explained by SIR ISAAC NEWTON. Thanks in part to Arago's belief in Fresnel's theories, they were finally proven to be correct.

In 1820 Arago became one of many scientists who became enthralled with the newly published theory of electromagnetism. Arago took simple steps to create an experiment in which he ran a current through a wire and attracted iron filings to the current, creating the first electromagnet. His next major interest was in sound waves. In 1822 Arago worked with Gaspard de Prony in a cannon-firing experiment to attempt to determine how fast sound traveled through air.

In 1824 Arago experimented with suspending a magnetized needle above a copper disk, and noticed that when the copper moved, the needle moved. This was the basis for Faraday's 1831 discovery of electromagnetic induction. The following year, in 1825, Arago was awarded the Royal Society Copley Medal, and his experimentation continued, this time taking him into the realm of compressing gases. His early work, along with that of Faraday and others, contributed to the foundation of work that continued until 1906, which identified the critical temperature that would liquefy each of the permanent gases.

Arago's active interest in scientific discovery had far-reaching consequences. In 1816 Nicéphor Niepce started work on a solution to making light-sensitive chemicals that would create a permanent photograph, which he originally wanted to use for lithography. Ten years later, Niepce created a solution that worked on an eight-hour exposure. He invited the French painter Louis Daguerre to join him, and when Niepce died in 1833, Daguerre continued the work. Eventually he began to use silver iodide and mercury vapor on the plate, with a hyposulphite of soda solution. Arago stepped in and insisted, in 1839, that the French government

pay Daguerre and Niepce's son for the rest of their lives to continue their work on the process. By 1874 the highly developed plates and solutions were capable of taking pictures at the very fast rate of 1/25 of a second, and astronomer Pierre-Jules-César Jannsen (1824–1907), codiscoverer of helium in 1868, took the first series of continuous photographs of the transit of Venus. Before the end of the century, cinematography was created from this same technology.

The discovery of Neptune, in 1846, is as tied to Arago as it is to JOHN COUCH ADAMS, Sir George Airy, and URBAIN-JEAN-JOSEPH LE VERRIER. Arago suspected in 1845 that the anomalies in Uranus's orbit were cause by an as-yet-undiscovered planet. Arago assigned the task of finding the planet to LeVerrier, who published his papers before Adams and won credit for the new planet, which Arago participated in naming. When the great scandal took place over the ownership of discovery, all of France, especially Arago, staunchly defended LeVerrier's claim.

Arago's work with light and optics contributed to the creation of instruments that enabled photographic study of planets and stars, and the ability to begin to measure their color and radiation. Measuring and comparing the brightness of stars through photometry was a theory first introduced in 1729 by Pierre Bouguer in his paper *Essai d'optique sur la gradation de la lumière*. SIR JOHN FREDERICK WILLIAM HERSCHEL was the first astronomer to create a photometer, in 1836, and although improvements were made during the next decade, many problems remained with the glass, or the prisms, and the methods used to work the photometer. In 1850 Arago's knowledge of polarization led to the solution when he suggested using a single object glass and two Nicol prisms as a basis for comparing brightness. Astronomers EDWARD CHARLES PICKERING (1846–1919) and Karl Friedrich Zöllner (1800–60) made a slight adjustment of the prisms and glass, and by 1860 the photometer was perfected. Zöllner then began using the instrument for a new purpose, to measure the light reflected by planets.

The matter of light waves was one that stayed with Arago throughout his lifetime. In 1838 he worked on the theory of an experiment involving the comparison of the speed of light through air versus water versus glass, but there were problems actually conducting the experiment. In 1850 experimentation could finally begin, but Arago's age and failing eyesight meant that his colleagues Léon Foucault and Hippolyte Fizeau would have to carry out the work. Based on his specifications, the two proved his theory of light, in which the velocity of light decreases as it passes through a denser medium. Arago died shortly thereafter.

During his lifetime, Arago became a political figure campaigning for liberal reform in the French government. He was appointed as minister of war and marine, and is credited for eliminating slavery throughout the colonies. He has been honored for his lifetime contribution to science in many ways. Several Parisian streets bare his name, and he is one of 72 scientists commemorated with a plaque in the Eiffel Tower. The Moon and Mars each have a crater named Arago.

⊠ **Aratus of Soli**
(ca. 315–ca. 245 B.C.E.)
Greek
Writer

Prior to the famous publication of *Phaenomena*, by Aratus of Soli, the texts of astronomy in ancient Greece were mostly concerned with explanations of the spherical nature of the homocentric system, as detailed by Eudoxus of Cnidos, who died approximately 35 years before Aratus was born. The ordering of the heavens and the nature of their movement in relationship to the Earth and Sun, the reasoning behind why the stars moved from east to west, month to month,

and year to year, plus the description of the paths that were traveled in orbit, were all explained mathematically in an elaborate scheme of spheres within spheres, each system of spheres independent of the others.

Aratus of Soli was not an astronomer or a scientist as we would classify these roles today, but his contribution to astronomy was one of the greatest from ancient times, and had an impact on the works of Hipparchus and, eventually, PTOLEMY, centuries later.

Aratus was a Stoic who lived at the court of the Stoic king, Antigonus Gonatus of Macedonia. It is estimated that sometime around 270 B.C.E., Aratus wrote the poem *Phaenomena*, which explained the work of Eudoxus, not in detailed mathematical terms, but rather in simple verse, making the information more accessible to everyone. This document gave the positions of the stars, their risings and settings, and descriptions of 44 constellations, all under the pretense of a poem.

Hipparchos was the next astronomer to work with Aratus's poem, in the first century B.C.E. The next great astronomer to come along, Ptolemy, also used the text as a basis for his work, and expanded the number of constellations to 48.

Phaenomena was considered such an important work that it transcended time. It was translated from Greek into Latin and remained one of the predominate textbooks on astronomy well into the 16th century.

⊠ **Argelander, Friedrich Wilhelm August**
(1799–1875)
German
Astronomer

This 19th-century German astronomer investigated variable stars and compiled a major telescopic (but prephotography) survey of all the stars in the Northern Hemisphere brighter than the ninth magnitude. From 1859 to 1862 Argelander published the four-volume star catalog entitled *Bonner Durchmusterung* (Bonn survey), an amazing compendium containing more than 324,000 stars.

Friedrich Argelander was born on March 22, 1799, in the Baltic port of Memel, East Prussia (now Klaipeda, Lithuania). His father was a wealthy Finnish merchant and his mother a German. He studied at the University of Königsberg in Prussia, where he was one of FRIEDRICH WILHELM BESSEL's most outstanding students. In 1820 Argelander decided to pursue astronomy as a career after becoming Bessel's assistant at the Königsberg Observatory.

Argelander received his Ph.D. in astronomy in 1822 from the University of Königsberg. His doctoral dissertation involved a critical review of the celestial observations made by John Flamsteed. Argelander's academic research interest in assessing the observational quality of earlier star catalogs so influenced his later professional activities that the hardworking astronomer would eventually develop *Bonner Durchmusterung*, his own great catalog of Northern Hemisphere stars.

In 1823 Bessel's letter of recommendation helped Argelander secure a position as an observer at the newly established Turku (Åbo) Observatory in southwestern Finland (then an autonomous grand duchy within the Imperial Russian Empire). Here, the young astronomer poured his energies into the study of stellar motions. Unfortunately, a great fire in September 1827 totally destroyed Turku, the former capital of Finland, halting Argelander's work at the observatory. Following this catastrophe, the entire university community moved from Turku to the new Finnish capital at Helsinki.

In 1828 the university promoted Argelander to the rank of professor of astronomy and also gave him the task of designing and constructing a new observatory. Argelander found a suitable site on a hill south of Helsinki and construction

was completed in 1832. This beautiful and versatile observatory served as the model for the Pulkovo Observatory constructed by Friedrich Georg Wilhelm von Struve (1793–1864) near St. Petersburg (Leningrad) for use as the major observatory of the Imperial Russian Empire.

Argelander summarized his work on stellar motions in the 1837 book *About the Proper Motion of the Solar System*. His work in Helsinki ended in 1837 when Bonn University in his native Prussia offered him a professorship in astronomy that he could not refuse. The offer included construction of a new observatory at Bonn, financed by the German crown prince Friedrich Wilhelm IV (1795–1861), who became king in 1840. Argelander was a personal friend of Friedrich Wilhelm IV, having offered the crown prince refuge in his own home in Memel, following Napoleon's defeat of the Prussian army in 1806.

The new observatory in Bonn was inaugurated in 1845. From then on, Argelander devoted himself to the development and publication of his famous star catalog, *Bonner Durchmusterung*, which was published between 1859 and 1862. Argelander and his assistants worked very hard measuring the position and brightness of 324,198 stars in order to compile the largest and most comprehensive star catalog ever produced without the assistance of photography. Argelander's enormous work listed all the stars observable in the Northern Hemisphere down to the 9th magnitude. In 1863 Argelander founded the Astronomische Gesellschaft (Astronomical Society), whose mission was to continue his work by developing a complete celestial survey using the cooperation of observers throughout Europe.

Friedrich Argelander died in Bonn on February 17, 1875. His assistant and successor, Eduard Schönfeld (1828–91), extended Argelander's astronomical legacy into the skies of the Southern Hemisphere by adding another 133,659 stars. Schönfeld's own efforts ended in 1886, but other astronomers continued Argelander's quest. In 1914 the *Cordoba Durchmusterung* (Cordoba Survey) appeared, containing the position of 578,802 Southern Hemisphere stars measured down to the 10th magnitude, as mapped from the Córdoba observatory in Argentina. This effort completed the huge systematic (pre-astrophotography) survey of stars begun by Bessel and his hardworking assistant Argelander nearly a century before.

⊠ Aristarchus of Samos
(ca. 320–ca. 250 B.C.E.)
Greek
Mathematician, Astronomer

In the mid-1600s, NICOLAS COPERNICUS revealed his revolutionary idea of a heliocentric universe. While Copernicus remains one of the most renowned astronomers of all time, it was the Greek astronomer and mathematician Aristarchus, born on the island of Samos, who first introduced the concept that the Earth revolved around the Sun, some 17 centuries before Copernicus was born.

Little has been found through the ages about Aristarchus's work. It is believed that he studied under Strato of Lampsacus at ARISTOTLE's Lyceum in Alexandria, sometime after 287 B.C.E. Several authors of other treatises have referred to his work, giving some historical perspective on his contribution to both astronomy and mathematics, as well as the opinions of his work during the time.

The only remaining text found to date written by Aristarchus is a document entitled *Treatise on the Sizes and Distances of the Sun and Moon*. The significance of this piece has nothing to do with whether the Sun or the Earth is the center of the universe, but rather with the fact that he reasoned, through mathematics and observation of reflected sunlight on the surface of the Moon, a method to determine the relationships between the Moon, Earth, and Sun

without the benefit of instruments or trigonometry. He calculated that the Sun was 20 times the size of the Moon, and 20 times as far from the Earth as the Moon. Through modern astronomy we know that his calculations were off by an order of magnitude, but the fact that he was even able to make this kind of measurement was of great importance to science.

Through the writings of the Roman architect Vitruvius (90–20 B.C.E.), we get some indication of Aristarchus's importance in this field of study. He mentions Aristarchus among others in acknowledging their contribution to mathematical principles and inventions (Aristarchus is also credited for inventing a bowl-shaped sundial).

It is through the works of Archimedes (287–12 B.C.E.) and Plutarch (45–125) that Aristarchus's theory of a heliocentric system can be found. While a critic of Aristarchus's notion that the Sun, not the Earth, is the center of the universe, Archimedes makes it clear that Aristarchus gets the credit for being the first to propose such an idea, as well as the first to suggest that the universe is enormous, much larger than ever believed. Plutarch's writings show that Aristarchus also believed that, despite how it might appear that the stars rotate around the Earth, the truth was actually that the Earth rotated on its axis.

Aristarchus's work was key in introducing the concept of mathematical astronomy. Like many astronomers in the centuries to come, this student of the cosmos suffered from criticism that his outrageous theories denied both mathematical and religious known "truths."

⊠ Aristotle
(384–322 B.C.E.)
Greek
Philosopher

Aristotle's approach to astronomy differed little from his approach to all of the other fields of knowledge he pursued. He looked at everything, including the universe, as an organism. His ideas on the order of the heavens were just a small part of the volumes he amassed for the purpose of teaching, and like the rest of Aristotle's philosophy, his views on astronomy went unchallenged for centuries.

The son of a medical doctor named Nicomachus and his wife Phaestis, Aristotle was born in 384 B.C.E. in the northern Greek city of Stagirus. As the son of a physician, Aristotle was destined to learn about organisms and how living beings worked, because the practice of medicine at the time was handed down from father to son as sacred, secret knowledge. But his parents died when he was about 10 years old, and his travels and instruction with his father, who was the personal physician to Amyntas III, the king of Macedonia, came to an end. Aristotle was thereafter raised and educated in rhetoric, poetry, writing, and Greek by Proxenus of Atarneus.

By the time Aristotle was 17, he was ready for higher education, and went to study at the institution that PLATO founded 20 years earlier, the Academy in Athens, named after the Greek landowner Academus. Eudoxus (ca. 400–347 B.C.E.) was acting head of the university when Plato traveled for political purposes, giving Aristotle direct exposure to Eudoxus's work. Aristotle's philosophy soon began to be noticed, and Plato consistently referred to his famous student as "the intelligence of the school."

Aristotle's primary philosophy was based in nature, and he believed that everything could easily, and logically, be explained through observation. It was this application of logic to his observations that made it possible for Aristotle to devise such convincing explanations that his ideas became intellectually irreproachable.

In 347 B.C.E., upon Plato's death, Aristotle was not named to succeed his teacher as the head of the Academy. Dedicated to learning, he began a personal journey of acquiring and collecting knowledge, writing down his observations and thoughts on topics ranging from biology and zoology, to logic and politics. He left Athens and traveled first to Assos, where he married Pythias, the niece and adopted daughter of the ruler Hermias of Atarneus, and had a daughter also named Pythias. He went on to Lesbos, then to Macedonia in 343, staying at the court of King Philip, son of Amyntas III. During his stay in Macedonia, Aristotle endured the death of his wife, and eventually met Herpyllis, with whom he had a son, Nicomachus. When internal political problems arose, Aristotle supported the views of Philip's son, Alexander, who soon became King Alexander the Great. Alexander supported the works of the university, and asked Aristotle to start another university of his own in Athens.

In 335 B.C.E. Aristotle founded the Lyceum, bringing with him all of the writing he had accumulated over the years. The Lyceum became famous for its teachings and discussions, which often took place between students and instructors as they walked around the school, causing it to be referred to as the Peripatetic school, meaning "walking about."

Aristotle believed that "a man could not claim to know a subject unless he was capable of transmitting his knowledge to others, and he regarded teaching as the proper manifestation of knowledge." Aristotle's knowledge, in addition to politics, psychology, economics, logic, zoology, ethics, poetry, rhetoric, and theology, included astronomy, meteorology, chemistry, geography, physics, and metaphysics, many of which did not exist as fields of study until he created them.

There were some underlying philosophical concepts that led to his overall view of how the universe worked; and the logic that he implemented and applied to all science had to follow. First, Aristotle was a theoretical scientist. His ideas were based on philosophical speculation, never on scientific measurement, and were somewhat guided by the knowledge and culture of the time. His place was to give these theories a sense of order. Additionally, Aristotle did not believe in the mathematical concept of "infinite." Logic prevailed, and he ordered astronomy around the fact that the universe was finite.

His works *On the Heavens* and *On Physics* were two of the main texts concerning Aristotle's order of the universe. Having studied under Eudoxus, Aristotle expounded on the work of his teacher regarding the spherical definition of the planets, Earth, Moon, Sun, and stars. He believed that all of these were shaped like spheres, and that all existed in spheres, because spheres are the perfect shape. It was in fact logical, and became common knowledge, that the Earth was a sphere, based on the arch of the shadow of a lunar eclipse, and the appearance of new stars in the horizon as one walked north.

Unlike Eudoxus, however, Aristotle believed that these spheres actually physically existed. The heavens, he said, were divided into two distinct parts. Everything below the Moon, in the subluminary sphere, contained the four elements of earth, air, fire, and water, and this was the part of the universe where all change occurred, accounting for the existence of growth, maturity, corruption, death, and decay.

Everything above the Moon, however, existed in a pure, unchanging, eternal quintessence, or fifth element, which he called aether. In these spheres, the heavens were in constant spherical movement, and they never stopped or changed because change did not exist. This caused two problems. First, additional spheres had to be added to make the whole system work, and second, there was an issue with the logic of motion.

Aristotle changed the spherical layout Eudoxus had devised by adding 22 more spheres to explain how the motion of some of the spheres worked in a way that would not interfere with the motion of others. With the Earth at the cen-

ter, there were now 55 concentric spheres, all attached, all rotating at different velocities, working in the complex way that an organism works. Aristotle also used this system to explain why stars twinkled. It was the movement of the spheres, he concluded, that caused friction and heated the air around a star, "particularly in the part where the sun is attached to it," and that friction explained why a star shines.

To address the issue of motion, Aristotle created another explanation. Since everything that is in motion has to be set in motion, Aristotle created the existence of a Prime Mover, responsible for the initial movement of all of nature. He argued that this Prime Mover exists in the outermost sphere of the universe, that it causes the circular motion, which is perfect because it has no beginning and no end, and that this activity because of its perfection is the highest form of joy. Since the universe is finite, nothing exists beyond the limits of the universe, and since everything is an organism, this highest degree of life in the cosmos must reside in the sphere of the aether.

In the 13th century, the medieval theologian Thomas Aquinas (1225–74) found translations of Aristotle's teaching notes on astronomy, and convinced the church to use this work as the basis for explaining Christianity. When the church adopted these philosophies as the foundation for the structure of the universe to explain heaven and God, it became the word of God, and questioning Aristotle was tantamount to questioning God. This is what got GALILEO GALILEI in trouble, and is, in part, what kept Aristotle's work unchallenged for 2,000 years.

Most of Aristotle's writings published during his lifetime are lost. Only quoted fragments remain in the work of others. The writings that do exist from Aristotle consist of his lecture notes from courses at the Lyceum, more than 2,000 pages, although it is generally believed that at least some of this material has been added to by other teachers, students, and translators over the centuries.

In 323 Alexander the Great died, and politics again caused Aristotle to leave Athens. He moved to his mother's family estate in Chalcis, and died the following year, 322, at the age of 62. Aristotle is now a crater on the Moon.

⊠ Arrhenius, Svante August
(1859–1927)
Swedish
Chemist, Exobiologist

Years ahead of his time, Svante August Arrhenius was the pioneering physical chemist who won the 1903 Nobel Prize in chemistry for a brilliant idea that his conservative doctoral dissertation committee barely approved in 1884. His wide-ranging talents anticipated such Space Age scientific disciplines as planetary science and exobiology. In 1895 Arrhenius became the first scientist to formally associate the presence of "heat trapping" gases, such as carbon dioxide, in a planet's atmosphere with the greenhouse effect. Then, early in the 20th century, he caused another scientific commotion when he boldly speculated about how life might spread from planet to planet and might even be abundant throughout the universe.

Arrhenius was born on February 19, 1859, in the town of Vik, Sweden, on the University of Uppsala's estates, to Carolina Christina Thunberg and Svante Gustaf Arrhenius, a land surveyor responsible for managing the estates. His uncle, Johan Arrhenius, was a well-respected professor of botany and rector of the Agricultural High School near Uppsala who also served as the secretary of the Swedish Academy of Agriculture.

In 1860 Arrhenius's family moved to Uppsala. While a student at the Cathedral School, he demonstrated his aptitude for arithmetical calculations and developed a great

In 1908 the Nobel laureate Svante August Arrhenius introduced the panspermia hypothesis when he published his book *Worlds in the Making*. His hypothesis was a bold speculation that life could spread through outer space from planet to planet by the diffusion of spores, bacteria, or other microorganisms. He was also one of the first scientists to anticipate global warming issues, when he presented a paper at the end of the 19th century that discussed the role of carbon dioxide as a heat-trapping gas in Earth's atmosphere. *(Elliot V. Fry, courtesy AIP Emilio Segrè Visual Archives)*

interest in mathematics and physics. Upon graduation in 1876, he entered the University of Uppsala, where he studied mathematics, chemistry, and physics. He earned his bachelor's degree from the university in 1878, and continued his studies there for an additional three years as a graduate student. However, Arrhenius encountered professors at the University of Uppsala who dwelled in what he felt was too conservative a technical environment and who would not support his innovative doctoral research topic involving the electrical conductivity of solutions. So in 1881 he went to Stockholm to perform his dissertation research in absentia under Professor Eric Edlund at the Physical Institute of the Swedish Academy of Sciences.

In this more favorable research environment, Arrhenius pursued his scientific quest to answer the mystery in chemistry of why a solution of saltwater conducts electricity when neither salt nor water do by themselves. His brilliant hunch was "ions"—that is, electrolytes that when dissolved in water split or dissociate into electrically opposite positive and negative ions. In 1884 he presented this scientific breakthrough in his thesis, *"Recherches sur la conductibilité galvanique des électrolytes"* ("Investigations on the galvanic conductivity of electrolytes"). But, the revolutionary nature of Arrhenius's ionic theory simply overwhelmed the orthodox thinkers on the doctoral committee at the University of Uppsala. They just barely passed him, giving his thesis the equivalent of a blackball fourth-class rank and declaring only that his work was "not without merit."

Undeterred, Arrhenius accepted his doctoral degree, continued to promote his new ionic theory, visited other innovative minds throughout Europe, and explored new areas of science that intrigued him. For example, in 1887 he worked with LUDWIG BOLTZMANN in Graz, Austria. In 1891 Arrhenius accepted a position as a lecturer in physics at the Stockholms Högskola, Stockholm's Technical University. In Stockholm in 1894, he married his first wife, Sofia Rudeck, his student and assistant. The following year, he received a promotion to professor in physics and the newly married couple had a son, Olev Wilhelm. But his first marriage was a brief one, ending in divorce in 1896.

The Nobel Prize committee viewed the quality of Arrhenius's pioneering work in ionic theory quite differently than did his doctoral committee. It awarded Arrhenius the 1903 Nobel Prize in chemistry "in recognition of the extraordinary services he has rendered to the advancement of chemistry by his electrolytic theory of dissociation." At the awards ceremony in Stockholm, he met another free-spirited genius, Marie Curie (1867–1934), who shared the 1903 Nobel Prize in physics with her husband, Pierre Curie (1859–1906), and A. Henri Becquerel (1852–1908) for the codiscovery of radioactivity. Certainly, 1903 was an interesting and challenging year for the members of the Nobel Prize selection committee. Arrhenius's great discovery shattered conventional wisdom in both physics and chemistry, while Marie Curie's pioneering radiochemistry discoveries forced the committee to include her as a recipient, making her the first woman to receive the prestigious award.

In 1905 Arrhenius retired from his professorship in physics and accepted a position as the director of the newly created Nobel Institute of Physical Chemistry in Stockholm—a position expressly tailored by the Swedish Academy of Sciences to accommodate his wide-ranging technical interests. That same year, he married his second wife, Maria Johansson, who bore him two daughters and a son.

Soon a large number of collaborators came to the Nobel Institute of Physical Chemistry from all over Sweden and numerous other countries. The institute's creative environment allowed Arrhenius to spread his many ideas far and wide. Throughout his life, he took a very lively interest in various branches of physics and chemistry and published many influential books, including *Textbook of Theoretical Electrochemistry* (1900), *Textbook of Cosmic Physics* (1903), *Theories of Chemistry* (1906), and *Theories of Solutions* (1912).

In 1895 Arrhenius boldly ventured into the fields of climatology, geophysics, and even planetary science, when he presented an interesting paper to the Stockholm Physical Society. Titled

"On the Influence of Carbonic Acid in the Air upon the Temperature of the Ground," the paper anticipated by decades contemporary concerns about the greenhouse effect and the rising carbon dioxide (carbonic acid) content in Earth's atmosphere. In the article, Arrhenius argued that variations in trace atmospheric constituents, especially carbon dioxide, could greatly influence Earth's overall heat (energy) budget.

During the next 10 years, Arrhenius continued his pioneering work on the effects of carbon dioxide on climate, including his concern about rising levels of anthropogenic (human-caused) carbon dioxide emissions. He summarized his major thoughts on the issue in his 1903 book *Lehrbuch der kosmichen Physik* (Textbook of cosmic physics)—an interesting work for planetary science and Earth system science, scientific disciplines that did not yet exist.

A few years later, in 1908, Arrhenius published the first of several of his popular technical books, *Worlds in the Making*. In this book, he describes the "hot-house theory" (now called the greenhouse effect) of the atmosphere. He was especially interested in explaining how high-latitude temperature changes could promote the onset of the ice ages and interglacial periods.

In *Worlds in the Making*, he also introduced his "panspermia" hypothesis—a bold speculation that life could be spread through outer space from planet to planet or even from star system to star system, by the diffusion of spores, bacteria, or other microorganisms.

In 1901 he was elected to the Swedish Academy of Sciences despite lingering academic opposition in Sweden to his internationally recognized achievements in physical chemistry. The Royal Society of London awarded him the Davy Medal in 1902, and in 1911 elected him as a foreign member. That same year, during a visit to the United States, he received the first Willard Gibbs Medal from the American Chemical Society. Finally, in 1914 the British Chemical Society presented him with its prestigious

Faraday Medal. He remained intellectually active as the director of the Nobel Institute of Physical Chemistry until his death in Stockholm on October 2, 1927.

⊠ Aryabhata (Aryabhata I, Aryabhata the Elder)
(476–550)
Indian
Mathematician, Astronomer

Born in India in 476, Aryabhata is considered to be one of the most brilliant original thinkers in mathematics and astronomy, making computations and explaining the nature of our solar system some 1,000 years before NICOLAS COPERNICUS suggested the heliocentric system.

The exact years of Aryabhata's birth and death are fixed from 476 to 550, but the location of his birthplace has never been determined with any consensus. Drawing conclusions from texts others have written about him, many experts believe he was born in Kerala or one of a number of other cities in southern India. Some believe he was born in Bengal in the northeast, while others suspect that it was Pataliputra in the north, or the more common assumption of Kusumapura, which is thought to be current-day Patna, but even the exact location of this ancient city is unclear.

It is agreed, however, that Aryabhata did most, if not all, of his work in Kusumapura, one of the mathematical capitals of India, and that around the year 499 he wrote the *Aryabhatiya*, the only surviving text of several he is believed to have written concerning astronomy. The *Aryabhatiya* is written in 121 verses, or couplets, on mathematics and astronomy. The first section is an introduction to some of the mathematics in the document. The next collection of 33 verses covers algebra, trigonometry, fractions, and equations, and includes the accurate estimate of the value of *pi*, which he concluded was 62732/20000, or 3.1416. The remainder of the text focuses on the heavens.

Aryabhata's astronomy is one of many firsts. He is the first to describe the Earth as a sphere, and as a planet that rotates about its axis. It is this rotation, he explains, that makes the night sky appear to move above us.

His work also looks at the relationships between the Sun, the Earth, and the Moon. He is the first to compute the ratio between lunar orbits and rotations of the Earth, and also to calculate the length of the solar orbit. In measuring time, Aryabhata determined that the length of a year is 365 days, six hours, 12 minutes, 30 seconds, an extremely close calculation to the modern standard of 365 days, six hours.

Much of his work concerns the Moon. Aryabhata explains the phases of the Moon, which he says result from the shadows of the Earth, and using this same concept of shadows and the proximity of the Earth, the Sun, and the Moon, he explains solar and lunar eclipses. He also determined that the Moon revolves around the Earth. All of Aryabhata's work was done before the invention of the telescope.

There is an ancient Sanskrit saying that there are suns in all directions. Another suggests that when our Sun "sinks below the horizon, a thousand suns take its place." It is likely that Aryabhata's original work led to this very early understanding by Indian astronomers that our Sun is in fact a star.

Aryabhata's work was originally written in Sanskrit, and *Aryabhatiya* was finally translated into Latin toward the end of the 13th century, making its way to Europe, where its biggest impact was on mathematicians. By this time, astronomers were already familiar with the work of Copernicus, so Aryabhata's text was not considered groundbreaking.

In honor of his magnificent original work in the field of astronomy, the first Indian satellite was named after him, and a crater on the Moon carries the name of Aryabhata.

B

Baade, Wilhelm Heinrich Walter
(1893–1960)
American
Astronomer

German-born American astronomer Walter Baade made a name for himself many times over for his significant contributions to the understanding of the universe. He collaborated with Swiss astronomer FRITZ ZWICKY on supernovae and neutron stars, and with Rudolf Leo Bernhard Minkowski on radio sources as they pertained to Cygnus A and others. But it was taking advantage of wartime blackouts and peering into Andromeda that led to his greatest discoveries, in which he determined through new classifications of stars that the galaxy was twice as big and twice as old as scientists had previously imagined.

Baade was born in Schröttinghausen, Germany, on March 24, 1893. He began his higher education at the University of Münster, and obtained his Ph.D. in astrophysics from the University of Göttingen in 1919. Through the 1920s and into the early 1930s, he worked on staff at the Hamburg Observatory where, despite the less-than-ideal instruments he had to work with, he began studying comets, star clusters and variable stars, minor planets, and galaxies.

Soon after beginning work at Hamburg, Baade made his first major discovery, that of an unusual asteroid he named Hidalgo, which has the largest known orbit of any asteroid. He had already started to get attention from the astronomical community, and was invited to the Mount Wilson Observatory in California for a one-year visit through a Rockefeller fellowship. Upon his return to Germany, he urged that studies should be done in the Southern Hemisphere rather than competing with the United States in Northern Hemisphere astronomy. He was particularly interested in studying variable stars.

When Bernhard V. Schmidt (1879–1935), an accomplished optical specialist who made telescopic mirrors, joined the staff at Hamburg in 1926, Baade made another contribution to the field. Schmidt joined him in a 1929 expedition to the Philippines, and Baade suggested that the astronomical community really needed a good aberration-free wide-field camera. The cameras used at the time were not precise enough for astronomers to make any accurate estimates, in Baade's opinion. Schmidt set upon the task immediately after returning to Hamburg, and created the Schmidt camera in 1930, which ultimately replaced all of the old-technology portrait lenses previously used in astronomy, providing astronomers with clear, sharp images for study.

By 1931 Baade made his next big move. He left Hamburg to immigrate to the United States, securing a position at Mount Wilson Observatory. This was the first Carnegie observatory built, and it housed the historic 60-inch reflecting telescope, which was completed in 1909, plus the world's largest telescope, the 100-inch, which was completed in 1918. EDWIN POWELL HUBBLE had used this telescope to make his discovery of the expanding universe. It was here that Baade peered into the night sky to study spiral galaxies, especially the Andromeda Galaxy (M31).

Baade's collaborators over the next several years included Swiss astronomer Fritz Zwicky. In 1934 the two collaborated on their theories and announced that neutron stars and cosmic rays were the result of supernovae.

Baade's interest in finding the center of the Milky Way Galaxy began in earnest in 1937. He expanded his work to include observations of the Sculptor and Fornax dwarf galaxies with Hubble in 1939, and searched for the central star of the Crab Nebula, which he ultimately identified as having resulted from the supernova of 1054.

By the 1940s Baade discovered a window of opportunity to gather data on our galaxy. The Milky Way had proved difficult to penetrate because it was so obscured by cosmic dust. But Baade made a discovery near the center of the Galaxy, finding an area, which he called a window, that contained relatively little opaque dust, allowing observers to peer inside. This region contains millions of visible stars, and is viewed through what is now called Baade's Window to study the composition and design of the Milky Way. Baade's interest in spiral galaxies turned toward the Andromeda Galaxy.

In 1941 the United States went to war. Many of Baade's colleagues were helping with the war effort, but Baade, a German-born immigrant who had intended to file for U.S. citizenship but with his busy schedule and his dislike of bureaucracy let his papers expire, found himself prohibited from getting involved in military research. During World War II, blackouts were frequent enough in Los Angeles that Baade was able to use the 100-inch Hooker telescope to focus on the center of the Andromeda Galaxy. Classified as M31, Andromeda was first recorded by the Persian astronomer Abd-al-Rahman al-Sufi (903–986) who called it "little cloud," but its discovery was credited to Simon Marius (1573–1624) in 1612 by the French astronomer CHARLES MESSIER, who was the first astronomer to view the galaxy through a telescope, and catalogued it as the 31st nonstar object in 1774. Baade became the first astronomer of record to study the galaxy.

During his observations, Baade focused his attention on variable stars, both eruptive variables, and a special kind of pulsating variable—the Cepheid variables. First discovered in 1784 by a 19-year-old English astronomer, John Goodricke (1764–86), Cepheid variable stars are yellow supergiants that expand and contract in a pulsating fashion. The pulsating is not only an expansion and contraction of the physical size of the star, but also of the star's brightness or luminosity. The time it takes to go through a pulsing cycle depends on the star's density—longer for a giant star with low density, and shorter for a smaller star with high density. The North Star, Polaris, is a Cepheid variable that takes a little less than four days to go through a pulsing cycle.

In 1912 an astronomer at Harvard, HENRIETTA SWAN LEAVITT, was examining 25 Cepheid variables that she discovered in the Small Magellanic Cloud, and found that there is a relationship between the stars' luminosity and its period of pulsation—stars that appear to be brighter have longer periods of light variation. During the same time that Leavitt made her discoveries, HARLOW SHAPLEY had concluded that Cepheid variables could be used to calculate distance. Based on his conclusions, Shapley was involved in a famous debate in 1920 with HEBER

DOUST CURTIS over the size of the galaxy, which at that time was synonymous with the size of the universe. The score on this issue was finally settled some years later by Hubble.

Baade's work in 1941 allowed him to make a remarkable discovery about the variable stars he saw and expand the ever-evolving ideas about the state of the universe. He realized that there were two kinds of Cepheid variable stars in Andromeda. The younger, whitish-blue stars found in the spiral arms he called Population I stars. The older, reddish stars, found in the core of the galaxy, became known as Population II stars. During the next decade, Baade was able to identify more than 300 Cepheid variables, and he categorized these stars into two types. This information was significant.

The Cepheid variables in the arms of the Andromeda Galaxy were found to be four times brighter than their counterparts. Hubble had previously used only the older stars to estimate the age of the universe based on the equipment he had at the time for observation, and he determined that Andromeda was 800,000 light-years away, and the universe was approximately 1 billion light-years in size. But with Baade's discovery of two kinds of Cepheid variables, with differing factors of brightness, using both Population I and Population II stars for computations allowed the astronomer to determine that Andromeda was actually 2 million light-years away. It followed that other galaxies were also farther away, twice as far as had been originally calculated. Suddenly, the size of the universe had increased by a magnitude of two. And so, too, the age of the galaxy doubled, to approximately 10 billion years old.

This had an impact back home in the Milky Way. Baade suggested that data about Andromeda's spiral nature could be applied to our own galaxy. And he suggested a search for new cluster variables in the Magellanic Clouds, which he urged to be studied from the new 1.5-meter reflector telescope in Argentina.

Another of Baade's major discoveries involved another type of variable star—the eruptive variable. This kind of star has an unpredictable outburst—or, more rarely, a decrease—in its luminosity. A nova is an eruptive variable. *Nova*, which means "new," describes a star that is very hot and very small. Because of its size, a nova has such a low luminosity that it cannot be seen by the naked eye. It suddenly flares up to thousands of times its luminosity, often resulting in exploding fragments of itself hurtling through space. When the star flares up to this kind of brightness, it appears to suddenly burst into the night sky. The ancient Chinese astronomers, called them "guest stars" because they suddenly just show up.

A nova can take years or decades to return to its original luminosity, and its course can be charted as a light curve. Another eruptive star is the supernova, and it is a far more rare phenomenon than the nova. The most famous are the Supernova of 1054, and Tycho's Star in 1572, viewed in the constellation Cassiopeia by the Renaissance astronomer TYCHO BRAHE, who named it Stella Nova. Baade took a look at another star that had appeared to have undergone a nova reaction in 1604 in Serpens, which JOHANNES KEPLER and GALILEO GALILEI had witnessed. Studying the star's light curve, and using the photographic capability of the 100-inch telescope, Baade set out searching for evidence and he soon found it in fragments of the star's explosion. The discovery made this the third famous supernova in recorded history.

The next big event came in 1944, again using the 100-inch reflector. Baade found the center of the Andromeda Nebula, and its two companion galaxies M32 and NGC205, two of the four small cluster galaxies in the spiral Andromeda Galaxy. This was a feat that many were anticipating with the upcoming completion of the 200-inch telescope, still under construction at Palomar, and one that had been considered impossible with the current equipment. It was a major victory for Baade.

Baade's career came full circle with the discovery of another asteroid in 1949. As unique as his first, this new asteroid was both an Earth-grazer, because of its close proximity of a mere 4,000,000 miles in its orbit from Earth, and an Apollo-object, because it passes the Sun by a distance of just 17,700,000 miles during its 1.12 year orbit. Baade named this asteroid after the Greek mythic character Icarus, whose wax wings melted and cost him his life when he flew too close to the Sun.

Prior to 1952, astronomers had argued that radio sources were coming from distant galaxies. In fact, they were right, but it took Baade, in collaboration with Minkowski, to prove them so. Centaurus A was discovered on August 4, 1826, by James Dunlop (1793–1848) and described in detail in 1849 by SIR JOHN FREDERICK WILLIAM HERSCHEL, but basically ignored by the astronomical community until 1949 when an 80-foot radio antenna erected in Dover Heights, Australia, was used by astronomers John Bolton, G. Stanley, and Bruce Slee, who became the first to identify it as a radio galaxy. By 1954 Baade and Minkowski had begun a new project of systematically surveying the positions of radio sources with the new 200-inch telescope at the Palomar Observatory. They needed optical identification, and they got it. The two visually confirmed that Centaurus A was truly a galaxy. In addition to making visual identifications of the radio source from this galaxy, they identified others as well, including Cassiopeia A, Cygnus A, Perseus A, and Virgo A.

Baade was awarded the Royal Astronomical Society Gold Medal in 1954, and the prestigious Bruce Medal in 1955. He continued his work at Mount Wilson until 1958, and said later that he "greatly regretted" his retirement. Baade's publications were few, considering the enormous contributions he made to science, and this fact was noted, and yet accepted, by his colleagues, one of whom, a former executive at Carnegie Institute, explained that while Baade had published very little in writing, "he 'publishes' his data by conversations in his office with the world's astronomers," adding that Baade "is one of the most prolific of our staff."

After his retirement, Baade went to Australia for six months and studied the center of the Milky Way on a 74-inch telescope at the Mount Stromlo Observatory in Canberra. The following year, he returned to Göttingen to accept a position as Gauss professor, bringing his astronomical career full circle. Plagued with hip trouble for many years, Baade underwent surgery to take care of the problem, and died of respiratory failure, a complication from the surgery, on June 25, 1960. His work, finally in preparation for publication, is now in the hands of the Mount Wilson, Palomar, and Leiden observatories. In honor of his contributions to astronomy, lunar crater Baade was named for him, and the Magellan I telescope has been renamed the Walter Baade Telescope.

⊠   **Barnard, Edward Emerson**
(1857–1923)
American
Astronomer

A Nashville, Tennessee, native son born into extreme poverty, Edward Emerson Barnard excelled in astronomy against all odds through his intelligence, perseverance, and dedication to the field, and became recognized as one of the greatest astronomers of the late 19th and early 20th centuries. Gifted with uncanny eyesight, great photographic skills, and exceptional powers of observation, Barnard discovered 16 comets, a moon, a couple hundred nebulae, some binary stars, and a namesake star, and was a pioneer in astrophotography, using wide-angle photography to record stars, comets, nebulae, and the Milky Way. For his extraordinary work and contributions, he gained both fortune and international fame during his lifetime.

Barnard was born in Nashville, Tennessee, when the Union army had control of the city during the Civil War. His father died before he was born, leaving Barnard's mother destitute. Unable to afford schooling for her son, she educated him herself until he was nine years old, and was finally able to scrimp together enough money to send him to school. But after just two months, the depth of their poverty hit home, and with no child labor laws in place, Barnard was forced to quit school and get a job to help support them. He found work at a photography gallery manning a solar enlarger, a device that tracks the sun to make photographic prints. His job was to manually position the enlarger to keep it pointed directly at the Sun on sunny days.

In 1876 the book *The Practical Astronomer* was given to him by a friend as a gift for a favor he had done, and Barnard was suddenly introduced to the field that would become his life's work. Aided by a coworker, James W. Braid, Barnard built his first telescope out of a cardboard tube and a broken glass lens they found on the street. He later recalled in his writings that looking through this homemade telescope "filled my soul with enthusiasm when I detected the larger lunar mountains and craters, and caught a glimpse of one of the moons of Jupiter." He was inspired enough that he decided to invest $380 the following year to buy a five-inch refracting telescope. This amounted to two-thirds of his entire salary for the year.

Armed with his new investment, Barnard was presented with another exciting opportunity in 1877, when the American Association for the Advancement of Science (AAAS) held its annual meeting in Nashville. Buoyed by his love for the field, an excited Barnard sought out the AAAS president, SIMON NEWCOMB, hoping to get some solid advice on how to become a professional astronomer. When Newcomb heard that Barnard had no formal education and no mathematical background, he allegedly told Barnard that there was no opportunity for him

to make astronomy a profession, and advised that he should just stick with it as a hobby. Many years later, Barnard confessed that he left the meeting with Newcomb, found a secluded spot behind the state capitol building, and cried.

Fortunately, he did not heed Newcomb's advice. He continued to peer into space, and began to record his thoughts about planetary observation in 1880 in a little book he wrote, but never published, about Mars. Then came news about a lucrative, although unconventional, way to generate income in astronomy. H. H. Warner, a wealthy astronomy patron from Rochester, New York, announced that he would give $200 to anyone who discovered a new comet. Barnard jumped at the opportunity, and made his first new comet discovery through his five-inch telescope in 1881. As a newlywed, Barnard used the money from the discovery of comet 1881 VI, as it was named, as a down payment to build a house for his bride, Rhoda Calvert.

Later that year he was offered a job at the Mount Hamilton Lick Observatory, near San Jose, California, taking inventory of the property. He worked during the day and anticipated the opportunity to make important observations with the big 36-inch telescope at night, for which he traveled west. But Edward Holden (1846–1914), the observatory's director, had grand ideas of his own, assigning himself two nights a week on the apparatus while giving two other astronomers, S. W. Burnham and James Keeler, two nights each, leaving one night open for the public and leaving Barnard out of the loop. If Barnard wanted to do any observations, it would have to be on the 12-inch refractor in the observatory's other dome, which he had to share with the other astronomers, or on his own telescope.

Holden's newfound desire for personal fame led him to decide to produce a pictorial atlas of the Moon, and he wanted Barnard to help develop his pictures. Upon discovering that

Holden's pictures were for the most part out of focus and of very poor quality, Barnard instead rigged up a camera on the observatory's inexpensive six-inch telescope and started tracking comets. His pictures were extremely well focused. In 1892 he made the first photographic discovery of a new comet. With such clear pictures, he began photographically recording the Milky Way and, as with his early days at the photography gallery using the solar enlarger, it demanded that he keep the telescope pointed exactly at its target for hours at a time while it tracked the night sky.

Barnard did get some use out of the 12-inch telescope at Lick, and began to observe Jupiter that same year, noting that the moon Io appeared to him darker at the poles and lighter at the equator. This took amazing eyesight and focused observation to see with such limited equipment, and it was not until bigger telescopes were developed in the 1900s that his observations about Io were confirmed.

From 1888 to 1892, Barnard did not get to use the 36-inch telescope. But that was about to change. Holden was alienating the staff, and Burnham decided to leave the observatory. Barnard jumped at the opening on the telescope, asking the regents to let him have the time previously allocated to Burnham. Within three short months, in September 1892 Barnard made a discovery that propelled him into instant fame. While looking at Jupiter's four moons, discovered by GALILEO GALILEI in 1610, he found that the planet actually had five moons. His discovery of Jupiter's fifth moon, Amalthea, on the 36-inch telescope, made him an instant celebrity. Today, Jupiter is known to have 60 moons and is predicted to have 100 or so, the most recent 43 of which were discovered at the University of Hawaii by physics professor David C. Jewitt and graduate student Scott S. Sheppard, using the 3.6-meter (approximately 140-inch) Canada-France-Hawaii telescope atop Mauna Kea—a far cry from Barnard's 36-inch variety.

Barnard's interest in Mars was never far from his mind, and in 1892 he had the opportunity to observe it at opposition. But the timing was cursed with poor viewing conditions that meant he could only use magnifications of 350x or less on the 36-inch telescope, making it nearly impossible to record any visual observations. Undaunted, he waited for the next opportunity in 1894, and he began making observations three months before opposition, in July. By September, he was using magnifications of 1000x, observing from sunset to sunrise, and recording in his notes that he "failed to see any of Schiaparelli's canals as straight narrow lines," adding that even "under the best conditions," they "could never be taken for the so-called canals." He was certain there was nothing artificial on the Martian landscape.

By 1895, despite his growing public acclaim, Barnard was getting nowhere with Holden. The president refused to publish any of Barnard's stunning Milky Way photographs, and when the disillusioned Barnard was presented with another opportunity to further his contributions to his field, he took it. The University of Chicago was surpassing the Lick Observatory by building a 40-inch refracting telescope at the new Yerkes Observatory, and they wanted Barnard to join them as a professor of practical astronomy.

True to form, Barnard and his wife moved to Chicago nearly a year before the Yerkes observatory was finished, and he began working at the Kenwood Observatory for GEORGE ELLERY HALE, who was designated to become the director at Yerkes. This time, Barnard also had his portfolio of photographs to work on from the Lick Observatory, which he was preparing to publish as an atlas, as well as his continued observations of the Milky Way, comets, and nebulae, while he waited for construction of Yerkes to be completed.

In 1897 Barnard and his wife moved to Yerkes in Williams Bay, Wisconsin, and he was joined on staff by his old colleague Burnham from the Lick Observatory, where the two worked with Hale to compare observations between the Yerkes 40-inch refractor and the 36-inch from Lick. During their tests, Barnard's observational skills, coupled with the great power of the new telescope, led to his discovery that Vega was actually a double star.

In May Barnard and Hale's assistant Ferdinand Ellerman were scheduled to observe nebulae one evening using the 40-inch refractor. At around 12:45 A.M., the men heard a noise when they raised the elevated floor, but they could not figure out the source of the sound. After two and a half hours of observation, Barnard uncharacteristically cut short his typical all-night observation. Soon after the two men left the building, the 37½-ton floor's supporting cable broke, and the floor crashed to the ground, exploding into a heap of rubble.

Burning with a passion to collect the best photographic data of the galaxy, Barnard solicited a $7,000 grant in 1897 from another New York astronomy patron, Catherine W. Bruce (1816–1900). With Bruce's gift, Barnard oversaw the creation of a new five-inch telescope on which he had mounted two photographic doublets of a 6¼-inch and 10-inch aperture. Barnard's unbending determination in getting the widest photographic field and the shortest relative focus caused multiple delays in the completion of the telescope, which took until 1904. In the meantime, he continued observations throughout the night at Yerkes, sometimes to his own detriment in the cold Wisconsin winters. Barnard was known for getting completely lost in his work, so it was not too surprising when one particularly cold Wisconsin night he did not realize that his nose had frozen stuck to the eyepiece of the telescope.

By the time the new Bruce five-inch photographic telescope was finally completed in 1904, a small wooden observatory with a 15-foot dome was built for it near Barnard's home. Soon, Hale wanted to put the telescope to use on solar

research he was overseeing at the Mount Wilson Observatory in California, so the telescope, and Barnard, were shipped to California for several months to gather visual data that was not possible at the altitudes at Yerkes. Here, Barnard continued his photographic discovery of the Milky Way for about seven months before returning to Wisconsin.

In May 1916, while looking at a new photographic plate he had taken, Barnard discovered a star that he could not find when he compared it to another plate he took in August of 1894. However, there was a star at a different location on the 1894 plate that was not on the 1916 plate. He compared the two to a third plate taken in 1904, and found that neither star was on the 1904 plate, but there was another star that was located about half the distance between the two. Barnard had discovered a new star. A red dwarf, Barnard's Star has the largest proper motion of any known star, at 10.3 arcsecond per year as compared to the typical rate of about 1 arcsecond, and is the fourth closest star to our system, after the three Alpha Centauri stars, at 1.821 parsecs.

Barnard spent a total of 28 years at Yerkes, lecturing, researching, photographing, and writing. In 1897 the Royal Astronomical Society of Great Britain awarded him the gold medal for his photographic work with the six-inch telescope at Lick. In 1917 he was awarded the Bruce Medal. He is the only faculty member at Vanderbilt University to have a building named after him, and after his death in 1923, the Barnard Astronomical Society was founded in his honor in Chattanooga, Tennessee. His home at Yerkes was deeded to the university upon his wife's death, and it remains today as the residence of the director of the observatory.

Barnard discovered nearly 200 nebulae in his lifetime, published more than 900 articles, and created more than 4,000 photographic plates of the Milky Way. He was the first to discover a comet with photography, and the last to discover a moon visually. His life's work, *Photographic Atlas of Selected Regions of the Milky Way*, was published after his death by the University of Chicago Press. His belief that sleep was a complete waste of time, along with his dedication to observation, earned him the title as the man who never sleeps, and he was universally recognized as one of the greatest and most prolific observers ever to work in the field of astronomy.

⊠ **al-Battani (Abu Abdullah Mohammad ibn Jabir ibn Sinan al-Raqqi al-Harrani al-Sabi al-Battani, Albategnius)**
(ca. 858–929)
Arab
Astronomer, Mathematician

Al-Battani (Latinized name: Albategnius) is regarded as the best and most famous astronomer of medieval Islam. As a skilled naked-eye observer he refined the sets of solar, lunar, and planetary motion data found in PTOLEMY's great work, *The Almagest*, with more accurate measurements. Centuries after his death, these improved observations, as contained in his compendium, *Kitab al-Zij*, worked their way into the Renaissance and exerted influence on many western European astronomers and astrologers alike. Al-Battani and his fellow Muslim stargazers preserved and refined the geocentric cosmology of the early Greeks.

Though al-Battani was a devout follower of Ptolemy's geocentric cosmology, his precise observational data encouraged NICOLAS COPERNICUS to pursue heliocentric cosmology—the stimulus for the great scientific revolution of the 16th and 17th centuries. Al-Battani was also an innovative mathematician who introduced the use of trigonometry in observational astronomy. In 880 he produced a major star catalog (the *Kitab al-Zij*) and refined the length of the year to approximately 365.24 days.

Al-Battani was born in about 858, in the ancient town of Harran, located in northern Mesopotamia some 44 kilometers southeast of the modern Turkish city of Anliufra. His full Arab name is Abu Abdullah Mohammad ibn Jabir ibn Sinan al-Raqqi al-Harrani al-Sabi al-Battani, which later European medieval scholars who wrote exclusively in Latin changed to Albategnius. His father, Jabir ibn Sinan al-Harrani, was an astronomical instrument maker and a member of the Sabian sect, a religious group of star worshippers in Harran. So al-Battani's lifelong interest in astronomy probably started at his father's knee in early childhood. Al-Battani's contributions to astronomy are best appreciated within the context of history. At this time, Islamic armies were spreading through Egypt and Syria and westward along the shores of the Mediterranean Sea. Muslim conquests in the eastern Mediterranean resulted in the capture of ancient libraries, such as the famous one in Alexandria, Egypt. Ruling caliphs began to recognize the value of many of these ancient manuscripts and so they ordered Arab scribes to translate any ancient document that came into their possession.

By fortunate circumstance, while the Roman Empire was starting to collapse in western Europe, Nestorian Christians busied themselves preserving and archiving Syrian-language translations of many early Greek books. Caliph Harun al-Rashid ordered his scribes to purchase all available translations of these Greek manuscripts. His son and successor, Caliph al-Ma'mun, who ruled between 813 and 833, went a step beyond. As part of his peace treaty with the Byzantine emperor, Caliph al-Ma'mun was to receive a number of early Greek manuscripts annually.

One of these tribute books turned out to be Ptolemy's great synthesis of ancient Greek astronomical learning. When translated, it became widely known by its Arabic name, *The Almagest* (or "the greatness"). So the legacy of geocentric cosmology from ancient Greece narrowly survived the collapse of the Roman Empire and the ensuing Dark Ages in western Europe by taking refuge in the great astronomical observatories of medieval Islam as found in Baghdad, Damascus, and elsewhere.

But Arab astronomers, like al-Battani, did not just accept the astronomical data presented by Ptolemy. They busied themselves checking these earlier observations and often made important refinements using improved instruments, including the astrolabe, and more sophisticated mathematics. History indicates that al-Battani was a far better observer than any of his contemporaries. Yet, collectively these Arab astronomers participated in a golden era of naked-eye astronomy enhanced by the influx of ancient Greek knowledge, mathematical concepts from India, and religious needs. For example, Arab astronomers refined solar methods of timekeeping so Islamic clergymen would know precisely when to call the faithful to daily prayer.

Al-Battani worked mainly in ar-Raqqah, and also in Antioch (both now located in modern Syria). Ar-Raqqah was an ancient Roman city along the Euphrates River, just west of where it joins the Balikh River at Harran. Like other Arab astronomers, he essentially followed the writings of Ptolemy and devoted himself to refining and improving the data contained in *The Almagest*. While following this approach, he made a very important discovery concerning the motion of the aphelion point—that is, the point at which Earth is farthest from the Sun in its annual orbit. He noticed that the position at which the apparent diameter of the Sun appeared smallest was no longer located where Ptolemy said it should be with respect to the fixed stars of the ancient Greek zodiac. Al-Battani's data were precise, so he definitely encountered a significant discrepancy with the observations of Ptolemy and early Greek astronomers. But neither al-Battani nor any other astronomer who adhered to Ptolemy's geocentric cosmology could

explain the physics behind this discrepancy. They would first need to use heliocentric cosmology to appreciate why the Sun's apparent (measured) diameter kept changing throughout the year as Earth traveled along its slightly eccentric orbit around the Sun. Second, the Sun's annual apparent journey through the signs of the zodiac corresponded to positions noted by the early Greek astronomers. But because of the phenomenon of precession (the subtle wobbling of Earth's spin axis), this correspondence no longer took place in al-Battani's time—nor does it today. As al-Battani looked out from Earth, he noticed that when the Sun's apparent diameter was smallest, it did not occur where the great star master Ptolemy said it was supposed to. But without the benefit of a wobbling Earth model (to explain precession) and heliocentric cosmology (to explain the variation in the Sun's apparent diameter) it was clearly impossible for him, or any other follower of Ptolemy, to appreciate the true scientific significance of what he discovered.

Al-Battani's precise observations also allowed him to improve Ptolemy's measurement of the obliquity of the ecliptic—that is, the angle between Earth's equator and the plane defined by the apparent annual path of the Sun against fixed star background. He also made careful measurements of when the vernal and autumnal equinoxes took place. These observations allowed him to determine the length of a year to be about 365.24 days. The accuracy of al-Battani's work helped the German Jesuit mathematician Christopher Clavius (1537–1612) reform the Julian calendar and allowed Pope Gregory XIII (1502–85) to replace it in 1582 with the new Gregorian calendar that now serves much of the world as the international civil calendar.

While al-Battani was making his astronomical observations between the years 878 and 918, he became interested in some of the new mathematical concepts other Arab scholars were discovering in India. Perhaps al-Battani's greatest contribution to Arab astronomy was his introduction of the use of trigonometry, especially spherical trigonometric functions, to replace Ptolemy's geometrical methods of calculating celestial positions. This contribution allowed Muslim astronomers to use some of the most complicated mathematics known in the world up to that time.

Al-Battani died in 929 in Qar al-Jiss (now in modern Iraq) on a homeward journey from Baghdad. His contributions to astronomy continued for many centuries after his death. For example, medieval astronomers became quite familiar with Albategnius, primarily through a 12th-century Latin translation of his *Kitab al-Zij*, renamed *De motu stellarum* (On stellar motion). The invention of the printing press in 1436 by Johann Gutenberg soon made it practical to publish and widely circulate the Latin translation. This occurred in 1537 in Nuremberg, Germany—just in time for the centuries-old precise observations of al-Battani to spark interest in some of the puzzling astronomical questions that spawned the scientific revolution.

⊠ Bayer, Johannes (Johann)
(1572–1625)
German
Lawyer, Astronomer (Amateur)

For a man whose work in astronomy was conducted purely as a hobby while he made his living as a lawyer, the Bavarian-born Johann Bayer made one of the most important contributions to the field of astronomy when he published his star atlas, *Uranometria*.

Little is known about Bayer's early life. His formal education began with an interest in philosophy at the university in Ingolstadt, Germany, but he switched schools and changed his focus to study law at Augsburg, where he

eventually became a lawyer and the city's legal adviser. As analytical as his chosen profession was, his interest in astronomy allowed him to integrate both his analytical mind and his aesthetic talents, as evidenced by his artistic and meticulously accurate book.

Printed in 1603 by the 31-year-old Bayer, *Uranometria* was considered the first modern atlas of the constellations. Devised to give astronomers a precise guide to the stars, his atlas was a combination of beautiful plates of the constellations plus technical data regarding their positions and brightness.

The atlas included many firsts. Instead of using PTOLEMY's original 48 constellations, Bayer used the star calculations compiled by TYCHO BRAHE and JOHANNES KEPLER, which they had published only one year prior to Bayer's book, and which included more than 700 stars. *Uranometria* was also the first book to include the 12 new Southern Hemisphere constellations of Apus, Chamaeleon, Dorado, Grus, Hydrus, Indus, Musca (also called Apis), Pavo, Phoenix, Triangulum Australe, Tucana, and Volans, originally discovered and catalogued by the 16th-century Dutch navigator Pietr Dirksz Keyser. The book took a new artistic path as well. Rather than using woodcuts, the star maps were done as large, detailed engravings. And despite the fact that they were presented as beautiful art, Bayer never lost sight of the fact that the plates and the accompanying tables were first and foremost meant to be accurate maps.

The most important feature of Bayer's atlas was the system he created to name and classify the stars, using lowercase Greek letters followed by the plural Latin name of the constellation. He designated the brightest star of a constellation with the first letter in the Greek alphabet, making it the alpha star. The remaining letters of the Greek alphabet were used in decreasing order to indicate each star's brightness within a constellation, making the brightest star alpha, the second brightest beta, and so on. Under this system, the brightest star in the constellation Centaurus became Alpha Centauri. This nomenclature was printed in tables, along with descriptions and positions of the stars, used in conjunction with the engraved charts.

Bayer's classifications were modified somewhat in 1725, when the British astronomer John Flamsteed created a numbering system that indicated a star's position from west to east. In 1774 CHARLES MESSIER expanded on and devised his own system when he began cataloging nebulae and galaxies. And in 1843 FRIEDRICH WILHELM AUGUST ARGELANDER released *Uranometria Nova*, a new, updated version of the atlas, in homage to the work done by Bayer nearly two and half centuries earlier.

The International Astronomical Union, founded in 1919, picked up where Bayer and his successors left off, regulating star classifications as more and more stars become visible with the aid of modern technology. For his contribution through *Uranometria*, Bayer's work became the foundation of today's star classifications, and the standard to which star atlases have been held for more than 400 years.

⊠ Beg, Muhammed Taragai Ulugh
(ca. 1393–1449)
Persian
Mathematician, Astronomer

The mathematician and astronomer Ulugh Beg was by birthright a Persian prince, son of Shah Rukh, and grandson of the Mongol conqueror Timur, who amassed a great empire, including parts of modern-day Iran, Iraq, Turkey, Syria, and India. Married around age 10, Beg tended to his princely duties under his father's rule, and became even more active in his role after his grandfather's death in battle in 1405. In 1409 Shah Rukh moved the seat of his empire to present-day western Afghanistan, and set out to make the chosen city of Herat a cultural, educational,

and trading capital. When Beg was 16, his father turned over leadership of the Mawaraunnahr region to him, giving him control of, among other cities, Samarkand.

The learned Beg was a gifted poet and historian, and a Hafiz, one who could recite the Qur'an by heart. But despite the legacy left by his grandfather, Beg was not afflicted with a desire to conquer. Instead, his intentions were to further the study of science, particularly mathematics and astronomy.

Beg's interest in astronomy reportedly came about from a visit to the ruins of the Maragheh observatory, the workplace of the famous astronomer Nas ir al-Din Tusi (1201–74), who authored nearly 150 books and translated the great works of the ancients, including Archimedes, Euclid, and PTOLEMY. In 1271 al-Tusi wrote the *Zij-i Ilkhani*, also known as the famous Ilkhan Tablets, a compilation of tables of planetary orbits.

In pursuit of his desire to turn Samarkand into a leading educational center, Beg ordered the building of a *madrasah*, an institution for higher learning, in 1417. When construction was completed in 1420, the facilities opened with approximately 70 of the leading scientists of the day participating in the most advanced research, observations, discussions, and lectures. Surrounded by great minds, and an accomplished astronomer in his own right, Beg's vision for furthering scientific study grew into a plan to build a massive observatory. In 1428 construction began on a great circular building that was three stories high and more than 50 meters in diameter. The combination of the huge observatory and the brilliant minds working at the *madrasah* on important observations made Samarkand the world capital for major astronomical study at the time.

Beg was the first in recorded history to develop the concept of permanently mounting astronomical instruments in a building. His observatory included an armillary sphere, a device that consisted of spherical rings used to model the shapes and positions of celestial spheres and to help identify the position of the stars; a marble sextant, known as the Fakhri sextant, for observing stars and planets and measuring their declinations; and a quadrant that measured stellar and planetary altitudes. The quadrant was so massive that the ground had to be excavated to bring the piece of equipment into the building for installation.

Inspired, perhaps, by the writings of al-Tusi, Beg is also credited with the creation of his own *zij*, which is an astronomical compilation containing tables for calculating current positions and predicting the future positions of celestial bodies. Beg's *Zij-i Sultani* (Catalog of the stars), was published in 1437, and included the names and positions of approximately 1,000 stars (the reported number varies from 992 to 1,018 to 1,022), plus calendar information, an explanation of the methods and uses of astronomical observations, a description of the movements of the Sun, the Moon, and the planets, a treatise on astrology, and trigonometry tables of sines and tangents that are accurate to eight decimal places. Beg calculated the length of a year as 365 days, five hours, 49 minutes, 15 seconds—less than one minute off today's measurement.

Beg's *Zij* was the first major star compilation since Ptolemy's work around the year 170. And despite the fact that it preceded TYCHO BRAHE's work, and the invention of the telescope, by approximately 200 years, it remained unknown in Europe until some 50 years after Brahe's publications, which were so accurate that they made the discovery of Beg's writings inconsequential.

Leadership of the family's empire defaulted to Beg in 1447 when his father died. In such a political position, the scholarly Beg soon became victim to another family member who had aspirations of conquest. Beg's son, 'Abd al-Latif, secured the services of an assassin and had his father murdered by beheading in 1449.

After Beg's death the city of Samarkand eventually lost its status as a leading educational center. The observatory fell to ruins and was left virtually untouched until excavated by Russian archaeologists in 1908. Beg's fully-clothed body was discovered in 1941 in a mausoleum in Samarkand, buried at the foot of his grandfather's tomb. The importance of the clothing indicates that Beg's people considered him a martyr.

Today, the observatory at Samarkand is an archaeological site consisting of the building's foundation and part of the sextant, plus a small museum in honor of Beg. Using the Latin spelling of his name, the lunar crater Ulugh Beigh has been named in his honor.

⊠ Bell Burnell, Susan Jocelyn
(1943–)
Irish
Radio Astronomer, Astrophysicist

Born and raised in Belfast, Northern Ireland, near the Armagh Observatory, Jocelyn Bell took an interest in astronomy during her early teenage years. As an accomplished astronomer, Bell is known for her discovery of "little green men" (or LGM, as she nicknamed them), more commonly known as pulsars.

Bell's family strongly encouraged her as a young girl to study her field of choice, and she had access to the observatory in part because of her proximity to it, and in part because her father, an architect, had designed it. She lived in a very literate household, and was serious about education, although she did not pass entrance exams to get into the British university she wanted to attend, and ended up attending a boarding school before starting her university education.

Encouraged by radio astronomer Professor Bernard Lovell, Bell took up physics and became the only woman in her class of 50 at the University of Glasgow, from which she graduated with a physics degree in 1965. She immediately went to Cambridge University to start working on her Ph.D., and she immersed herself in building an 81.5-megahertz radio telescope designed to track quasi-stellar radio sources, or quasars. The intent of the project was to significantly improve research in the field of radio astronomy, which originally began as a field of study in the 1950s. To that end, this new radio telescope was massive, using more than four and a half acres of cabling and wires to connect thousands of nine-foot poles.

With the telescope's construction completed in July 1967, Bell's task turned toward that of a research student, operating the telescope and analyzing the data at the university's Mullard Radio Astronomy Laboratory. The radio waves, which turned into signals once they hit the telescope, were recorded on chart paper every four hours, spewing out 121.8 meters' worth of paper, nearly 400 feet, every four days, all of which had to be carefully studied and analyzed by hand. Within a few months, the November analysis showed something different. Bell noticed some glitches in a length of data that took up just 2.5 centimeters, or just under an inch. Uncertain about what she was seeing, she thought the unusual signal might be "scruff," as she called it, some sort of cosmic interference. She made notes and waited.

The signals occurred again a few weeks later, and it soon became clear that she was seeing some very regular patterns: a series of fast pulses occurring exactly every 1.337 seconds, originating from outside of our solar system. Excitement over the discovery fueled the imagination of the entire research group, and soon speculation was rampant throughout the astronomical community that the signal might be from beacons developed by extraterrestrials, instantly giving teeth to the theory of intelligent life elsewhere in the universe. Bell named the signal Little Green Men (LGM) and waited some more.

As the weeks went on, more signals occurred, this time from other areas in the galaxy,

and the theory about extraterrestrial life fizzled based on the probability of the science. Odds were decidedly against multiple civilizations sending the same signal at the same frequency at the same time to the same planet. Plus, when Bell recognized that the source changed at a rate very similar to the rate of the movement of the stars, it became clear that these signals were not artificially manufactured; in fact, they had to be naturally occurring phenomena.

Further research proved that the radio waves emulated from a special kind of rapidly spinning neutron star, the dense result of an exploding star. The stars, named pulsars, for "pulsating radio stars," emit signals much like beams of light from a lighthouse beacon sweeping across the night sky. Only in this case, they are electromagnetic radiation combined with radio waves pulsing through space in regular bursts as they are thrown off the rapidly spinning stars. The original pulsar, dubbed LGM, was officially named CP (Cambridge Pulsar) 1919. Due to naming standards, the opportunity to designate this newly discovered star as Bell's Star, as had been the precedent in previous centuries, was completely out of the question. Additional pulsars were soon discovered by Bell, but the first one was, of course, the most important discovery in this newly expanded field of radio astronomy.

For Bell's work, her thesis adviser at Cambridge University, ANTONY HEWISH, received the Nobel Prize in physics in 1974, the first time the award was given for a discovery in the field of astronomy. Controversy was on the heels of the Nobel announcement, since the discovery had clearly been made by the hard work and dedication of Jocelyn Bell. But research students typically do not get Nobel prizes—their professors do. Bell graciously accepted her fate, stating that giving it to a student could "demean" the prize.

Bell received her Ph.D. from Cambridge in 1968, married Martin Burnell that same year, and moved to the University of Southampton to begin work on gamma-ray research, where she stayed until 1973. During her time at Southampton, she was elected to the Royal Astronomical Society in 1969, and eventually served as vice president of the organization. In 1973 she moved to the Mullard Space Science Laboratory at University College in London to work on X-ray astronomy. Nearly a decade later, in 1982, she took a position to focus on infrared and optical astronomy, heading the James Clerk Maxwell Telescope project in Hawaii for the Royal Observatory in Edinburgh.

Despite the Nobel Prize incident in 1974, Bell Burnell received many awards for her accomplishments in later years. She was awarded the Beatrice M. Tinsley Prize in 1987 by the American Astronomical Society and the Herschel Medal from the Royal Astronomical Society in 1989. She redirected her career in 1991 to become a physics professor and department head at Open University in Milton Keynes, England. She won the Edinburgh Medal in 1999, and took a leave from Open University to work as a distinguished visiting professor at Princeton University in 1999 and 2000. In October 2001 she accepted the position of dean of science at the University of Bath, England. In addition to her other kudos, Bell Burnell is the recipient of the Oppenheimer Prize, the Michelson Medal, the Tinsley Prize, and the Megallanic Premium Award. She has received honorary doctorates from universities all over the world, and dedicates a portion of her time to furthering women's study of physics.

⊠ **Bessel, Friedrich Wilhelm**
(Wilhelm Bessel)
(1784–1846)
German
Astronomer

It is often difficult to know the exact event that causes someone to head his life in a certain direction. But in the case of Friedrich Wilhelm Bessel, the credit clearly goes to physician HEINRICH WILHELM MATTHÄUS OLBERS, an amateur astronomer who encouraged the young Bessel to develop a career devoted to astronomy. The resulting good fortune for the field of astronomy was the introduction of Bessel functions, the discovery of the proper motion and location of more than 50,000 stars, and the first accurate measurement of the distance to the stars.

Bessel began his life in Minden, Westphalia (now part of Germany), living a modest life as one of nine children of Carl Friedrich Bessel, a government employee, and Friederike Ernestine Schrader, a pastor's daughter. His childhood education was a brief one at the Gymnasium in Minden, where after four years, at the age of 14, the uninspired student quit school to work as an unpaid accounting apprentice for an import-export business in the town of Bremen. This first job proved to be valuable for Bessel because it introduced him to the world at large, and as his interest in other countries grew, so did his motivation to learn about them. Within a few short years, Bessel taught himself geography, English, Latin, Spanish, and mathematics, and he began to study astronomy as a by-product of his interest in navigation. At some point during his studies, Bessel ran across calculations on Halley's Comet made by the British mathematician and astronomer Thomas Harriot (1560–1621), who observed the comet on September 17, 1607—just seven days after JOHANNES KEPLER made his observation. Using Harriot's data, Bessel recalculated the comet's orbit and sent his information to Olbers, who was considered something of an expert on comets despite his amateur status in the field.

Olbers responded with praise and encouragement, asking Bessel to do more—make observations, do more detailed calculations—gently pushing the 20-year-old to create a high-quality paper worthy of publishing. He also put Bessel in communication with the noted mathematician

Johann Gauss (1777–1855), and the two developed a relationship that lasted throughout Bessel's life. When Bessel finally sent Olbers his completed work, Olbers recognized that the impressive paper would have earned Bessel a Ph.D. if he were in a formal university setting, and he advised Bessel to pursue astronomy as a profession.

The opportunity arose for Bessel to do just that in 1806, when a job was offered to him as assistant director at the private Lilenthal Observatory near Bremen, owned by German astronomer Johann Schroter (1745–1816). But seven years with the import-export firm had given him a sense of security, not to mention a decent salary, and Bessel had to weigh giving up his well-paying job for one that paid next to nothing, yet was in the field he loved. Whether it was his youth or his passion that swayed his decision is unknown, but Bessel left the import-export business and never looked back.

Bessel's first order of business in his new career was to learn the art of observation, which he did through an expertly crafted seven-foot telescope built by SIR WILLIAM HERSCHEL. He was given the task, and the opportunity, to focus his observations on the planet, rings, and satellites of Saturn, and he continued observing comets, which always fascinated him. While advancing his observational skills, he devoted time to furthering his knowledge of celestial mechanics, a field derived from SIR ISAAC NEWTON's laws of motion.

In 1807 Bessell began delving into the work of the third Astronomer Royal, Reverend JAMES BRADLEY, whose more than 60,000 observations of heavenly bodies, made between 1748 and 1762, were published in two volumes in 1798 and 1805, over a quarter of a century after his death. Star by star, Bessell worked on identifying the precise locations of more than 3,000 stars as they appeared in 1755. As word got out about his work, he was offered positions at both the Greifswald and Leipzig ob-

servatories, but he chose to stay at Lilenthal until an offer came around that he could not refuse.

In 1809 Prussia's King Frederick William III (1770–1840) ordered the building of a new observatory in Königsberg (now Kaliningrad, Russia), and the position of director was offered to the 26-year-old Bessel, along with the role of teaching astronomy at Albertus University. However, the uneducated Bessel needed a doctorate to teach at the university level, so his famous friend and colleague Gauss offered his whole-hearted recommendation, and Bessel was granted a Ph.D. on the basis of his previous work from the university at Göttingen. With the observatory still under construction, Bessel moved to Königsberg in May 1810 and his career in his new home was officially launched when he began teaching classes that summer. Bessel stayed in Königsberg for the rest of his life.

The year 1812 marked two milestones in Bessel's life—he married Johanna Hangen, with whom he would have four children, and he was elected to the Berlin Academy of Sciences. His monumental task of working on Bradley's observations took a great deal of time, but the following year he would become internationally renowned for undertaking such an extraordinary venture.

In 1813 the observatory was finally ready, and Bessel began his job as director. His timing to leave the Lilenthal Observatory proved to be very fortunate. King Frederick William III's decision in 1806 to start a war with France, invoking the wrath of the emperor Napoleon Bonaparte (1769–1821), resulted in an attack that divided his country. In the fray, the French marched into Schroter's observatory and completely destroyed it and everything inside. Stricken by the loss, Schroter died of apparent agony two years later, in 1815—the same year Bessel was honored for his work by the Berlin Academy.

In 1817 Bessel started working on the problem of planetary perturbations—changes in their orbits—and devised a new class of mathematical functions called cylindrical functions, more commonly known as Bessel functions. This type of analysis ultimately had a wider application than working on planetary perturbations, and is now used in hydrodynamics and wave theory, as well as in applied mathematics and engineering.

The overwhelming task of defining Bradley's work was finally realized in 1818, when Bessel published *Fundamentae Astronomiae pro Anno MDCCLV deducta ex Observationibus viri incomparabilis James Bradley* (Foundations of astronomy for the year 1755, deduced from the observations of the incomparable man James Bradley). Written in Latin, the work was a thorough compilation of the positions of 3,222 stars, including proper motion and spherical astronomy theory.

Bessel seemed to enjoy delving into the work of the royal astronomers. Perhaps it was his induction into the Royal Society in 1825 that sparked an interest in the work of the Astronomer Royal Nevil Maskelyne (1732–1811), and his "fundamental stars." In 1830 Bessel published his work on 38 stars, of which 36 were fundamental stars and the remaining two were polar stars, giving their mean and apparent positions as they appeared between 1750 and 1850. While Bessel worked on this production, he became particularly interested in two of Maskelyne's stars, Sirius and Procyon. He would return to these two stars in another 14 years and achieve another milestone in his field.

But first, he returned to one of his favorite subjects—comets. After all, it was now 1835, and Halley's Comet, which was responsible for getting him started in astronomy in the first place, was swinging back around in its 76-year orbit, and Bessel would have the good fortune of being able to observe it. In 1836 he published his observations in *Physical Theory of Comets*.

The first official star catalog, credited to PTOLEMY in the second century, consisted of more than 1,000 stars and was used until the 17th century. Bessel's undertaking was somewhat larger than Ptolemy's, creating a star catalog that would have the proper motions and positions of more than 50,000 stars. The resulting catalog of 75,011 stars would be completed by Bessel's student FRIEDRICH WILHELM AUGUST ARGELANDER, and named *Bonner Durchmusterung*.

While working on this extraordinary task of cataloging, Bessel focused his attention on 61 Cygni. Since this star had the greatest proper motion of any star he knew, Bessel (correctly) thought it might mean that the star was relatively close to the planet. He wanted to know exactly how close.

Accurately measuring the distance to a star was something that had never been accomplished, and was in fact a task that seemed to baffle the scientific community. In order to measure the distance, the star needed to be observed from two different locations, and a triangle formed between the two points of measurement and the star. The problem had been that there were no two places on Earth that were far enough apart to set up a usable triangulation to the star. Bessel came up with the brilliant idea of observing the star from one place on Earth when the Earth was in two different locations—along its path as it rotated around the Sun—taking calculations of the stars position every six months, which he did for three years.

This measurement of looking at an object from two different views, and creating a triangle connecting the points of view and the object, is called parallax. In 1838 Bessel announced that he had determined the distance to 61 Cygni using parallax, becoming the first astronomer to measure the parallax of a star, to see the motion of a star due to its parallax, and to calculate the distance to a star. The star 61 Cygni, located in the constellation Cygnus the Swan, was declared to be 10.3 light-years away—

remarkably close to current calculations of 11.2 light-years, or 219,072,000,000 miles, from Earth. In 1841 his accomplishment was officially recognized, and SIR JOHN FREDERICK WILLIAM HERSCHEL congratulated him on "the greatest and most glorious triumph which practical astronomy has ever witnessed." Bessel was consequently awarded the Gold Medal by the Royal Society.

Now it was time to return his attention to Sirius. Bessel noticed that there were inequalities in the orbits of the two Dog Stars, Sirius and Procyon. After studying them at great length, he felt he had figured out the reason why. In 1844 he announced that their proper motions were off because they each revolved around another object that could not be seen but which had a mass similar to that of the Sun. This hypothesis became known as "astronomy of the invisible," and was an important break from traditional astronomy, which relied upon visible observations. Bessel was proved correct in his theory when companion stars were discovered for Sirius in 1682 by ALVAN GRAHAM CLARK while he tested an 18-inch refractor telescope, and for Procyon in 1896, by John Schaeberle (1853–1924) using the 36-inch Lick refractor. This gave Bessel the distinction of being the first astronomer to predict the existence of "dark stars," which in this case turned out to be white dwarf stars.

Bessel was dedicated to the Königsberg observatory from its inception through the end of his life, forsaking all offers from competitive observatories, including the University of Berlin. He left on a scientific trip in 1842 to attend the Congress of the British Association in England, where he at long last met many of the most famous scientists of his time. The trip inspired him to do more publishing, and he published *Astronomische Untersuchungen* (Astronomical observations) that same year. After six years of declining health, Bessel died of cancer in 1846.

Bessel made numerous contributions to astronomy and science, and was deservedly awarded memberships in a variety of prestigious organizations, including the Royal Astronomical Society, the Royal Meteorological Society, the Berlin Academy, Palermo Academy, Stockholm Academy, and others. A star atlas, *Akademische Sternkarten* (Academic star maps), inspired by Bessel, was completed 13 years after his death, in 1859, through the collaboration of a number of observatories. In 1935 a lunar crater was named Bessel in his honor, and in 1938 an asteroid was given his name.

⊠ Bethe, Hans Albrecht
(1906–)
American
Astrophysicist

The scientist who answered the question "Where do stars get their energy?" is Hans Albrecht Bethe, born on July 2, 1906, in Strassburg, Germany (now Strasbourg, Alsace-Lorraine, France). The son of a psychologist, Bethe started school at the age of nine, attending the Gymnasium in Frankfurt until 1924, then into higher education at the University of Frankfurt, where he started his studies in physics. After two years, Bethe transferred to Munich, and in July 1928 earned his Ph.D. in theoretical physics under the expert tutelage of Arnold Sommerfeld, a good friend of ALBERT EINSTEIN.

Bethe immediately began teaching after Munich—first at the University of Frankfurt, where he taught for one term, then on to the University of Stuttgart, also for just one term until he obtained a teaching position in fall 1929 at the University of Munich. By the following May, Bethe was promoted to a *Privatdozent* (lecturer), and given a fellowship to travel by the International Education Board. He wasted no time in setting out for Cambridge in the fall term, and completed his fellowship travels with spring terms in Rome the following two years.

Bethe advanced his teaching career in 1932, when he became a faculty member at the highly regarded Tübingen University in Germany, founded in 1477 and modeled after the Renaissance methods of learning in Italy. But his assistant professorship came to a sudden end when Adolf Hitler (1889–1945) came into power and summarily fired all of the Jews in academic positions. Bethe, who had a Jewish mother but did not consider himself Jewish, found out about his dismissal from a student who saw it printed in the local newspaper. Neither Sommerfeld nor his student Bethe could find another position in Munich, so Bethe went back to England, this time to lecture at the University of Manchester for a year, then accepting a fellowship for the 1934 fall term at the University of Bristol. While he was at Bristol, Cornell University, located in Ithaca, New York, contacted Bethe and offered him a position that became the career move of a lifetime. Bethe packed his bags and moved to the United States to become an assistant professor at Cornell. The following year, 1935, Bethe started the project that would earn him a permanent place in the history of astronomy.

About 20 years earlier, the British astrophysicist SIR ARTHUR STANLEY EDDINGTON devoted a great deal of his efforts to uncover the internal structure of stars. At the time, the theoretical physicist HENRY NORRIS RUSSELL held the leading theory that a star's energy was generated through gravitational contraction. But Eddington thought it had something to do with radiation pressure, espousing the theory that the actual energy source was from a "transmutation of elements," in other words, some sort of conversion of hydrogen into helium that resulted in an energy release. But no definitive answer was found.

Around 1930, the collaborating team of Austrian physicist Fritz Houtermans (1903–66)

and British physicist Robert d'Escourt Atkinson (1893–1981) published a paper using the newly accepted theory of relativity and quantum mechanics explaining that fusion occurs in a star's center. Along came Bethe. In 1938, at a summit in Washington, D.C., leading physicists and astronomers were introduced to Bethe's work on nuclear reactions.

Stars with a mass the size of the Sun or smaller generate heat up to 16 million kelvins, burning hydrogen via the process known as the proton-proton chain—the fusion of hydrogen into helium. This happens in several steps. First, two hydrogen protons combine—one proton turns into a neutron and in the process releases a positron (an electron with a positive charge), and a neutrino (which has no charge), creating a hydrogen isotope called deuterium. The neutrino escapes, and the positron collides with an electron, creating gamma rays. Next, a third hydrogen proton combines with the deuterium and produces a lightweight isotope of helium—the helium nucleus now contains two protons and one neutron, and more high-radiation gamma rays are released. Finally, two helium nuclei combine to produce an ordinary nucleus of helium, and releases two protons—hydrogen. The end result is that six hydrogens combined to form one helium and two hydrogens. The mass that is lost becomes energy, which is proved by ALBERT EINSTEIN's equation $E = mc^2$.

For stars more massive than the Sun with temperatures exceeding 16 million kelvins, another process occurs. In this case, carbon serves as a catalyst for burning hydrogen. The fusion begins with a hydrogen and carbon nucleus forming nitrogen. After a much more complicated combination, the carbon is converted into nitrogen and oxygen, and in the end releases one helium and one carbon nucleus. Since there is the same amount of carbon in the end of the process as the beginning, the cycle stars again. This is known as the CNO (carbon-nitrogen-oxygen) cycle.

Bethe's work on nuclear reactions from 1935 through 1938 became his most famous work, and is considered one of the greatest accomplishments in science in the 20th century, with far-reaching effects for other scientists. FRITZ ZWICKY ran experiments that confirmed Bethe's theories. SUBRAHMANYAN CHANDRASEKHAR created the Chandra limit and opened the doors to exploring the collapse of stars, which was subsequently picked up by Roger Penrose (1931–) and SIR STEPHEN WILLIAM HAWKING. RALPH ASHER ALPHER, working under his doctoral adviser GEORGE GAMOW, created the Big Bang theory, determining when the universe was created and the elements that were present in the beginning—hydrogen, helium, and lithium. Fred Hoyle (born 1915), WILLIAM ALFRED FOWLER, and Geoffrey and ELEANOR MARGARET PEACHEY BURBIDGE showed how the nuclear fusion in stars could create the heavier elements, and how they are dispersed throughout the galaxy through such events as supernovae. In 1967 Bethe received the acknowledgment of a lifetime when he won the Nobel Prize in physics for his work on solar and stellar energy.

While Bethe was creating his amazing work, he was promoted to full professor of physics at Cornell, and he settled in to continue on his path of research and discovery. Then came World War II.

Bethe was eager to contribute to the war effort to help defeat Hitler. He left Cornell and traveled to the Massachusetts Institute of Technology (MIT), in Cambridge, Massachusetts, where he joined fellow scientists at the radiation laboratory working on microwave radar. But in summer 1942 he was invited to take part in a project that he initially wanted nothing to do with because he felt it was entirely too implausible. Robert Oppenheimer (1904–67) was assembling the most brilliant scientific minds in the United States to take part in the Manhattan Project—an effort to design and build the first atomic bomb. Bethe's participation was requested

and, after some consideration, he decided to join the team. Between 1943 and 1946 Bethe was the director of the theoretical physics division at the Los Alamos National Laboratory in Los Alamos, New Mexico. After the war, Bethe returned to Cornell and resumed his teaching and research.

In 1947 Bethe explained an occurrence in a hydrogen atom called the Lamb shift, which helped create the new field of quantum electrodynamics. He also did work involving the production via electromagnetic radiation of particles known as pi mesons—the lightest particles consisting of a quark and an antiquark. The following year found Bethe as an unwilling participant in the science news event of the century—the announcement of the big bang theory. Gamow, who wanted to create the play on words of "alpha, beta, gamma," added Bethe's name to Alpher's paper on the big bang, saying it was authored by Alpher, Bethe, and Gamow. Bethe refused to take credit for work that was not his own.

Bethe's old friend and colleague Edward Teller (1908–2003) from the Manhattan Project, contacted Bethe in 1949, hoping to persuade him to return to Los Alamos and work on the next big project—the hydrogen bomb. Bethe declined on the basis that a war fought with a hydrogen bomb would result in humankind losing "the things we were fighting for." He did consult with the Los Alamos staff for about six months in 1952, and has ever since been a vocal opponent to the H-bomb, working throughout his life for disarmament.

In the mid-1950s Bethe began to look once again at the physics of atoms and collision theory, a subject he had worked on in the earlier days of his career. He became the president of the President's Science Advisory Committee from 1956 to 1959, working on the failed negotiations with the Soviet Union to ban nuclear testing. In 1995 the 88-year-old professor was still working for the good of the planet when he wrote an open letter calling on all of the scientists of the world to "cease and desist from work creating, developing, improving, and manufacturing further nuclear weapons . . ." adding that, as one of the few original Manhattan Project scientists still alive, he viewed with "horror that tens of thousands of such weapons have been built since that time, one hundred times more than any of us at Los Alamos could ever have imagined."

In the 1990s Bethe worked on calculating the maximum mass of neutron stars, and in 1994 he was the guest of honor at a Cornell physics symposium entitled "Celebrating 60 Years at Cornell with Hans Bethe," where he was described as the "father of nuclear physics," and the "social conscience of physics." Bethe's career has continued into the 21st century. He won the prestigious Bruce Gold Medal from the Astronomical Society of the Pacific in 2001, and he remains, at the age of 98, a professor at Cornell University in New York. Between 1928 and 1996 Bethe published 290 papers, and he continues to publish. He is the author of the book *The Road from Los Alamos* (1991), and the 1997 book *Intermediate Quantum Mechanics,* cowritten with R. W. Jackiw. With the exception of his time at Los Alamos and a few sabbaticals to Columbia University, University of Cambridge, and Copenhagen, Bethe has never left his work at Cornell.

⊠   **Bode, Johann Elert**
(1747–1826)
German
Astronomer

The self-taught German astronomer Johann Bode is best known for the unusual mathematical formula that bears his name and explains the relationship of planetary distances from the Sun. But his lifelong dedication to astronomy also includes the discovery of five deep-sky objects, the creation of several publications that were unlike any ever done in the field, and 43 years of service at Berlin's Academy of Science.

Bode's scientific career began in 1768 when, at age 21, he published the first of many editions to come of his textbook *Anleitung zur Kenntnis des gestirnten Himmels* (Instruction for the knowledge of starry heavens). In 1772 Bode became an astronomer at Berlin's Academy of Science, where he stayed throughout his entire career.

Viewing the heavens and publishing about them were the themes for Bode's working life. In 1774 he teamed up with mathematician and publisher Johann Lambert (1728–77), famous for proving that pi is immeasurable, and perhaps more important to Bode, the editor of an early astronomy-based almanac. The pair created one of the first European publications dedicated specifically to scientific research, an ephemeris—a publication that lists data on the locations of stars and other heavenly bodies for each day of the year. Originally named the *Astronomisches Jahrbuch oder Ephemeriden* (Astronomical yearbook and ephemeris), it was soon shortened to the *Astronomisches Jahrbuch*, and finally to *Berliner Astronomisches Jahrbuch.* It became an annual publication that was widely used by astronomers and navigators as an accurate assessment of the night skies.

With Bode's publishing goals firmly established, he aimed his sights at the night skies and began searching for nebulae. In 1774 he made his first two deep-sky discoveries, each of which he referred to as a "nebulas patch" in the Ursa Major constellation: the Spiral Galaxy M81, which he described as "mostly round [with] a dense nucleus in the middle," and the Irregular Galaxy M82, also known as the Cigar Galaxy, which he considered to be "very pale" and "elongated" in shape. Both are at a distance of 12 million light-years (12,000 kly), and both were found on December 31, 1774. Less than six weeks later, on February 3, 1775, he discovered the Globular Cluster M53. At nearly 60 thousand light-years (60 kly), it appeared to Bode in the "northern wing of Virgo," looking "rather vivid and of round shape." In 1776 Bode became the director of the Berlin Observatory.

Publishing gave Bode his next big career boost, when he popularized a mathematical phenomenon that ultimately became known as Bode's Law. The origin of the formula came from David Gregory (1659–1708), a University of Edinburgh mathematics and astronomy professor, who introduced the concept in his book *Astronomiae physicae et geometricae elementa* (The elements of astronomy). Printed first in Latin in 1702, then in English in 1715, with a second printing in each language in 1726, Gregory's book suggested that the distances between the planets and the Sun could be assigned a simple mathematical value: "supposing the distance of the Earth from the Sun to be divided into ten equal parts, of these the distance of Mercury will be about four, of Venus seven, of Mars fifteen, of Jupiter fifty two, and that of Saturn ninety five[.]"

This is essentially the idea of the astronomical unit (AU), in which the distance from the Earth to the Sun is 1 AU, and the distances of the remaining planets are expressed as a ratio.

The idea resurfaced in 1724, when the German mathematician, philosopher, and prolific writer Christian Wolff (1679–1754) mentioned the relationships in his book *Vernünfftige Gedanken von den Absichten der natürlichen Dinge*. It faded into obscurity for another 40 years, but emerged again thanks to Swiss philosopher Charles Bonnet (1720–93), who became famous as the first to propose a scientific theory of evolution. Bonnet wrote about the ratios in his 1764 book, *Contemplation de la Nature* (The contemplation of nature), which was subsequently translated into several languages.

In 1766 Johann Titius (1729–96), astronomer, mathematician, and professor at the University

of Wittenberg, Germany, became the German translator of Bonnet's popular book. But when he got to the text explaining the relationship between the distances of the planets, Titius added his own formula to his translation. First, a set of numbers is created starting with 0, then 3, then doubling each subsequent number, creating the series of numbers 0, 3, 6, 12, 24, 48, 96, 192, 384, and so on. The next step is to add 4 to each number, creating the new series 4, 7, 10, 16, 28, 52, 100, 196, 388, and so on. The final step is to divide this new group of numbers by 10, which results in .4, .7, 1.0, 1.6, 2.8, 5.2, 10.0, 19.6, 38.8, etc. This last group of numbers gives the location of the known planets by their distances from the Sun in AU. Titius failed to note that this was his own work, and it was not until his second translation of Bonnet's book was printed two years later, in 1768, that he pointed this out in an added footnote, clearing things up. Not that it mattered. No one noticed the concept.

But that same year Bode's *Anleitung zur Kenntniss des gestirnten Himmels* was printed, and it became a well-received and widely read astronomy textbook—so much so that he revised it and reprinted it in 1772, and in this edition he added Titius's formula. As with his predecessors, Bode included the information in his own book without mentioning the origin of the work. But unlike the earlier authors, Bode was a much more public figure and a highly regarded astronomer. So when the concept was published in his book, it was immediately accepted and became known as Bode's Law.

Since Bode's Law gave the nearly perfect locations of the then-known planets, Mercury at .4 AU (actual distance is 0.39), Venus at .7 (actual 0.72), Earth at 1.0, Mars at 1.6 (actual 1.52), Jupiter at 5.2, and Saturn at 10.0 (actual 9.54), speculation was high that there were more planets waiting to be discovered. The search was on, with a sense of certainty that Bode's Law would lead to new discoveries.

The next several years proved fruitful for Bode. His publishing goals were ambitious. In 1777 he created a compilation of all of the known nebulous stars and clusters and published it in *Astronomisches Jahrbuch* for the 1779 calendar year. Using data recorded by Johannes Hevelius (1611–87), along with the catalog compiled in a two-year journey to the Cape of Good Hope by French astronomer Nicolas-Louis de Lacaille (1713–62), and CHARLES MESSIER's 1771 deep-sky catalog, Bode created the *Complete Catalogue of Hitherto Observed Nebulous Stars and Star Clusters*, consisting of 75 entries, many of which, because of their creators' mistakes, did not actually exist. Not one to skimp on his own observational tasks, Bode dutifully ended the year by discovering the Globular Cluster M92 at approximately 26 kly, on December 27, 1777, in the Hercules constellation.

The *Complete Catalogue of Hitherto Observed Nebulous Stars and Star Clusters* grew by two objects in its 1779 printing. Bode added his discovery of M92, which he described as ". . . a mostly round figure with a pale glimmer of light," and his newest nebula, M64, which he found on April 4, 1779.

But on March 23 the Welsh astronomer Edward Pigott (1753–1825) had also found the nebula, just 12 days before Bode's discovery, although his discovery was not published until 1781. Messier found it as well, on March 1, 1780, and published his finding in the summer of that year. For more than two centuries, most forgot that Pigott had rightful title to the discovery, until April 2002, when it was uncovered by British astrophysicist Bryn Jones. Pigott and Bode are now considered codiscoverers of the Spiral Galaxy M64.

Bode's discovery of a comet in 1779—named C/1779 A1, 1779 Bode—led to a slew of new discoveries by other astronomers. Six new nebula were found, by astronomers Antoine Darquier de Pellepoix (1718–1802), who discovered the Ring Nebula M57 (2.3 kly), Johann

Gottfried Koehler, who discovered the Elliptical Galaxy M59 (60,000 kly) and the Spiral Galaxy M60 (60,000 kly), Barnabas Oriani (1752–1832), who discovered the Spiral Galaxy M61 (60,000 kly) and Messier, who discovered the Globular Cluster M56 (nearly 33 kly) and the Spiral Galaxy M58 (60,000 kly). All were discovered while observing Bode's comet.

Finally, in 1781 the solar system's seventh planet was found in the constellation Gemini, by the amateur astronomer SIR WILLIAM HERSCHEL, located at 19.18 AU, near the 19.6 AU location predicted by Bode's Law.

In honor of Herschel's discovery, Bode created a new constellation that he named Herschels Teleskop, depicting the instrument Herschel used in his momentous discovery, and placed next to the constellation Auriga.

As astronomers around the world observed the new planet, Bode was more than happy to collect and print their data. During his research Bode found evidence that it had actually been seen at least twice before; first in 1690 by John Flamsteed (1646–1719), who designated it in his catalog as star 34 Tauri, and again in 1756 by the German mathematician and astronomer Tobias Mayer (1723–62).

In 1782 Bode turned his attention to a new publishing format, creating a star atlas he named *Die Gestirne*. It was the third in a succession of atlases that began in 1792, when Flamsteed printed his famous *Atlas coelestis* in response to what he felt were problematic constellations in JOHANNES BAYER's *Uranometria*. Next in line came *Atlas Céleste de Flamstéed*, (Flamsteed's Celestial Atlas), printed in 1776, by the French astronomer Jean Fortin (ca. 1750–1831) in homage to Flamsteed's atlas. Bode's was the last to use Flamsteed's data, although Bode did add a nebulae he apparently discovered, named Bode I, which today is believed to possibly be the open cluster IC1434.

With the discovery of Herschel's planet in 1781, excitement about Bode's Law was rekindled, and astronomers became even more convinced that a planet had to exist in the gap between Mars and Jupiter at 2.8 AU. A worldwide team of astronomers cooperated to systematically divide the night sky and search for the missing planet. In 1801 the Sicilian monk GIUSEPPE PIAZZI, astronomer and director of the Palermo Observatory, found the next best thing to a missing planet—the minor planet or asteroid Ceres, which he named Cerere di Ferdinando for the goddess of grain and the patron goddess of Sicily and for King Ferdinand, Piazzi's patron. This was the first asteroid ever discovered, and remains the largest known, with a diameter of 623 miles, compared with the smallest planet, Pluto, which has a diameter of 1,457 miles, Earth, which is 7,926 miles, and the largest planet, Jupiter, which rolls in at 88,700 miles. At 2.77 AU, it was nearly exactly where Bode's Law predicted a planet should be.

Never far from publishing, Bode returned again to the monumental task of creating a new atlas, this time designing the biggest star atlas ever made. Bode did not research the validity of any information, and he did not edit anything out, but rather made a compilation of all known star atlases and catalogs. The book consists of 20 double-page plates, including all of the constellations that were ever suggested, with Bode's new constellations Apparatus Chemica, Custos Messium, Felis, Globus Aerostaticus, Honores Frederici, and Officina Typographica, none of which still exist, plus new figures for old constellations, more than 17,000 stars, and approximately 2,000 nebulae discovered by Herschel. The constellations were represented by beautiful, stylistically new artwork, making it unlike any atlas ever published. Bode's *Uranographia* is considered the book that marked the end of an era in star atlases, giving way to maps that simply connected the stars with lines, rather than depicting the constellations with artistic figures.

Bode revised and reprinted *Die Gestirne* in 1805, publishing it as *Vorstellung der Gestirne* (Introduction of the Luminaries). In addition to the stars he added, Bode customarily revised the artwork of the constellations. This became known as the Bode-Flamsteed atlas, and includes 34 maps featuring 26 constellations that Bode could see from his observatory in Germany, plus planispheres, hemispheres, star clusters, and nebulas.

A long and full career at the Berlin Observatory ended after nearly 40 years when Bode retired from his position as director in 1825. He continued to publish *Berliner Astronomisches Jahrbuch* until his death in November of 1826, just one year after his retirement.

Bode's Law was confirmed by the discovery of Uranus in 1781 and the Ceres asteroid and subsequent asteroid belt in 1801. But when Neptune was discovered in 1846 at 30.06 AU, where no planet should have been, the idea fell from grace with the astronomical community. Not until Pluto's discovery in 1930, at 39.44 AU, near the predicted 38.8 distance of Bode's Law, did it come back into favor. To this date, there remains no explanation of why Bode's Law works.

The Spiral Galaxy M81, discovered in 1774 by the 25-year-old Bode, is known as Bode's Galaxy, and M82 and M81 are often referred to as Bode's Nebulae or Bode's Galaxies. The asteroid first discovered on August 6, 1923, by Karl Wilhelm Reinmuth (1892–1979) in Heidelberg, Germany, was named Asteroid Bodea, and in 1935, a crater on the Moon was named in Bode's honor.

Bok, Bartholomeus Jan (Bart Bok)
(1906–1983)
Dutch/American
Astronomer

In the 1930s and 1940s this Dutch astronomer conducted detailed investigations of the star-forming regions of the Milky Way Galaxy. He

The Dutch-American astronomer Bart Bok at the University of Arizona's Steward Observatory in Tucson, a major astronomical facility under his direction between 1966 and 1970. Earlier in his career, he discovered the small, cool (~10 kelvins) dark nebulas now called "Bok globules" in his honor. (*AIP Emilio Segrè Visual Archives, Physics Today Collection*)

made careful studies of interstellar dust and gas and their relation with respect to star formation. In 1947 he discovered the small, near-spherical, dense dark nebulas, now called "Bok globules," that contain enough interstellar material to eventually condense into star clusters.

Bartholomeus (Bart) Jan Bok was born on April 28, 1906, in Hoorn, the Netherlands, to Sergeant Major Jan Bok and Gesina Annetta van der Lee Bok. The family moved to The Hague in 1918, where Bok attended high school. At age 13 Bok knew he wanted to be an astronomer, having been strongly influenced in science by one of his schoolteachers. Throughout his high

school years he was an active amateur astronomer and became an admirer of HARLOW SHAPLEY. Upon completing high school with high honors, he entered the University of Leiden in 1924 and remained there until 1927. Leiden proved to be a stimulating environment for this young astronomer, where one of his classmates was GERARD PETER KUIPER and among his professors were ENJAR HERTZSPRUNG and JAN HENDRIK OORT. His studies at the University of Leiden provided Bok with a strong foundation in classical astronomy. In 1927 Bok went to the University of Groningen to pursue his doctoral degree.

In summer 1928 Bok attended the Third General Assembly of the International Astronomical Union in Leiden, where two special things occurred that would greatly influence the rest of his life. First, he met Harlow Shapley, who was then director of the Harvard College Observatory. Shapley was impressed with the young Dutch astronomer and invited Bok to come the following year to do his research on the Milky Way at the Harvard College Observatory. Second, Bok met and immediately fell in love with the American astronomer Priscilla Fairfield (1896–1975). Despite the fact she was 10 years his senior, Bok decided to marry Fairfield, whom he called "the love of his life." She eventually accepted his proposal of matrimony and they were wed in Troy, New York, on September 9, 1929—just two days after Bok arrived in the United States to work at Harvard College Observatory.

Bok's marriage provided him with a foundation upon which he could build his career as a world-class astronomer. As an astronomer herself, Priscilla Fairfield Bok was an eager collaborator on her husband's astronomical projects. As a devoted wife, she was always there to provide the gentle, introspective judgment needed to balance his often spontaneous and unrestrained approach to new opportunities and ideas. Bok found the Harvard Observatory a most exciting place to do research. The couple's first child, a son, was born in August 1930. After conducting research at the Harvard College Observatory, Bok successfully defended his doctoral dissertation, "A Study of the Eta Carinae Region," at the University of Groningen on July 6, 1932. He became an assistant professor at Harvard in 1933. Later that year, the couple's second child, a daughter, arrived.

In the 1930s he performed detailed studies of the Milky Way and published an important monograph, *The Distribution of the Stars in Space*, in 1937. The following year he became a naturalized citizen of the United States, and in 1939 he earned a promotion to associate professor of astronomy at Harvard College Observatory. Starting in about 1937, Bok collaborated with his wife to write a book called *The Milky Way*. Five years later, the published book served as a much-needed comprehensive treatment of galactic astronomy and became a very popular work that went through five editions. As a result, this collaborative writing effort attracted many bright young minds to the field of astronomy. In 1946 Harlow Shapley appointed Bok as the associate director of the Harvard College Observatory—a position he kept until Shapley retired in 1952.

Bart Bok became a full professor of astronomy at the Harvard College Observatory in 1947. That same year he also made his most famous astronomical discovery—the small, dark, cool (about 10 kelvins), almost circular, interstellar clouds observable against the background of stars or luminous gas regions in the Milky Way. Because they contain enough interstellar material, these protostar regions, or "Bok globules," can be thought of as candidate stellar nurseries for some of the Galaxy's lower-mass stars.

Starting in the late 1940s Bok became involved in the process of establishing new observatories in different parts of the world. For example, in February 1950 he and his family traveled to South Africa and set up a new telescope at the Harvard Boyden Station. Bok especially enjoyed this opportunity to view regions of the Milky Way and the Magellanic Clouds from Earth's Southern Hemisphere.

Several years after Shapley retired as director of the Harvard College Observatory, Bok resigned his professorship and departed the institution he served for so many years. In January 1957 he and his wife went to Australia, where he accepted a position as the director of the Mount Stromlo Observatory, near Canberra. He also became a professor of astronomy at the Australian National University. Just before returning to the United States in 1966, Bok helped establish Australia's Siding Spring Observatory.

When the couple returned to the United States, Bok become director of the University of Arizona's Steward Observatory in Tucson and kept that position until he retired in 1970. He suffered a crushing blow when his wife died on November 19, 1975. He remained withdrawn for several years, emerging from his self-imposed solitude in 1977 to receive the prestigious Bruce Medal of the Astronomical Society of the Pacific. He resumed some of his astronomical activities in the early 1980s, but died suddenly on August 5, 1983, while making plans to attend an upcoming astronomical conference.

⊠ **Boltzmann, Ludwig**
(1844–1906)
Austrian
Physicist

This brilliant, though troubled, Austrian physicist developed statistical mechanics and the kinetic theory of gases. His pioneering work affected many areas of science, including thermodynamics and astronomy. In the 1870s and 1880s Boltzmann collaborated with JOSEF STEFAN in creating an important physical principle, now called the Stefan-Boltzmann Law,

In the 1870s and 1880s Ludwig Boltzmann, the brilliant Austrian physicist, collaborated with his mentor, Josef Stefan, to create the Stefan-Boltzmann law—an important physical principle now used by astronomers and astrophysicists to relate the total radiant energy output (or luminosity) of a star to the fourth power of its absolute temperature. *(University of Vienna, courtesy of AIP Emilio Segrè Visual Archives)*

that relates the total radiant energy output (or luminosity) of a star to the fourth power of its absolute temperature.

Ludwig Boltzmann was born on February 20, 1844, in Vienna, Austria. His father was a government official involved in taxation. He studied at the University of Vienna, earning a doctoral degree in physics in 1866. His dissertation was supervised by Stefan and involved the kinetic theory of gases. Following graduation, Boltzmann joined his academic adviser as an assistant at the university.

Starting in 1869 he held a series of professorships in mathematics or physics at a number of European universities. Boltzmann's physical restlessness mirrored the hidden torment of his mercurial personality that would swing him suddenly from intellectual contentment into a deep depression. His academic appointments included the University of Graz (1869–73; 1876–79); the University of Vienna (1873–76; 1894–1900; 1902–06); the University of Munich (1889–93), and the University of Leipzig (1900–02). On the positive side, his movement from institution to institution brought him into contact with many great 19th-century scientists, including ROBERT WILHELM BUNSEN, RUDOLF JULIUS EMMANUEL CLAUSIUS, and GUSTAF ROBERT KIRCHHOFF. Boltzmann was a popular lecturer and entertained diverse academic interests that encompassed physics, mathematics, chemistry, and philosophy.

As a physicist, he is best remembered for his invention of statistical mechanics and its pioneering application in thermodynamics. His theoretical work helped scientists connect the properties and behavior of individual atoms and molecules (viewed statistically on the microscopic level) to the bulk properties and physical behavior (such as temperature and pressure) that a substance displayed when examined on the familiar macroscopic level used in classical thermodynamics. One key to this development was Boltzmann's equipartition of energy principle. It stated that the total energy of an atom or molecule is equally distributed, on an average basis, over its various translational kinetic energy modes. Boltzmann postulated that each energy mode had a corresponding degree of freedom. He further theorized that the average translational energy of a particle in an ideal gas was proportional to the absolute temperature of the gas. This principle provided an important connection between the microscopic behavior of an incredibly large number of atoms or molecules and macroscopic physical behavior (such as temperature) that physicists could easily measure. The constant of proportionality in this relationship is now called Boltzmann's constant, for which the symbol "k" is usually assigned.

Boltzmann developed his kinetic theory of gases independent of the work of the great Scottish physicist James Clerk Maxwell (1831–79). Their complementary activities resulted in the Maxwell-Boltzmann distribution—a mathematical description of the most probable velocity of a gas molecule or atom as a function of the absolute temperature of the gas. The greater the absolute temperature of the gas, the greater will be the average velocity (or kinetic energy) of individual atoms or molecules. Considering a container filled with oxygen (O_2) gas at a temperature of 300 kelvins (K), for example, the Maxwell-Boltzmann distribution predicts that the most probable velocity of an oxygen molecule in that container is 395 meters per second.

In the late 1870s and early 1880s Boltzmann collaborated with Stefan in developing the very important physical principle that describes the amount of thermal energy radiated per unit time by a blackbody. In physics, a blackbody is defined as a perfect emitter and perfect absorber of electromagnetic radiation. All objects emit thermal radiation by virtue of their temperature. The hotter the body, the more radiant energy it emits. By 1884 Boltzmann had finished his theoretical work in support of Stefan's observations of the thermal radiation emitted by blackbody radiators at various temperatures. The result of their collaboration was the famous Stefan-Boltzmann law of thermal radiation. A physical principle of great importance to both physicists and astronomers, this law states that the luminosity of a blackbody is proportional to the fourth power of the body's absolute temperature. The constant of proportionality for this relationship is called the Stefan-Boltzmann constant, and scientists usually assign this constant the symbol "σ"—a lowercase sigma in the Greek alphabet. The Stefan-Boltzmann law tells scientists that if the absolute temperature of a blackbody doubles, for example, its luminosity will increase by a factor of 16.

The Sun and other stars closely approximate the thermal radiation behavior of blackbodies, so astronomers often use the Stefan-Boltzmann law to approximate the radiant energy output (or luminosity) of a stellar object. A visible star's apparent temperature is also related to its color. The coolest red stars (called stellar spectral class "M" stars), such as Betelgeuse, have a typical surface temperature of less than 3,500 K. However, very large hot blue stars (called spectral class "B" stars), like Rigel, have surface temperatures ranging from 11,000 to 28,000 K.

In the mid-1890s Boltzmann related the classical thermodynamic concept of entropy (symbol S), recently introduced by Rudolf Clausius (1822–88), to a probabilistic measurement of disorder. In its simplest form, $S = k \ln \Omega$. Here, Boltzmann boldly defined entropy (S) as a natural logarithmic function of the probability of a particular energy state (Ω). The symbol k again represents Boltzmann's constant. This important equation is even engraved on his tombstone.

Toward the end of his life, Boltzmann encountered very strong academic and personal opposition to his atomistic views. Many eminent European scientists of this period could not grasp the true significance of the statistical nature of his reasoning. One of his most bitter professional and personal opponents was the Austrian physicist Ernest Mach (1838–1916), who as chair of history and philosophy of science at the University of Vienna essentially forced the brilliant but depressed Boltzmann to leave that institution and move to the University of Leipzig in 1900.

Boltzmann continued to defend his statistical approach to thermodynamics and his belief in the atomic structure of matter, but he could not handle having to continually defend his theories against mounting opposition in certain academic circles. So, while on holiday with his wife and daughter at the Bay of Duino, near

Trieste, Austria (now part of Italy), Boltzmann hanged himself in his hotel room on October 5, 1906, as his wife and child enjoyed a swim. Was the tragic suicide of this brilliant physicist the result of a lack of professional acceptance of his work or the self-destructive climax of his life-long battle with bipolar disorder? No one can say for sure. Sadly, at the time of his death, other great physicists were performing key experiments that would soon prove Boltzmann's atomistic philosophy and life's work correct.

⊠ **Bradley, James**
(1693–1762)
British
Astronomer

James Bradley was the 18th-century English minister turned astronomer who made two important astronomical discoveries. His first was the aberration of starlight, announced in 1728. It represented the first observational proof that Earth moves in space, confirming the Copernican hypothesis. His second discovery was that of nutation—a small wobbling variation in procession of Earth's rotational axis, reported in 1748, after 19 years of careful observation. Upon the death of the legendary Sir Edmund Halley in 1742, Bradley was appointed as the third Astronomer Royal of England.

He was born sometime in March 1693 (the precise date is not recorded) in Sherborne, England. As a young child, Bradley was being prepared for a life in the ministry when he suffered an attack of smallpox. His uncle, the Reverend James Pound, introduced Bradley to astronomy while nursing him back to health. The Reverend Pound was an avid amateur astronomer and friend of Halley. Before long, Bradley began to demonstrate his great observational skills as he practiced amateur astronomy while studying to become a vicar in the Church of England. By 1718 Bradley's talents sufficiently impressed

England's second Astronomer Royal, and he encouraged the young man to pursue astronomy as a career. Halley also recommended that Bradley be elected as a fellow of the Royal Society in London.

As a result of this encouragement, Bradley abandoned any thoughts of an ecclesiastical career. In 1721 he accepted an appointment to the Savilian Chair of Astronomy at the University of Oxford. Throughout his career as an astronomer, Bradley concentrated his efforts on making improvements in the precise observation of stellar positions. For example, in 1725 he started on his personal quest to be the first astronomer to measure stellar parallax—the very small apparent change in the position of a star when viewed by an observer on Earth as the planet reaches opposing extremes of its elliptical orbit around the Sun. However, the difficult measurement of stellar parallax managed to baffle and elude all astronomers, including Bradley, until Friedrich Wilhelm Bessel accomplished the task in 1838. Nevertheless, while searching for stellar parallax, Bradley accidentally discovered another very important physical phenomenon that he called the aberration of starlight.

The aberration of starlight is the tiny apparent displacement of the position of a star from its true position due to a combination of the finite velocity of light and the motion of an observer across the path of the incident starlight. For example, an observer on Earth would have Earth's orbital velocity around the Sun, approximately 30 kilometers per second. As a result of this orbital motion effect, during the course of a year the light from a "fixed" star appears to move in a small ellipse around its mean position on the celestial sphere.

By a stroke of good fortune, on December 3, 1725, Bradley began observing the star Gamma Draconis, which was within the field of view of his large telescope pointed at zenith. He was forced to look at the stars overhead because his refractor telescope was assembled in a long chimney, with its telescopic lens above rooftop and its objective (viewing lens) below, in the unused fireplace. Bradley was actually attempting to become the first astronomer to experimentally detect stellar parallax, and he used this relatively immobile telescope pointed at zenith to minimize atmospheric refraction. He viewed Gamma Draconis again on December 17 and was quite surprised to find that the position of the star had shifted a tiny bit. But the apparent shift was not by the amount or in the direction expected, if he was observing the phenomenon of parallax. He continued these careful observations over the course of an entire year and found that the position of Gamma Draconis actually traced out a very small ellipse.

Bradley puzzled over the Gamma Draconis problem for about three years, but still could not satisfactorily explain the cause of the displacement. Then, in 1728, while he was sailing on the Thames River, inspiration suddenly struck as he watched a small wind-direction flag on a mast change directions. Curious, Bradley asked a sailor whether the wind had changed directions. The sailor informed him that the direction of wind had not changed. So Bradley realized that the changing position of the small flag was due to the combined motion of the boat and of the wind. This connection helped him solve the Gamma Draconis displacement mystery.

Based on the previous attempts by the Danish astronomer Olaus Roemer (1644–1710) to measure the velocity of light, Bradley hypothesized correctly that light must travel at a very rapid, finite velocity. He also further speculated that a ray of starlight traveling from the top of his tall telescope to the bottom would have its image displaced in the objective ever so slightly by Earth's orbital motion. Bradley finally recognized that the mysterious annual shift of the apparent position of Gamma Draconis, first in one direction and then in another, was due to the combined motion of starlight and Earth's annual orbital motion around the Sun. So it happened that early in his career as an astronomer, Bradley made a major discovery—the aberration of starlight. This discovery was the first observational proof that Earth moved in space, confirming the Copernican hypothesis.

Bradley soon constructed a new telescope, one not placed in a chimney, so he could make precise observations of other stars. Starting in 1728 and continuing for many years, Bradley precisely recorded many stellar positions. His efforts greatly improved upon the earlier work of many notable astronomers, including England's first Astronomer Royal, John Flamsteed (1646–1719). His ability to make precise measurements led him to a second important astronomical discovery. He found that Earth's axis experienced small periodic shifts that he called "nutation" after the Latin word *nutare*, "to nod." The new phenomenon, characterized as an irregular wobbling superimposed upon the overall precession of Earth's rotational axis, is primarily caused by the gravitational influence of the Moon, as it travels along its somewhat irregular orbit around Earth. Bradley waited almost two decades, until 1748, to publish this discovery, because he wanted to carefully study and confirm the very subtle shifts in stellar positions before announcing his findings.

When Halley died in 1742, Bradley was appointed as England's third Astronomer Royal. He most competently served in that distinguished post until his death on July 13, 1762, in Chalford, England.

⊠   **Brahe, Tycho (Tÿge Brahe)**
(1546–1601)
Danish
*Astronomer, Mathematician,
Cartographer*

History has recorded that if any of the great scientists had a nose for astronomy, it was Tycho

Brahe. Born December 14, 1546, to an aristocratic family in Knudstrup, Denmark, Tÿge Brage spent the better part of his 54 years becoming both a master of science and a masterful entrepreneur. (He adopted the Latin form of Tÿge, Tycho, when he was about 15.)

Brahe had the opportunity to study the sciences thanks in large part to his father's wealth and power. Otte Brahe was a member of Denmark's elite ruling oligarchy, Rigsraad, and was the first in an impressive line of patrons for the studious young Brahe. On April 19, 1559, at age 12, Tycho began his lessons in earnest at the University of Copenhagen. It was here, on August 21, 1560, that he experienced a solar eclipse that, according to philosopher Pierre Gassendi, turned Brahe's interest toward astronomy.

After three years in Copenhagen, Brahe left to attend the University of Leipzig, where for the next three years he focused on humanities and astronomy. With his father's funds available to pay for travel and continued studies, Brahe explored several universities over the following two years, including Wittenberg, which he left because of an outbreak of the plague, Rostock, and finally Basel.

It was while at Rostock that the hotheaded young Brahe took on fellow noble scholar, Manderup Parsbjerg, in a heated debate over which of the two was the better mathematician. This led to a duel with swords, and on December 29, 1566, by the time their score was settled the 20-year-old Tycho had forever lost the end of his nose.

Otte Brahe's death in 1571 left Tycho a substantial estate, consisting of assets including more than 200 farms. Even after dividing the assets with his brother, the 25-year-old Brahe had enough funds to live comfortably for the rest of his life.

Truly a scholar of science, Brahe's breadth of expertise included astronomy, astrology, meteorology, iatrochemistry, cartography, instrumentation, pharmacology, and mathematics. His first major astronomical discovery occurred in 1572, when he found a new star in the constellation Cassiopeia,

which he named Stella Nova. In truth, this was a supernova, one of three famous supernova events in the history of astronomy—one occurring in 1054; and another in 1604 observed by GALILEO GALILEI and JOHANNES KEPLER as a nova, then finally determined to be a supernova some 350 years later by WILHELM HEINRICH WALTER BAADE.

By mid-1573 Brahe met Kirsten Jörgensdatter, a priest's daughter, and the two married and eventually had a family of eight children. By 1574 it was clear that Brahe was ready to make himself, and his science, known to the world. He presented his first lecture, on mathematics, at the University of Copenhagen. By 1577, at age 30, he was offered what many would have considered the enviable position of rector at the university.

But six months earlier Brahe had received a notice from a new patron, King Frederick II, offering him an island, Hven, on which Tycho could build an observatory and leave his legacy. The castle Uranienborg was meant to be Brahe's astronomical haven. All expenses would be provided by the king to build, maintain, and prosper, for the rest of Brahe's life. The foundation was started with a helping hand from the French ambassador, Charles Dançay, who placed the first stone in the castle's wall.

Over the next few years, Tycho participated in more than stargazing. He continued amassing assets and significant revenue in an effort to secure his status as nobility in his own right. He was granted farms, a fiefdom, and the position of canon in Roskilde cathedral in 1579, which he held for nearly 20 years.

But in 1597 Brahe's time at Hven came to an end, and on April 22, after losing royal support for his work, the astronomer left his haven of 21 years, writing, "boring weather on a boring day." After a trip to Wandsbech, near Hamburg, Brahe settled in Prague. The following year, Brahe wrote *Mechanica*, in which he described many of the extraordinary instruments he had designed over the past 20 years. In an attempt to elicit support for his work, Brahe dedicated the piece to the Holy Roman Emperor Rudolf II. And it worked. The emperor spent an estimated 10,000 guldens on a house for Brahe in Prague, and in 1599 appointed him imperial mathematician.

It was during the following year that Brahe, who was a patron himself, employed Kepler. Brahe's system of explaining the Earth and its place in the universe was flawed, but it gave Kepler an extraordinary opportunity that ultimately concluded in proving the correct nature of our solar system.

Ever concerned with his status, Brahe applied for the rank of nobility in February 1601. Several months later, while having dinner with Baron Peter Vok Rosenberg in Prague, Brahe became quite ill. Over the next few days, his condition worsened with fever, and on October 24, 1601, the noted scientist and mathematician uttered his last words to Kepler: "Don't let me seem to have lived in vain." On November 11, 1601, Brahe was buried in Prague as nobility.

Kepler's work ensured that Brahe got his final wish. While Kepler fundamentally supported, and helped prove, the Copernican theory of a heliocentric universe, disproving Brahe's system of the planets revolving around the Sun and all revolving around the Earth, it was Brahe's extremely accurate observations of Mars that led to Kepler's conclusions and the creation of the laws of planetary motion.

Braun, Wernher Magnus von
(1912–1977)
German/American
Rocket Engineer

The brilliant rocket engineer Wernher von Braun turned the impossible dream of interplanetary space travel into a reality. He started by developing the first modern ballistic missile, the liquid-fueled V-2 rocket, for the German army. He then assisted the United States by developing a family of ballistic missiles for the U.S. Army

and later a family of powerful space launch vehicles for the National Aeronautics and Space Administration (NASA). One of his giant *Saturn V* rockets successfully sent the first human explorers to the Moon's surface in July 1969, as part of NASA's Apollo Project.

Braun was born on March 23, 1912, in Wirsitz, Germany (now Wyrzsk, Poland). He was the second of three sons of Baron Magnus von Braun and Baroness Emmy von Quistorp. As a

The brilliant rocket engineer Wernher von Braun poses with a Saturn IB launch vehicle perched on its pad at the Kennedy Space Center in 1968. Von Braun devoted his professional life to creating ever more powerful liquid-propellant rockets. In 1958 he quickly adapted a U.S. Army rocket to carry the first U.S. satellite (*Explorer 1*) into orbit around Earth. Then in the 1960s, he developed the family of powerful Saturn rockets for NASA. Starting in July 1969, his giant *Saturn V* rockets rumbled skyward, successfully sending teams of human explorers to the Moon's surface in timely fulfillment of President John F. Kennedy's bold lunar exploration commitment. *(Courtesy of NASA/Marshall Space Flight Center)*

young man, he enjoyed the science fiction novels of Jules Verne and H. G. Wells, which kindled his lifelong interest in space exploration. HERMANN JULIUS OBERTH's book *The Rocket into Interplanetary Space* (1923) introduced him to rockets. In 1929 Braun received additional inspiration from Oberth's realistic "cinematic" rocket that appeared in Fritz Lang's (1890–1976) motion picture *Die Frau im Mond* (The woman in the moon). That same year, he became a founding member of *Verein für Raumschiffahrt*, or VFR, the German Society for Space Travel. Through the VFR, he came into contact with Oberth and other German rocket enthusiasts. He also used the VFR to carry out liquid propellant rocket experiments near Berlin in the early 1930s.

He enrolled at the Berlin Institute of Technology in 1930. Two years later, at age 20, he received his bachelor's degree in mechanical engineering. In 1932 he entered the University of Berlin, where as a graduate student he received a modest rocket research grant from a German army officer, Artillery Captain Walter Dornberger (1895–1980). This effort ended in disappointment when Braun's new rocket failed to function before an audience of military officials in 1932. But the young graduate so impressed Dornberger that he hired him as a civilian engineer to lead the German army's new rocket artillery unit. In 1934 Braun received his Ph.D. degree in physics from the University of Berlin. A portion of his doctoral research involved testing small liquid-propellant rockets.

By 1934 Braun and Dornberger had assembled a rocket development team totaling 80 engineers at Kummersdorf, a test facility located some 100 kilometers south of Berlin. That year, the Kummersdorf team successfully launched two new rockets, nicknamed Max and Moritz after German fairy tale characters. Max was a modest liquid-fueled rocket, and Moritz flew to altitudes of about two kilometers and carried a gyroscopic guidance system. These successes earned the group additional research work from

the German military and a larger, more remote test facility near the small coastal village of Peenemünde, on the Baltic Sea.

The Kummersdorf team moved to Peenemünde in April 1937. Soon after relocation, von Braun's team began test firing the A-3 rocket, which was plagued with many problems. As World War II approached, the German military directed Braun to develop long-range military rockets and he accelerated design efforts for the much larger A-4. At the same time, an extensive redesign of the troublesome A-3 rocket resulted in the A-5 rocket.

When World War II started in 1939 Braun's team began launching the A-5. The results of these tests gave Braun the technical confidence he needed to pursue final development of the A-4 rocket—the world's first long-range, liquid-fueled military ballistic missile.

The A-4's state-of-the-art liquid-propellant engine burned alcohol and liquid oxygen, and was approximately 14 meters long and 1.66 meters in diameter. These dimensions were intentionally selected to produce the largest possible mobile military rocket capable of passing through the railway tunnels in Europe. Built within these design constraints, the A-4 rocket had an operational range of up to 275 kilometers.

Braun's first attempt to launch the A-4 rocket at Peenemünde occurred in September 1942 and ended in disaster. As its engine came to life, the test vehicle slowly rose from the launch pad for about one second. Then the engine's thrust suddenly tapered off, causing the rocket to slip back to Earth. As its guidance fins crumpled, the doomed rocket toppled over and destroyed itself in a spectacular explosion. But success finally occurred on October 3, 1942. In its third flight test, the experimental A-4 vehicle roared off the launch pad and reached an altitude of 85 kilometers while traveling a total distance of 190 kilometers downrange from Peenemünde. The successful flight marks the birth of the modern military ballistic missile.

Impressed, senior German military officials immediately ordered the still-imperfect A-4 rocket into mass production. By 1943 the war was going badly for Nazi Germany, so Adolf Hitler decided to use the A-4 military rocket as a vengeance weapon against Allied population centers. He named this long-range rocket the Vergeltungwaffe-Zwei (Vengeance Weapon Two), or V-2. In September 1944 the German army started launching V-2 rockets armed with high-explosive warheads (about one metric ton) against London and southern England. Another key target was the Belgian city of Antwerp, which served as a major supply port for the advancing Allied forces. More than 1,500 V-2 rockets hit England, killing about 2,500 people and causing extensive property damage. Another 1,600 V-2 rockets attacked Antwerp and other continental targets. Although modest in size and payload capacity when compared to modern military rockets, Braun's V-2 significantly advanced the state of rocketry and was a formidable, unstoppable weapon that struck without warning. Fortunately, the V-2 rocket arrived too late to change the outcome of World War II in Europe.

In May 1945 Braun, along with some 500 of his colleagues, fled westward from Peenemünde to escape the rapidly approaching Russian forces. Bringing along numerous plans, test vehicles, and important documents, he surrendered to U.S. forces at Reutte, Germany. Over the next few months, U.S. intelligence teams, under Operation Paperclip, interrogated many German rocket personnel and sorted through boxes of captured documents.

Selected German engineers and scientists accompanied Braun and resettled in the United States to continue their rocketry work. At the end of the war, American troops also captured hundreds of intact V-2 rockets. After Germany's defeat, Braun and his colleagues arrived at Fort Bliss, Texas, to help the U.S. Army (under Project Hermes) reassemble and launch captured German V-2 rockets from the White Sands Proving Ground in southern New Mexico. During this period, Braun also married Maria von Quistorp on March 1, 1947. Relocated German engineers and captured V-2 rockets provided a common technical heritage for the major military missiles initially developed by both the United States and the Soviet Union during the cold war.

In 1950 the U.S. Army moved Braun and his team to the Redstone Arsenal near Huntsville, Alabama. At the Redstone Arsenal, Braun supervised development of early ballistic missiles, such as the Redstone and the Jupiter, for the U.S. Army. These missiles descended directly from the technology of the V-2 rocket. As the cold war arms race heated up, the U.S. Army made Braun chief of its ballistic weapons program.

Starting in fall 1952, Braun also provided technical support for the production of a beautifully illustrated series of visionary space travel articles that appeared in *Collier* magazine. The series caught the eye of the American entertainment genius Walt Disney (1901–66). By the mid-1950s Braun became a nationally recognized space travel advocate through his frequent appearances on television. Along with Walt Disney, Braun served as a host for an inspiring three-part television series on human space flight and space exploration. Thanks to Braun's influence, when the Disneyland theme park opened in southern California in 1955 (the year Braun became a U.S. citizen), its "Tomorrowland" section featured a "Space Station X-1" exhibit and a simulated rocket ride to the Moon. A giant, 25-meter tall, needle-nosed rocket ship (personally designed by Willy Ley [1906–69] and Braun) also greeted visitors to the Moon ride. During the mid-1950s, the Disney–Braun relationship introduced millions of Americans to the possibility of the Space Age, which was actually less than two years away.

On October 4, 1957, the Space Age arrived when the former Soviet Union launched *Sputnik 1*, Earth's first artificial satellite. Following the successful Soviet launches of the *Sputnik 1* and

Sputnik 2 satellites and the disastrous failure of the first American Vanguard satellite mission in late 1957, Braun's rocket team at Huntsville was given less than 90 days to develop and launch the first U.S. satellite. On January 31, 1958, a modified U.S. Army Jupiter C missile, called the *Juno 1* launch vehicle, rumbled into orbit from Cape Canaveral Air Force Station. Under Braun's direction, this hastily converted military rocket successfully propelled *Explorer 1* into Earth orbit. The *Explorer 1* satellite was a highly accelerated joint project of the Army Ballistic Missile Agency (AMBA) and the Jet Propulsion Laboratory (JPL) in Pasadena, California. The spacecraft carried an instrument package provided by JAMES ALFRED VAN ALLEN, who discovered Earth's trapped radiation belts.

In 1960 the U.S. government transferred administration of Braun's rocket development center at Huntsville from the U.S. Army to a newly created civilian space agency called the National Aeronautics and Space Administration (NASA). Within a year, President John F. Kennedy made a bold decision to put U.S. astronauts on the Moon within a decade. The president's decision gave Braun the high-level mandate he needed to build giant new rockets. On July 20, 1969, two Apollo astronauts, Neil Armstrong (born 1930) and Edwin (Buzz) Aldrin (born 1930), became the first human beings to walk on the Moon. They had gotten to the lunar surface because of Braun's flawlessly performing *Saturn V* launch vehicle.

As director of NASA's Marshall Space Flight Center in Huntsville, Alabama, Braun supervised development of the Saturn family of large, powerful rocket vehicles. The *Saturn I, Saturn IB,* and *Saturn V* vehicles were well-engineered expendable rockets that successfully carried teams of U.S. explorers to the Moon between 1968 and 1972 and performed other important human space flight missions through 1975.

But just after the first human landings on the Moon in 1969, NASA's leadership decided to move Braun from Huntsville and assigned him to a Washington, D.C., headquarters "staff position" in which he could "perform strategic planning" for the agency. Braun, now an internationally popular rocket scientist, had openly expressed desires to press on beyond the Moon mission and send human explorers to Mars. This made him a large political liability in a civilian space agency that was now faced with shrinking budgets. In less than two years at NASA headquarters, Braun decided to resign from the civilian space agency. By 1972 the rapidly declining U.S. government interest in human space exploration clearly disappointed him. He then worked as a vice president of Fairchild Industries in Germantown, Maryland, until illness forced him to retire on December 31, 1976. He died of a progressive cancer in Alexandria, Virginia, on June 16, 1977.

Braun's aerospace engineering skills, leadership abilities, and technical vision formed a unique bridge between the dreams of the founding fathers of astronautics: KONSTANTIN EDVARDOVICH TSIOLKOVSKY, ROBERT HUTCHINGS GODDARD, and Oberth and the first golden age of space exploration, which took place from approximately 1960 to 1989. In this unique period, human explorers walked on the Moon and robot spacecraft visited all the planets in the solar system (except tiny Pluto) for the first time. Much of this was accomplished through Braun's rocket engineering genius, which helped turn theory into powerful, well-engineered rocket vehicles that shook the ground as they broke the bonds of Earth on their journey to other worlds.

⊠ **Bruno, Giordano (Filippo Bruno)**
(1548–1600)
Italian
Philosopher

While many would argue that Giordano Bruno was one of the great philosophical influences of all time, it cannot be overlooked that his beliefs about the universe and our place within it were in great part responsible for his execution at the hands of the Roman Inquisition.

In 1548, just five years after the death of NICOLAS COPERNICUS, Filippo Bruno was born in Nola, Italy, just outside Naples. Very little is known about his early childhood. His father was a soldier in Naples and the family was presumably poor, yet Bruno managed to begin his education in logic in Naples at about age 11. But it was his studies in theology and philosophy, which he started at age 15, that began his life of piety, and eventually led to discord with the church.

From 1563 to 1576 Bruno was a member of the St. Dominico monastery in Naples. While there, he changed his name to Giordano and began studying both ancient and current philosophy. It was also during this time that he learned of the works of Copernicus. Outspoken and perhaps dangerously frank, Bruno had the innate ability to easily offend those around him. There is little doubt that he started exhibiting these characteristics while in the monastery.

By the time he was 28, he had angered the church on more than one occasion. His first offense involved giving away images of saints. Charges were brought against him, most likely as a disciplinary tactic to keep him in line (they were dropped almost immediately), but in 1576 they were brought back up in conjunction with a new charge of heresy for reading philosophical texts that were out of favor with the church. Threatened with the possibility of proceedings against him in Rome as a heretic, Bruno fled to avoid persecution, giving up all affiliation with the Dominicans.

Over the next 16 years, Bruno led the life of a Renaissance man, wandering through Europe, espousing his philosophy, and gaining favor with royalty for his unique ways of thinking. France's King Henri III was so intrigued with Bruno's teachings in mnemonics that he helped the young philosopher obtain a university lec-

turing position. Bruno also began publishing books, in which he wove his philosophy through dialogue and questioned the then-known realities of the church, God, and the universe.

Of particular importance is his book *The Ash Wednesday Supper* (1584), which is typically cited as Bruno's most significant perspective on astronomy. Following Copernicus's lead, Bruno defied the church's teachings, based purely on ARISTOTLE, and supported the theory that the Earth is not the center of the universe. Bruno went even further to argue that the universe is infinite, and imagined that within this infinite universe there exists an infinite number of suns and earths. He believed that our Sun and Earth, when looked at from other vantage points, would look like any other stars. Using logic, Bruno threatened current knowledge. Unfortunately, this was tantamount to challenging the foundation of the church, thus defying God.

Bruno's grace with the French aristocracy helped him in his subsequent travels from 1583 to 1592. His audience with royalty, including Queen Elizabeth, and his nearly 20 books and plays contributed significantly to the acceptance of debating his philosophical viewpoints in intellectual circles. In March 1592 he agreed to return to Venice by invitation of another aristocrat, Zuane Mocenigo, who proclaimed interest in learning about mnemonics.

Apparently, Mocenigo was hoping mnemonics would also involve black magic. When Bruno was both unwilling and unable to deliver the secrets of the occult, Mocenigo turned him over to the Venetian Inquisition for blasphemy and heresy, on May 23, 1592. The Roman Inquisition took custody of Bruno the following January 27, and carried out its mandate to deliver truth through interrogation and torture for the next seven years. Unwilling to recant his beliefs, Bruno was accused of heresy on January 20, 1600, by the Holy Inquisition. All of his books and writings were proclaimed "heretical and erroneous" and placed on the church's Index of

Forbidden Books, and Bruno was sentenced to death. Upon receiving his verdict, it is said that he replied, "Perhaps your fear in passing judgment on me is greater than mine in receiving it."

Giordano Bruno is known as a wanderer, an independent thinker, a rationalist, and a martyr. For the crime of freethinking, on February 17, 1600, he was led from his dungeon in Rome to a stake in the Campo de' Fiori square. His mouth was bound with an iron gag made with spikes to pierce his tongue and palette, and he was burned alive, becoming the first martyr of modern science.

⊠   Bunsen, Robert Wilhelm
(1811–1899)
German
Chemist, Spectroscopist

This innovative German chemist collaborated with GUSTAV ROBERT KIRCHHOFF in 1859 to develop spectroscopy. Their innovative efforts entirely revolutionized astronomy by offering scientists a new and unique way to determine the chemical composition of distant celestial bodies. An accomplished experimentalist, Bunsen also made numerous contributions to the science of chemistry, including improvement of the popular laboratory gas burner that carries his name.

Robert Bunsen was born in Göttingen, Germany, on March 31, 1811, the youngest child in a family of four sons. The fact that his father was an academician and served as a professor of linguistics at the University of Göttingen allowed Bunsen to experience the academic environment at an early age and to select a similar professional path. He attended preparatory school in Holzminden and, upon graduation, enrolled at the University of Göttingen as a chemistry major. An excellent student, Bunsen graduated in 1830 with his doctoral degree in chemistry at age 19. After graduation, the young scholar traveled for an extensive period through-

The German chemist Robert Bunsen's collaboration with Gustav Kirchhoff in 1859 revolutionized astronomy by allowing scientists to determine the chemical composition of distant luminous celestial bodies through spectroscopy. Despite his very important role in the development of spectroscopy as a means of identifying individual chemical elements, most science students usually remember him as the person who developed the popular gas burner that bears his name. *(AIP Emilio Segrè Visual Archives, E. Scott Barr Collection)*

out Germany and other countries in Europe, engaging in scientific discussions and visiting many laboratories. He remained in Vienna from 1830 to 1833. During these travels, Bunsen established a network of professional contacts that would prove very valuable during his long and illustrious career.

Upon his return to Germany, Bunsen accepted a position as a lecturer in chemistry at the University of Göttingen. It was here that he also began an interesting set of experiments investigating the insolubility of metal salts of arsenious acid. As part of this pioneering chemical research at Göttingen, the young chemist discovered what is still the best antidote against arsenic poisoning.

In 1836 Bunsen accepted a position at the University of Kassel, but departed within two years to work at the University of Marsburg. While continuing his arsenic chemistry research, he experienced a near fatal laboratory explosion in 1836 that cost him an eye. After that accident, his chemistry research still remained quite hazardous, because twice he nearly died from arsenic poisoning. An intrepid and intelligent experimenter, Bunsen was elected as a member of the Chemical Society of London in 1842.

He accepted a position at the University of Heidelberg in 1852 and he remained with this university until he retired in 1889. Bunsen was an outstanding teacher who never married and instead focused all his time and energy on his laboratory and his students. He eagerly taught the introductory chemistry courses normally shunned by his academic colleagues. His lectures emphasized experimentation and the value of independent inquiry. A superb mentor, Bunsen taught many famous chemists, including the Russian Dmitri Mendeleev (1834–1907), who developed the periodic table.

Bunsen's continuing contributions to chemistry were recognized by his induction to the Académie des Sciences in 1853 and by his election as a foreign fellow of the Royal Society of London. Despite his numerous achievements in the field of chemistry, he is often best remembered for his improvement of a simple laboratory burner. Although the "Bunsen burner" was actually invented by a technician, Peter Desaga, at the University of Heidelberg, Bunsen greatly improved the original device through his concept of premixing the gas and air prior to combustion in order to provide a burner that yielded a high-temperature, nonluminous flame.

In 1859 Bunsen began his historic collaboration with the German physicist Gustav Kirchhoff (1824–87)—a productive technical association that changed the world of chemical analysis and astronomy. While JOSEPH VON FRAUNHOFER's earlier work laid the foundations for the science of spectroscopy, Bunsen and Kirchhoff provided the key discovery that unleashed the great power of spectroscopy. Their classic 1860 paper, "Chemical Analysis through Observation of the Spectrum," revealed the important fact that each element has its own characteristic spectrum—much as fingerprints can help identify an individual human being. In a landmark experiment, they sent a beam of sunlight through a sodium flame. With the help of their primitive spectroscope (a prism, a cigar box, and optical components scavenged from two abandoned telescopes) they observed two dark lines on a brightly colored background just where the sodium D-lines of the Sun's spectrum occurred. Bunsen and Kirchhoff immediately concluded that gases in the sodium flame were absorbing the D-line radiation from the Sun, creating an absorption spectrum. The sodium D-line appears in emission as a bright, closely spaced double yellow line, and in absorption as a dark double line against the yellow portion of the visible spectrum. The wavelengths of these double bright emission lines and dark absorption lines are identical and occur at 5889.96 angstroms and 5895.93 angstroms, respectively.

After some additional experimentation, they realized that the Fraunhofer lines were actually absorption lines—that is, gases present in the Sun's outer atmosphere were absorbing some of the visible radiation coming from the solar interior. By comparing the solar lines with the spectra of known chemical elements, Bunsen and Kirchhoff detected a number of elements present in the Sun, the most abundant of which was hydrogen. Their pioneering discovery paved the way for other astronomers to examine the elemental composition of the Sun and other stars and made spectroscopy one of modern astronomy's most important tools. Just a few years later, in 1868, the British astronomer SIR JOSEPH

NORMAN LOCKYER attributed an unusual bright line that he detected in the solar spectrum to an entirely new element then unknown on Earth. He named the new element "helium," after *helios*, the ancient Greek word for the Sun.

Bunsen and Kirchhoff then applied their spectroscope to resolving elemental mysteries right here on Earth. In 1861, for example, they discovered the fourth and fifth alkali metals, which they named cesium (from the Latin *caesium*, "sky blue") and rubidium (from the Latin *rubidus*, "darkest red"). Today scientists use spectroscopy to identify individual chemical elements from the light each emits or absorbs when heated to incandescence. Modern spectral analyses trace their heritage directly back to the pioneering work of Bunsen and Kirchhoff.

The scientific community bestowed many awards on Bunsen. In 1860 he received the Royal Society's Copley Medal; he shared the first Davy Medal with Kirchhoff in 1877; and received the Albert Medal in 1898. After a lifetime of technical accomplishment, Bunsen died peacefully in his sleep at his home in Heidelberg on August 16, 1899.

Burbidge, Eleanor Margaret Peachey
(1919–)
American
Astronomer, Astrophysicist

With a lot of science genes going for her, Margaret Burbidge was born Eleanor Margaret Peachey on August 12, 1919, in Davenport, Cheshire, England, to parents who both studied chemistry at the Manchester School of Technology. She recalled in a biography that her interest in astronomy was kindled during a boat crossing from England to France when she was four years old and, having become seasick and confined to her bunk, spent many hours gazing at the brilliant stars through her porthole.

After excelling in mathematics at private primary and secondary schools, Peachey was excited to discover that the University of London offered a degree in astronomy, unlike most colleges at the time. She eventually earned her B.S. degree, the only woman in her class majoring in astronomy, and thanks to the frequent blackouts during German air raids was able to observe through the observatory's one functional telescope without the visual pollution of city lights. (A second telescope was rendered useless by a buzz bomb.)

She remained at the university's observatory after the war, where she observed for several years until she received her Ph.D. in astronomy, eventually becoming assistant director in the astronomy department. While in that position she applied for a Carnegie Fellowship at the Mount Wilson Observatory, in the San Gabriel Mountains above Pasadena, California; she received her first taste of gender discrimination when told in their response letter that women were not accepted. The excuse was that the observatory had only one toilet!

Two lifetime milestones in one year do not occur very often, but after receiving her Ph.D. in 1948, she also married her lifetime partner and co-researcher Geoffrey Burbidge, a theoretical physicist and her former physics teacher.

Burbidge's first significant appointment in the United States was as a research fellow at the University of Chicago's Yerkes Observatory, where she worked from 1951 to 1953. Geoffrey also did research at the university and, indeed, throughout Burbidge's career her husband would accompany her in associated academic positions. She then returned to England to do research at the Cavendish Laboratory, which is in the Department of Physics at Cambridge University. It was here that the couple met the renowned British astronomer Sir Fred Hoyle (born 1915) and the American nuclear physicist WILLIAM ALFRED FOWLER.

The English astrophysicist Eleanor Margaret Burbidge reviews images of quasars and various spiral galaxies in 1980. Collaborating with her husband and other physicists in 1957, she published an important scientific paper that described how nucleosynthesis creates elements of higher mass in the interior of stars. She also coauthored a fundamental book on quasars in 1967 and became the first female director of the Royal Greenwich Observatory, serving in that post from 1972 to 1973. *(AIP Emilio Segrè Visual Archives, Physics Today Collection)*

Burbidge returned to the United States in 1955 when her husband was awarded the Carnegie Fellowship for astronomical research at the Mount Wilson Observatory, the position for which Burbidge had previously been turned down ostensibly because of the solitary toilet. When she won a fellowship from California Institute of Technology (Caltech), which also was affiliated with the then-famous Mount Wilson telescope, she soon discovered that she not only was barred from the dormitory at the observatory, but also that women were not allowed to use any of the telescopes. However, she convinced administrators that she was Geoffrey's "assistant." The pair, along with other coworkers, protested the discrimination, but it would be 10 years before the "no women" policy was overturned.

In 1957 Burbidge returned to Chicago when Geoffrey received a position as assistant professor of astronomy at the Yerkes Observatory in Wisconsin. To avoid charges of nepotism, which she stated "are always used against the wife," she had to settle for a position as an associate professor of astronomy at the university.

It was during this tenure that she coformulated what is known as the "B²FH Theory," named for the principal authors: she, her husband, Fowler, and Hoyle. Based on the concept that there is no simple way by which the elements could have formed, since space matter is subjected to different conditions, the theory presented the then-revolutionary explanation that the elements were formed by nuclear reactions inside stars. In 1959 the American Astronomical Society awarded the Burbidge husband-and-wife team the Helen B. Warner Prize for Astronomy, recognizing the importance of their theory to the new field of nuclear astrophysics.

Yet another nepotism issue arose in 1962, when Geoffrey was given a position as professor in the Department of Physics at the University of California at San Diego (UCSD). Since UCSD did not have a separate astronomy department at the time, and the husband and wife were prohibited from working together, Burbidge had to accept a position in the chemistry department. Two years later, the university abolished its nepotism rule and she joined Geoffrey as a full professor of physics.

During these years Burbidge had been pioneering investigations into the nature of quasars, or quasi-stellar radio sources, which, in simple terms, are galaxies billions of light-years away that can be detected only with radio telescopes. In 1967 she and Geoffrey published the book *Quasi-Stellar Objects*, which even today remains a classic in the field of quasar research and radioastronomy. She also delved into the nature of galaxies and, using the telescope at the MacDonald Observatory operated by the University of Texas at Austin, she pioneered work in measuring the spectra of more than 50 spiral galaxies, determining their rotations, masses, and chemical compositions.

Burbidge took a leave from UCSD and returned to England in 1972 to take the position as director of the Royal Greenwich Observatory, the oldest scientific institution in Great Britain. However, Geoffrey could not find a similar position, so he stayed at UCSD. After a year had passed, she both missed her husband and grew frustrated with bureaucratic stumbling blocks. She resigned her position and returned to UCSD, where she eventually became the director of the university's Center for Astrophysics and Space Sciences. In what must have been an emotionally satisfying event after so many years of gender discrimination and nepotism rules, in 1976 she was named the first female president of the American Astronomical Society. One of her first accomplishments was to get the organization to refuse to hold meetings in states that had not ratified the Equal Rights Amendment. Soon after that, in 1978, she was elected to the National Academy of Sciences.

For years an idea had been rolling around the back of Burbidge's mind: Why not try to station a large telescope in space? When the concept of the *Hubble Space Telescope* was put forth by NASA, Burbidge helped design the faint object spectrograph (FOS), a high-resolution instrument aboard the *Hubble Space Telescope* that records the spectra of galaxies. Beginning in 1990, she was the coprincipal investigator of data emanating from the FOS for the next six years.

One of the great pioneers in the field of astronomy in the 20th century, Burbidge at age 82 was still professor emeritus of astronomy and research professor at UCSD. She has won dozens of professional honors, prizes, and honorary degrees, and is noted as having turned down the Annie J. Cannon Award in Astronomy from the American Association of University Women, a prestigious prize offered only to women astronomers, on the ground that special honors and discrimination for women should be abolished.

In 1982 she became the first woman in the 84-year history of the Astronomy Society of the Pacific to win the Catherine Wolfe Bruce Medal, one of the highest honors in the field of astronomy, for "a lifetime of distinguished service and achievement." In 1985 President Reagan presented her with the National Medal of Science, and in 1988 she won the Albert Einstein World Award of Science Medal. The minor planet 5490 has been named Burbidge in her honor.

C

Campbell, William Wallace
(1862–1938)
American
Astronomer

Early in the 20th century, the American astronomer William Campbell made pioneering spectroscopic measurements of stellar motions that helped astronomers better understand the motion of our solar system within the Milky Way Galaxy and the overall phenomenon of galactic rotation. He perfected the technique of photographically measuring radial velocities of stars, and discovered numerous spectroscopic binaries. Campbell also provided a more accurate confirmation of ALBERT EINSTEIN's theory of general relativity when he carefully measured the subtle deflection of a beam of starlight during his solar eclipse expedition to Australia in 1922.

Campbell was born on April 11, 1862, in Hancock County, Ohio. He experienced a childhood of poverty and hardship on his family's farm. He originally enrolled in the civil engineering program at the University of Michigan. However, his encounter with SIMON NEWCOMB's work *Popular Astronomy* (1878) convinced him to become an astronomer. After graduating from the University of Michigan in 1886, he accepted a position in the mathematics department at the University of Colorado. After working for two

years in Colorado, Campbell returned to the University of Michigan in 1888 and taught astronomy there until 1891. While teaching in Colorado, he also met the student who became his wife. Betsey (Bess) Campbell proved to be a very important humanizing influence on her husband, who was known for being domineering and inflexible, personal traits that made him extremely difficult to work with or for.

In 1890 he traveled west from Michigan to become a summer volunteer at the Lick Observatory of the University of California, located on Mount Hamilton near San Jose. His hard work and observing skills earned him a permanent position as astronomer on the staff of the observatory the following year. On January 1, 1901, he became director of the Lick Observatory and ruled that facility in such an autocratic manner that his subordinates often privately called him the czar of Mount Hamilton. In 1923 he accepted an appointment as the president of the University of California (in Berkeley), but he also retained his former position as the director of the Lick Observatory.

Wallace Campbell's permanent move from the University of Michigan to California in 1891 proved to be a major turning point in his career as an astronomer. Until then, he had devoted his observational efforts to the motion of comets. At the Lick Observatory, he brought about a

American astronomer William Wallace Campbell in 1893, standing beside the 36-inch refractor telescope with its specially designed spectrograph at the Lick Observatory, on Mount Hamilton near San Jose, California. Campbell used this equipment to make pioneering measurements of the radial (line of sight) velocities of stars by measuring the Doppler shift of their spectral lines. *(AIP Emilio Segrè Visual Archives)*

revolution in astronomical spectroscopy when he attached a specially designed spectrograph to the Lick Observatory's 36-inch refractor telescope—then the world's largest. He used this powerful instrument to measure stellar radial (that is, line of sight) velocities, by carefully examining the telltale Doppler shifts of stellar spectral lines. He noted that when a star's spectrum was "blueshifted," it was approaching Earth; when "redshifted," it was receding. He perfected his spectroscopic techniques and in 1913 published *Stellar Motions*, his classic textbook on the subject. In that same year, Campbell also prepared a major catalog, listing the radial velocities of more than 900 stars. He later expanded the catalog in 1928 to include 3,000 stars. His precise work greatly improved knowledge of the Sun's motion in the Milky Way and the rotation of the Galaxy.

Campbell also used his spectroscopic skills to show that the Martian atmosphere contained very little water vapor—a finding that put him in direct conflict with other well-known astronomers at the end of the 19th century. Using the huge telescope and clear-sky advantages of the Lick Observatory, Campbell, who was never diplomatic, quickly disputed the plentiful water vapor position of SIR WILLIAM HUGGINS, SIR JOSEPH NORMAN LOCKYER, and other famous European astronomers. While Campbell was indeed correct, the abrupt and argumentative way he presented his data won him few personal friends within the international astronomical community.

He enjoyed participating in international solar eclipse expeditions. His wife joined him on these trips, and she was often put in charge of the expeditions provisions. During a 1922 solar eclipse expedition in Australia, Campbell measured the subtle deflection of a beam of starlight just as it grazed the Sun's surface. His precise measurements helped confirm Albert Einstein's theory of general relativity and reinforced the previous, but less precise, measurements made by

SIR ARTHUR STANLEY EDDINGTON during the 1919 solar eclipse.

Campbell retired as the president of the University of California (Berkeley) and as the director of the Lick Observatory in 1930. He and his wife then moved to Washington, D.C., where he served as the president of the National Academy of Sciences from 1931 to 1935. Returning to California, he spent the last three years of his life in San Francisco. On June 14, 1938, he died by his own hand—driven to suicide by declining health and approaching total blindness. William Campbell's contributions to astronomy were acknowledged through several prestigious awards. The French Academy presented him with its Lalande Medal in 1903 and its Janssen Medal in 1910, the British Royal Astronomical Society awarded him its Gold Medal in 1906, and the National Academy of Sciences of the United States gave him the Henry Draper Medal in 1906. Finally, in 1915, the Astronomical Society of the Pacific awarded Campbell its Bruce Gold Medal.

⊠ Cannon, Annie Jump
(1863–1941)
American
Astronomer

At the turn of the 20th century, the director of the Harvard College Observatory hired several women as data analysts, not paying them very much but nevertheless remunerating them for reducing and filing complex data accumulated from stellar observations. Many of these women became respected members of the astronomical community. The most famous among them was Annie Jump Cannon, who became the most prolific and studious star spectrum classifier in history, and eventually the world's foremost authority on classifying stars.

Born in Dover, Delaware, Cannon was only 16 years old when she became one of the first

American astronomers Annie Jump Cannon (left) and Henrietta Swan Leavitt (right) outside the main entrance to the Harvard College Observatory, where they worked under the supervision of Edward Charles Pickering to publish the *Henry Draper Catalogue*—a nine-volume document that contained the spectral classifications of more than 225,000 stars. Cannon was the driving force behind the development of a refined system for classifying stellar spectra that became known as the Harvard Classification System. *(Courtesy AIP Emilio Segrè Visual Archives, Shapley Collection)*

women from that state to be admitted to college. Her father was Wilson Cannon, a shipbuilder and state senator, and her mother, Wilson's second wife, Mary Jump, taught her the constellations from the family's "star book." As a child, Cannon was fascinated by how crystal pendants on the candelabra separated sunlight into colorful rainbow patterns. These two interests led her to study physics and astronomy at Wellesley

College, where she learned how to make spectroscopic measurements. After graduating in 1884, and taking up the newly developed science of photography as an avocation, she traveled extensively with her camera.

During her travels she contracted scarlet fever and was struck deaf, having to use a hearing aid for the rest of her life. However, this affliction is said to have honed her great powers of concentration, responsible for propelling her into the highest ranks of astronomers of her time.

After the death of her mother in 1894, Cannon returned to Wellesley to teach physics, and did further studies in astronomy at Radcliffe. In 1896 she became one of "Pickering's women," taking part in some of the first X-ray experiments, and on May 14 she made her first observation of a star's spectrum. At the time, the accumulation of the massive stellar data was funded by Anna Draper, widow of the wealthy physician and amateur astronomer HENRY DRAPER. Pickering set up the Henry Draper Memorial as a long-term project to gather the optical spectra of as many stars as possible. This had begun in 1886 and was carried on by several women, each using her own system of classification.

Eventually the work fell to Cannon, who threw herself into the project with an intensity and devotion unparalleled in the history of astronomy. The speed with which she worked was almost superhuman. It is said that Cannon classified 5,000 stars a month between 1911 and 1915, and one report has it that before she was through she had classified the spectra of more than 400,000 stars.

She began by developing a classification system that utilized the best parts of those systems developed before her, and assigned a division of stars into O, B, A, F, G, K, M categories, leading to the famous "Oh, Be a Fine Girl, Kiss Me" mnemonic memorized by every astronomy student since. The classification catalog she ultimately developed, called the HD Catalog, is still in use today as the cornerstone of modern spectroscopic astronomy. She also published several catalogs of variable stars, 300 of which she discovered herself.

Cannon went on to win practically every award and honor the scientific community could bestow. In 1911 she became curator of the observatory; in 1923 she was reportedly declared to be one of the "12 greatest living American women." In 1925 she was the first woman to receive an honorary doctorate degree from Oxford University, and in 1931 she received the coveted Draper Gold Medal from the National Academy of Sciences, again the first woman so honored. The American Association of University Women named its annual award to the foremost woman of science the Annie J. Cannon Award. In all, she received a total of six honorary doctoral degrees.

Annie Jump Cannon died on April 13, 1941, after 42 years of dedication at Harvard. She is one of only 20 women who have a lunar crater named in her honor.

⊠   **Cassini, Giovanni Domenico
(Gian Domenico Cassini, Jean-
Dominique Cassini, Cassini I)**
(1625–1712)
Italian (naturalized French)
Astronomer, Mathematician

Born in what is now Italy, but eventually becoming a naturalized French citizen, Cassini is known as the "Father of French Astronomy." He was born on June 8, 1625, to Julia Crovesi and Jacopo Cassini in the city of Perinaldo, which was either in the French county of Nice or the Republic of Genoa, depending on the source, and is now a part of Italy. One of the great astronomers of his time, Cassini was also an expert in hydraulics and engineering, specializing in river management and flood control.

Cassini reportedly studied for two years at Vallebone before becoming a student, first at the Jesuit college in Genoa and then at the abbey of San Fructuoso, where he became very interested in astrology. But after casually perusing a book on astronomy, which by this time was a separate discipline, he decided to drop astrology and focus on learning astronomy instead. However, his technical background served him well when he was consulted about the flooding of the river Po, and he worked for Pope Alexander VII to negotiate arrangements for the navigation of the rivers Po and Remo. Later he was named superintendent of the fortifications of Fort Urban and the fortress of Perugia, and was appointed to solve problems surrounding the course of the river Chiana, as well as supervising effluent control of the rivers Tiber and Arno, which was causing friction between the pope and the grand duke of Tuscany.

The extensive knowledge he had of both astronomy and astrology, with the necessary mathematical ability to work in both fields, gave Cassini the foundation he needed for a new position offered to him in 1644, when he was invited by a powerful political figure in Bologna, the marquis Cornelio Malvasia, to come to the university and work in the new observatory. It was an opportunity Cassini could not, and did not, pass up, and it was the beginning of a 20-year relationship with the university.

By 1648 the new Panzano Observatory was finally built, and Cassini was able to start his observations, which were initially aimed at the Sun. Within two years he was also a professor of astronomy and mathematics. By 1653 he was occupied with a task that seemed nearly impossible to carry out—he wanted to build a gnomon similar to the one he had seen at a church in San Petrino, which had been built between the mid-15th and early 16th centuries. The gnomon would be used to calculate the altitude of the Sun as the Sun cast a shadow over the sundial-like apparatus, but the one in San Petrino had been obstructed and was worthless, and Cassini wanted to build a much bigger device.

Cassini designed the new instrument and oversaw its completion. His calculations had to be precise for it to work, and when it was finally installed, it was perfect. This accomplishment gave Cassini instant success and, more important, allowed him to start making observations with this new tool. He published his work in a book entitled *Specimen observationum bononiensium* in 1656 and dedicated it to the exiled Queen Christina of Sweden.

The period between 1656 and 1659 made Cassini's name in the world of astronomical science. In these few years he calculated a new value of the inclination of the ecliptic as 230°29′15″, the most precise of his era; he drew up a table of the atmospheric refractions, which were the most accurate for a century after his death; and he proved that the orbital velocity of the Earth was not constant.

In 1661, even while doing his hydraulic work on the flooding rivers, he worked out a way to determine longitudes by observing solar eclipses, but the Inquisitor of Modena refused to allow publication of his paper. It would be published 10 years later in France.

Cassini obtained one of the first long focal lenses made by the lensmaker Giuseppe Campani, and began to study the planets. With these lenses he observed the shadows of Jupiter's moons on the surface of the planet and calculated their rotational period, and he observed Jupiter's Great Red Spot, which in turn allowed him to calculate Jupiter's rotational period. At this time he also evaluated Mars's rotational period.

All of this innovational astronomy gave Cassini a reputation that went beyond the boundaries of Italy. In France Louis XIV was deciding that France should become as prominent in the arts and sciences as it was in warfare and weaponry, and charged one of his ministers,

Colbert, with inviting Cassini to come to France and join France's Académie Royale des Sciences, giving him an opportunity to work at the new observatory then being built. However, this required the permission of Cassini's new boss, Pope Clement IX, who generously gave his approval. If Italy wanted to keep Cassini for itself, this proved a costly mistake, for when Cassini left, he only returned to Italy to visit relatives or make observations.

Cassini reached Paris in 1669, and went to live at the new observatory, where he was provided with a sizable salary, a travel allowance, and free accommodations. In 1673 he requested French nationality, changed his name from the Italian Giovanni Domenico to the French Jean-Dominique, married Genevieve de Laistre, daughter of a high-ranking military officer, and moved into the castle of Thury, near Beauvais, as a family residence—which, coincidentally, was crossed by the Paris meridian.

At the new observatory, Cassini started a series of observations of the lunar surface, which resulted in his writing a lunar atlas, a large map, and to propose a theory of the oscillation of the Moon. He also used his tables of refractions to calculate the Mars parallax and that of the Sun. This information greatly increased the scale of the solar system at the time.

Of all the objects in the heavens, it is the planet Saturn that is most closely associated with Cassini's name. While SIR WILLIAM HUYGENS had discovered Titan, Saturn's first satellite in 1655, Cassini, working at the Paris Observatory, discovered four others: Iapetus in 1671, Rhea in 1672, and Dione and Thetys in 1684. In 1675 Cassini discovered that Saturn's famous ring was actually two rings separated by a dark space, now called Cassini's Division. He also recorded the flatness of the ring, and observed a cloudy strip parallel to Saturn's equator. Eventually he suggested that the ring was not solid, but composed of tiny "satellites" that could not be resolved with currently available lenses, a thesis borne out only recently by today's observational technology.

Cassini also made important measurements of the Earth, and undertook an important geographical work. He extended the meridian line established from Paris to Amiens by Jean Picard (1620–82) and made great determinations of longitudes. As young Jesuit priests studied under him at the Paris Observatory and then went on their missionary assignments, Cassini began to receive results of their observations from China, Africa, and America. He marked these on a large planisphere, 7.8 meters in diameter, and made a pen-and-ink sketch on the ground at the observatory. In this manner, the longitudes of distant countries could be corrected more accurately than ever before.

Cassini carried on his research until 1711, when the strain of a lifetime of squinting through lenses, often at the Sun, rendered him totally blind. He was always described as a man of kind and gentle demeanor, with a pleasant nature and deeply religious, which allowed him to bear his blindness with good cheer and stoicism. He died in the Paris Observatory on September 14, 1712. A spacecraft launched in 1997 was sent on a seven-year voyage that successfully reached the Saturn system in July 2004, where it would gather data and pictures never before available to astronomers. It is named Mission Cassini.

⊠ Cavendish, Henry
(1731–1810)
British
Chemist, Physicist

The reclusive Henry Cavendish was perhaps one of the oddest, and yet one of the most important, scientists of the 18th century. He was known as an expert at scientific experimentation, and dedicated his life completely to the study of science, resulting in the discovery of hydrogen, determining the composition of air, and calculating the mean density of the Earth. But he was considered as eccentric as he was brilliant, and led a life in which he almost never spoke, not even to his household staff.

Cavendish was born to English nobility on October 10, 1731, in Nice, France, where his family lived temporarily because of his mother's health. He was the oldest son of Lord Charles Cavendish and Lady Anne Gray. The duke of Devonshire was his paternal grandfather and the duke of Kent was his maternal grandfather. While little is known about his early education, it is safe to say that his family's status and wealth probably afforded him private tutors in his early childhood. When he turned 11 years old, Cavendish was sent to study at Dr. Newcome's School in Hackney, England. At 18 he was off to Cambridge, where he enrolled in St. Peter's College.

After dropping out of Cambridge during his fourth year and taking a traditional tour of Europe with his brother, he inherited a fortune that made him one of the wealthiest men in England. However, instead of behaving as one might expect after such good luck, Cavendish became a virtual recluse for the rest of his life, living frugally in London and immersing himself in scientific studies.

As strange and solitary as Cavendish was, he was an extraordinary scientist. One of his first accomplishments occurred in 1766 with the discovery of hydrogen. Known as "flammable air" and studied by others for a century before him, this gas was often also called "phlogiston," an ancient term for a nonexistent element that, prior to the discovery of oxygen, was thought to be released as a product of combustion. But Cavendish was the first to measure its specific gravity and posited that hydrogen was a different gaseous substance from ordinary air, whose components he subsequently analyzed as composed of nitrogen and oxygen in an approximate 4:1 ratio.

The discovery of water, or rather its composition, is also due to Cavendish's experiments. In 1781 he discovered that water was composed of hydrogen, his "flammable air" discovery, and oxygen, which he determined by burning the two chemicals together, in the proper ratio now known as H_2O. He was the first person to hear the famous "POP!" that high school chemistry students experience when they hold a Bunsen burner to the mouth of the bottle of collected hydrogen.

Cavendish wrote, ". . . on applying the lighted paper to the mouth of the bottle . . . with 3 parts of flammable air to 7 of common air, there was a very loud noise." Then, when he combined the hydrogen with the oxygen, he produced water, leading him to reverse the process and prove once and for all that water was composed of "flammable air" and oxygen. As Cavendish described it, "By the experiments . . . it appeared that when flammable and common air are exploded in a proper proportion, almost all of the flammable air and near one-fifth of the common air lose their elasticity and are condensed into dew."

While he was at it, Cavendish also noted that the water's weight was equal to the weight of the gases. French chemist Antoine-Laurent de Lavoisier (1743–94) named the new gas hydrogen—Greek for "water-former." Cavendish eventually calculated the composition of air to be 79.167 percent "phlogisticated air" (now known as nitrogen) and 20.833 percent "dephlogisticated air" (oxygen), but also established that $1/120$ was a mysterious third gas, which 100 years later was discovered to be argon in 1894 by the Scotsman Sir William Ramsey (1852–1916) at University College, London.

Such chemistry discoveries ultimately supported his work in astronomy. Cavendish was the first to determine SIR ISAAC NEWTON's "gravitational constant," which he published in 1798 as "Experiments to Determine the Density of the Earth," one of his few published papers in his lifetime. His work over a two-year period resulted in accurately measuring the mass and density of the earth as 5.48 times that of water, stunningly close to the 5.5268 calculated by modern techniques. This constant of proportionality used in Newton's Law of Universal Gravitation, or G, was found to be 6.67×10^{-11} newtons between two objects with a mass of one kilogram each at a distance from each other of one meter. Cavendish calculated this constant measuring the gravitational force between two metal spheres. Knowing G permitted the determination of the Earth's mass, and subsequently the mass of the Sun and planets.

Cavendish also dabbled in electrical measurements, sometimes measuring the strength of an electrical current by shocking himself and then comparing the levels of pain he experienced. Some of his experiments investigated the phenomenon of capacitance, and he came very close to formulating what was to become Ohm's Law. Cavendish's work in electricity was rediscovered and published after his death by James Clerk Maxwell (1831–79), the scientist known for discovering electromagnetism.

Although he himself admitted to having "a singular love of solitariness," Cavendish did have one social function in his life, albeit a scientific social function, that he religiously attended—he was a member of the Royal Society and is reported to have rarely missed the weekly dinner meetings. But while his scientific discoveries were highly regarded by his contemporaries, even they considered him an extremely odd man. It could have been due to the way he dressed. Although a man of amazing wealth, Cavendish wore the same clothes nearly every day—an outfit that consisted of "a crumpled violet suit" that was in style during the previous century, with ruffled cuffs and a high ruffled collar, topped off by his equally-out-of-style trademark three-cornered hat. Or perhaps it was the way he talked, or more accurately did not talk, that put off his contemporaries. When he spoke, which was rare even at the Royal Society meetings, his voice was described as shrill and high, and he was so shy that he stammered and communicated with great difficulty. Completely inept in dealing with women, Cavendish communicated with his household staff strictly using handwritten notes, and the women who worked for him were ordered to stay completely out of his sight under threat of being fired.

Social skills aside, his brilliance was still able to shine despite his lack of verbal and written communication with his colleagues. Cavendish published fewer than 20 papers in his 50-year career, but his first, titled "Factitious Airs" and published in 1766, earned him the Royal Society's Copley Medal. He was a member of the Royal Society from 1760 until his death. Never married, Cavendish's sole social contact was dining with other Royal Society members. Lord Henry Broughman (1778–1868), a fellow member who eventually became Lord Chancellor of England, once commented that Cavendish "probably uttered fewer words in the course of his life than any man who ever lived to fourscore years, not at all excepting the monks of La Trappe." Cavendish was a scientist to the end. According to Broughman, Cavendish wanted to be left alone on his deathbed, because he did not want any interruptions while he observed and recorded the progress of the disease that was running through his body. Cavendish died alone at home, except for his staff-in-hiding, on March 10, 1810. The famous Cavendish Laboratory at Cambridge University is named in his honor, as is a crater on the Moon.

⊠ **Chandrasekhar, Subrahmanyan (Chandra)**
(1910–1995)
Indian, Naturalized American
Astrophysicist

Subrahmanyan Chandrasekhar was known for most of his scientific life simply as "Chandra."

He was born in Lahore, southern India, then part of British colonial India, to an educated family. His uncle was a Nobel laureate and his father was an executive for the Indian railway system who moved the family to Madras when Chandra was six.

Chandra was an exceptionally brilliant student from the start. At the tender age of 15, he was admitted to the most prestigious college in Madras, the Presidency College, where he enrolled in the school's physics honors program, graduating with a bachelor's degree in theoretical physics at the top of his class. During this program he read Ralph H. Fowler's (1889–1944) early work at Cambridge University's Trinity College on "white dwarfs," stars that have burned off their internal supply of hydrogen and have collapsed into themselves as intensely hot, highly massive, but very small, remnant stars. This subject fascinated him, and at 18 years old Chandra wrote his first scientific paper and sent it to Fowler, who liked it so much he sent it to the Royal Society. It was published as "Compton Scattering and the New Statistics" in the *Proceedings of the Royal Society*. It was because of this paper that upon graduation Chandra was accepted as a research student at Cambridge after winning an Indian government scholarship.

When Chandra left Bombay on the voyage to England, he became seasick and, confined to his stateroom, worked out new calculations based on Fowler's work and his own training in relativity theory. By the time the ship docked two weeks later, Chandra had worked out his first suspicion that there was an upper limit to the mass of a white dwarf, a concept that went completely against the prevailing theory at the time. When he put forth his new theory at Cambridge, the country's two leading astrophysicists, SIR ARTHUR STANLEY EDDINGTON and Edward A. Milne (1896–1950), dismissed his proposition as impossible, and Eddington in particular publicly rejected Chandra's conclusions.

Chandra's theory was essentially this: During its evolution, a star goes through a stage known as a "red giant," in which its radius may be hundreds of times its original size; after this the star gradually implodes, even if it were once a 7- or 10-solar mass, into a mass lower than Chandra's 1.4 solar mass limit. At this stage no more nuclear energy can be gained, and when the new dwarf star's iron core grows to the 1.4 solar mass number, it collapses by gravitation again.

Chandra continued to work on his theory and, after receiving his Ph.D., was elected a Fellow of Trinity College. After undertaking more complete calculations, he confirmed his earlier result and concluded that the mass of a white dwarf had an upper limit of 1.4 solar mass. At age 25, he was then invited to present his results in a lecture in 1935 at the Royal Astronomical Society. When he did, Eddington again publicly rejected Chandra's results, and one report quotes him as saying, "I think there should be a law of nature to prevent a star from behaving in this absurd way." Scientific observers of the day realized that Eddington's life's work had been to prove that every star, regardless of its mass, had a stable configuration and Chandra's contention, if valid, would destroy that proof.

Nevertheless Chandra was devastated. He finally appealed to some respected physicists he knew—Léon Rosenfeld (1904–74), Neils Bohr (1885–1962), and Wolfgang Pauli (1900–58)—who unanimously made it known to the scientific community that they heartily agreed with Chandra's conclusions. His upper limit of 1.4 solar masses is known today as "Chandrasekhar's Limit." However, it is generally agreed in scientific circles that the monkey wrench thrown into the discussion by Eddington and Milne was responsible for Chandra having to wait 50 years before winning the Nobel Prize in physics in 1983. It apparently made it difficult for Chandra to acquire any kind of prestigious position in England, and the beginnings of

political unrest in his native India made returning home imprudent. Thus, he accepted an offer from Otto Struve (1897–1963) to join the faculty at the Yerkes Observatory of the University of Chicago in 1937. Before reporting, however, he journeyed back to India to marry Lalitha Doraiswamy, a former schoolmate in the physics program at Presidency College in Madras. Lalitha was then working in the Bangalore laboratory of Chandra's uncle, the physics Nobel laureate Chandrasekhara Venkata Raman (1888–1970).

Chandra would remain at the University of Chicago for the rest of his life, becoming Morton D. Hall Distinguished Service Professor in Astronomy and Astrophysics in 1952. While at Chicago, he delved into theoretical work on stellar interiors, and became keenly interested in the gravitational frictional drag on a star passing through a tenuous cloud of stars, using such "drag" to determine the ages of globular clusters. He also developed equations for interacting gravitational waves, and is responsible for developing the post-Newtonian approximation that has become the standard formal approach to calculating the gravitational waves from dynamic systems of massive particles.

Chandra was a great supporter of Mohandas Gandhi, but when World War II broke out he felt that helping the cause of victory over Hitler's Nazi Germany was his most important duty. His contribution to the war effort was working half time at the Aberdeen Proving Ground on shock waves, and only lengthy clearance problems prevented him from accepting an invitation to join the Manhattan Project at Los Alamos. He became an American citizen in 1953.

While at Chicago, Chandra published six books and numerous scientific papers, each of which is considered definitive in its field: *An Introduction to the Study of Stellar Structure* (1939); *Principles of Stellar Dynamics* (1943); *Radiative Transfer* (1950); *Hydrodynamic and Hydromagnetic Stability* (1961); *Ellipsoidal Figures of Equilibrium* (1968); and *The Mathematical Theory of Black Holes* (1983).

He and Eddington eventually patched up their differences, and Eddington was instrumental in Chandra's 1944 election to the Royal Society. Eddington died that year, and in an obituary speech, Chandra ranked Eddington next to KARL SCHWARZSCHILD as the greatest astronomer of his time. In 1962 Chandra received the Royal Society's coveted Royal Medal. In 1966 he received the National Medal of Science. He was also editor of the *Astrophysical Journal* from 1952 to 1971, taking it from a relatively small and unnoticed university journal to the internationally respected journal of the American Astronomical Society.

Chandra also developed an intense interest in literature and music, and was considered by many authorities to be a master of the English language. His 1987 book, *Truth and Beauty*, is a collection of essays exploring the similarities in motivation and aesthetic rewards of artists and scientists.

At his death in 1995, he was praised by students, associates, and world scientific leaders as possibly the greatest astrophysicist of the 20th century. NASA renamed its Advanced X-ray Astrophysics Facility (AXAF) the Chandra X-ray Observatory in honor of him, and launched it in the space shuttle *Columbia* in 1999 to study faint sources in X-rays in crowded fields from an elliptical high-earth orbit.

⊠ **Charlier, Carl Vilhelm Ludvig**
(1862–1934)
Swedish
Astronomer, Mathematician

Carl Vilhelm Ludvig Charlier was born April 1, 1862, in Ostersund, the capital of Jamtland, a province in northern Sweden. He was the son of Emanuel Charlier, a government official, and Aurora Kristina Hollstein. He attended the local high school and in 1881 entered the prestigious University of Uppsala, the oldest university in the Nordic countries, founded in 1477. There, he came under the influence of Professor Herman Shultz, then director of the Uppsala Observatory, becoming an assistant in his third year, and he studied theoretical mathematics and astronomy using the observatory's 9-inch refractor.

Charlier received his Ph.D. in 1887 with a dissertation on Jupiter's effect on the orbit of minor planet 17 Thetis. A year later he became an assistant at nearby Stockholm Observatory, under the direction of Johan A. H. Gylden, a leader in celestial mechanics research. However, Charlier brought his mathematical expertise to more practical problems, such as the new application of photography to astronomical observations. His initial task was to develop a practical method for photographic photometry. At Uppsala he had gained some experience in the use of a visual photometer, and at the Stockholm Observatory he used a small photographic telescope to observe the Pleiades group of stars. Using a micrometer, he measured the diameters of the various star disks on his plates and established a formula relating the image diameters to the photometric magnitudes of the brighter Pleiades stars. This formula enabled him to become the first observer to determine the magnitudes of the fainter Pleiades stars. At the time, his paper describing his Pleiades work was considered a groundbreaking contribution and was selected by the Astronomische Gesellschaft for dedication of the Pulkowa Observatory on its 50th anniversary.

After two years at Stockholm Observatory, Charlier went back to Uppsala, where from 1890 to 1897 he was chief assistant. During this time a new photographic refractor was installed and Charlier became interested in the theory of astronomical and photographic lens combinations. He published many papers on the development of an achromatic objective lens comprised of two lenses of the same glass material instead of the current standard of flint and crown glass.

In 1897, at age 35, Charlier received the highest position in Swedish astronomy: professor of astronomy and director of the observatory at the University of Lund, a position he held for 30 years. One of his accomplishments during this tenure was to establish two series of observatory publications, published under the name *Meddelanden*. More than 150 issues were printed under his direction, comprising 10 volumes. He continued his work in celestial mechanics at Lund. It is noteworthy that, while Charlier's early works were published in Swedish, he eventually also wrote his technical papers and books in German, French, and English. Writing scientific papers in four different languages is quite possibly a feat achieved by no other astronomer in history.

The contributions Charlier made to theoretical astronomy and statistical theory are many. Chief among them are a study on the effect on asteroid orbits of secular perturbations, that is, the slow progressive changes in planetary orbits, and a study of the rotation of planets around their axes. Also while at Lund, Charlier brought his knowledge of mathematical statistics to the field of astronomy. Charlier demonstrated that all distribution laws in nature can be represented by one of two classes of error functions. He was called upon to test this application in other fields as well, such as biology, medicine, and population statistics. The courses he taught at the University of Lund led to the establishment of a separate chair for mathematical statistics.

Charlier also made extensive statistical studies of the distributions and motions of stars near the Sun, and showed that hotter stars and galaxies form a flattened system. His hierarchical model of the universe proposed that it is made up of a successive series of higher and higher systems.

Charlier became a member of the Astronomische Gesellschaft in 1887, and in 1904 was elected to the board of directors, on which he served for 13 years. He was awarded the Watson Gold Medal of the National Academy of Sciences in 1924; the Memorial Medal of the Royal Physiographic Society at Lund, also in 1924; an honorary membership in the American Astronomical Society in 1921; and in 1933 was awarded the prestigious Bruce Gold Medal of the Astronomical Society of the Pacific. In his retirement he translated Sir Isaac Newton's *Principia Mathematica Philosophiae Naturalis* into Swedish, and he wrote a French textbook on the applications of statistics to astronomy.

Charlier was one of the most influential teachers of astronomers and statisticians, and elevated the University of Lund to a prominent position in the world astronomical community. Among his students were Gustav Stromberg (1882–1962), who did important spectroscopic research at the Lick Observatory on Mount Wilson, and Karl Malmquist (1893–1982), who described analysis pitfalls of brighter object bias in astronomical surveys.

Charlier was stricken by an apoplectic event in 1932, and he died at Lund on November 5, 1934, leaving his widow Siri Dorotea Leissner and five daughters.

⊠ Clark, Alvan Graham
(1832–1897)
American
Astronomer, Lensmaker

Alvan Graham Clark made several of the finest telescope lenses ever produced. Clark was born in Fall River, Massachusetts, in 1832. His father, Alvan Clark, made instruments for a living. Alvan Senior had very definite plans about his business life, and in 1846 he opened a company, Alvan Clark & Sons, that specialized in making lenses for telescopes. Alvan Junior wanted to be an artist, specifically a portrait painter, but at the age of 20 he joined his father's firm in Cambridgeport and learned the lens-grinding art, eventually becoming the firm's chief optician.

It was while Clark was testing one of his creations, a new 18-inch lens in a telescope he had built for the Dearborn Observatory in Chicago (now Northwestern University's observatory), that he made his first astronomical discovery. He turned the instrument toward Sirius, the Dog Star, and observed for the first time a companion star. This star had been predicted 18 years earlier by astronomer FRIEDRICH WILHELM BESSEL, who had observed Sirius for more than 10 years and concluded that it had an unseen companion, but now Clark confirmed its existence. Sirius B is a white dwarf star—an extremely hot but very small star—the first white dwarf to ever be discovered. For his achievement, Clark was honored with the Lalande Prize of the French Academy of Science. As a companion star to the Dog Star, Sirius B is nicknamed "the Pup."

But Clark's most significant contributions to astronomy were the famous and coveted lenses he built for the major observatories and astronomers of the time. He built five significant lenses in his lifetime, and each of these important lenses, bigger than the previous, became the largest refractor available, setting new world records when they were installed.

In 1896 Clark built the 24-inch lens for the Lowell Observatory in Flagstaff, Arizona. This telescope is still on display in the "wedding cake" dome at the observatory. PERCIVAL LOWELL (1855–1916), who was convinced that there was a planet beyond Neptune, funded three projects for the search. During the third project, on February 18, 1930, a young astronomer named CLYDE WILLIAM TOMBAUGH discovered the planet on his photographic plates. He is the only American to discover a planet. His plates of Pluto are on display at the observatory near the telescope.

In 1861 SIMON NEWCOMB was appointed to the Naval Observatory, which had been founded in 1830 as the Depot of Charts and Instruments, in Washington, D.C. He supervised Clark in building a new 26-inch refractor for the observatory, the largest refractor ever made. In 1877 ASAPH HALL used this telescope to discover Deimos and Phobos, the two moons of Mars. The telescope is still in use today to monitor double stars, and the observatory reports that its optics are "phenomenally good."

In 1878 a new 30-inch refractor was built for Polkovo Observatory near Leningrad, Russia, which was used some years later by the Russian astrophysicist Gavriil Adrianovich Tikhov (1875–1960), the founder of astrobiology, to study the Martian terrain. The telescope was destroyed during World War II.

The 36-inch refractor built for the Lick Observatory in Mount Hamilton, California, cost approximately $180,000 when it was installed in 1888 in the large dome at the observatory, considered the most advanced at the time. This was the first time an observatory had been built for its favorable location to observe rather than its proximity to a university. EDWARD EMERSON BARNARD made many of his famous observations on this telescope, including his discovery of the fifth moon of Jupiter. The telescope is still in operation today.

The 40-inch refracting telescope at the University of Chicago's Yerkes Observatory in Williams Bay, Wisconsin, remains the largest refractor ever made. The lens was installed into the 63-foot (19.2 m) telescope on May 20, 1897, and operated in a small but significant ceremony the next evening by GEORGE ELLERY HALE, who pointed the telescope at Jupiter and showed guests what the planet looks like at 400x the magnification of the naked eye. Some of the most famous astronomers have used the 40-inch Yerkes telescope to make their observations and discoveries, including Barnard, Hale, GERARD PETER KUIPER, and

SUBRAHMANYAN CHANDRASEKHAR. This was the last lens made by Clark, just before his death.

Because there is an optical limit to using lenses larger than 40 inches, scientists in that era began using mirrors in reflecting telescopes to achieve greater magnifications, which is why the Yerkes lens is still, more than a century later, the largest refracting telescope in the world.

Clark discovered 16 more double stars before he died in 1897, and most of his discoveries came by testing lenses he had made for other people or institutions. Eventually, refracting telescopes incorporating lenses made by Clark would make some of the major astronomical discoveries of the late 19th and 20th centuries.

Through his work on the second law of thermodynamics and his development of the concept of entropy, the German theoretical physicist Rudolf Clausius had a major impact on 19th-century cosmology. He concluded that the entropy of the universe was striving to achieve a maximum value under the laws of thermodynamics. The end condition would be what he called "the heat death" of the universe—a state of complete temperature equilibrium with no energy available to perform any useful work. *(AIP Emilio Segrè Visual Archives)*

⊠ Clausius, Rudolf Julius Emmanuel
(1822–1888)
German
Physicist

By introducing the concept of entropy in 1865, Rudolf Clausius completed his development of the first comprehensive understanding of the second law of thermodynamics—a scientific feat that allowed 19th-century cosmologists to postulate a future condition (end state) they called "heat death of the universe."

A German theoretical physicist, Clausius was born on January 2, 1822, in Köslin, Prussia (now Koszalin, Poland). His father, the Reverend C. E. G. Clausius, served as a member of the Royal Government School Board and taught at a small private school where Clausius attended primary school. In 1840 he entered the University of Berlin as a history major, but emerged in 1844 with his degree in mathematics and physics. After graduation Clausius taught mathematics and physics at the Frederic-Werder Gymnasium (secondary school), while pursuing his graduate degree. In 1848 he received his doctorate in mathematical physics from the University of Halle. In 1850 he accepted a position as a professor at the Royal Artillery and Engineering School in Berlin.

In 1850 Clausius also published his first and what is now regarded as his most important paper discussing the mechanical theory of heat. The famous paper, entitled *"Uber die bewegende Kraft der Wärmer"* ("On the Motive Force of Heat"), provided the first unambiguous statement of the Second Law of Thermodynamics. In this paper, read at the Berlin Academy on February 18, 1850, and then published in its *Annalen der Physik* (Annals of physics) later that year, Clausius built upon the heat engine theory of the French engineer and physicist, Sadi Carnot (1796–1832), and introduced his version of second law of thermodynamics in a simple, yet important statement: "Heat does not spontaneously flow from a colder body to a warmer body." While there is one basic statement of the First Law of Thermodynamics, namely, the conservation-of-energy principle, there are quite literally several hundred different, yet equally appropriate, statements of the second law. Clausius's perceptive statements and insights, as expressed in 1850, lead the long and interesting parade of second law statements. This particular paper is often regarded as the foundation of classical thermodynamics.

However, in the 19th century many physicists, including the famous British natural philosopher William Thomson Kelvin, first baron of Largs (1824–1907), were also grappling with the nature of heat and developing basic mathematical relationships to describe and predict how thermal energy flows through the universe and is transformed into mechanical work. Clausius is generally given credit as the person who most clearly described the nature, role, and importance of the second law—although others, like Lord Kelvin, were simultaneously involved in the overall quest for understanding the mechanics of thermophysics.

Between 1850 and 1865 Clausius published a number of other papers dealing with mathematical statements of the second law of thermodynamics. This effort climaxed in 1865, when he introduced the concept of entropy as a thermodynamic property to describe the availability of thermal energy (heat) for performing mechanical work. In his paper of 1865, Clausius considered the universe to be a closed system (that is, a system that has neither energy nor mass flowing across its boundary) and then succinctly tied together the first and second laws of thermodynamics with the following elegant statement: "The energy of the universe is constant (first law principle); the entropy of the universe strives to reach a maximum value (second law principle)."

By introducing the concept of entropy in 1865, Clausius provided scientists with a convenient mathematical way to understand and express the second law of thermodynamics. In addition to its tremendous impact on classical thermodynamics, his pioneering work also had an enormous impact on 19th-century cosmology. As previously stated, Clausius assumed that the total energy of the universe (taken as a closed system) was constant, so the entropy of the universe must then strive to achieve a maximum value in accordance with the laws of thermodynamics. According to this model, the end state of the universe is one of complete temperature equilibrium, with no energy available to perform any useful work. Cosmologists call this condition the heat death of the universe.

In 1855 Clausius accepted a position at the University of Zurich and in 1859 married his first wife, Adelheid Rimpam. She bore him six children, but died in 1875 giving birth to the couple's last child. While he enjoyed teaching in Zurich, he longed for Germany and returned to his homeland in 1867 to accept a professorship at the University of Würzburg. He moved on to become a professor of physics at the University of Bonn in 1869, where he taught for the rest of his life. At almost 50 years of age, he organized some of his students into a volunteer ambulance corps for duty in the Franco-Prussian War (1870–71). Clausius was wounded in action and received the Iron Cross in 1871 for his services to the German army. He remarried in 1886, and his second wife, Sophie Stack, bore him a son. He continued to teach until his death on August 24, 1888, in Bonn, Germany.

Despite a certain amount of nationalistically inspired international controversy over who deserved credit for developing the second law of thermodynamics, Clausius left a great legacy in theoretical physics and earned the recognition of his fellow scientists from throughout Europe. He was elected as a foreign fellow of the Royal Society of London in 1868 and received that society's prestigious Copley Medal in 1879. He also was awarded the Huygens Medal in 1870 by the Holland Academy of Sciences and the Poncelet Prize in 1883 by the French Academy of Sciences. The University of Würzburg bestowed an honorary doctorate upon him in 1882.

⊗ Compton, Arthur Holly
(1892–1962)
American
Physicist

Arthur Holly Compton was one of the pioneers of high-energy physics. In 1927 he shared the Nobel Prize in physics for his investigation of the scattering of high-energy photons by electrons—an important phenomenon called the "Compton effect." His research efforts in 1923 provided the first experimental evidence that electromagnetic radiation possessed both particle-like and wavelike properties. Compton's important discovery made quantum physics credible. In the 1990s the National Aeronautics and Space Administration (NASA) named its advanced high-energy astrophysics spacecraft the *Compton Gamma Ray Observatory* (CGRO) after him.

A. Compton was born on September 10, 1892, into a distinguished intellectual family in Wooster, Ohio. Elias Compton, his father, was a professor at Wooster College, and his older brother, Karl, studied physics and went on to become the president of the Massachusetts Institute of Technology. As a youth, Compton experienced two very strong influences from his family environment: a deep sense of religious service and the noble nature of intellectual work. While he was an undergraduate at Wooster College, he seriously considered becoming a Christian missionary. But his father convinced

The Nobel laureate physicist Arthur Holly Compton identified a special high-energy photon scattering phenomenon. This phenomenon, now called Compton scattering, has become the foundational principle behind many of the gamma ray detection techniques used in modern high-energy astrophysics. Compton shared the Nobel Prize in physics in 1927 for his discovery. To further honor his achievements, in 1991 NASA named its large astrophysics laboratory the Compton Gamma Ray Observatory (CGRO). *(Photograph by Moffett Studio, AIP Emilio Segrè Visual Archives)*

him that, because of his talent and intellect, he could be of far greater service to the human race as an outstanding scientist. His older brother also helped persuade him to study science, and in the process changed the course of modern physics by introducing him to the study of X-rays.

Compton carefully weighed his career options and then followed his family's advice by selecting a career of service in physics. He completed his undergraduate degree at Wooster

College in 1913 and joined his older brother at Princeton. He received his master of arts degree in 1914 and his Ph.D. degree in 1916 from Princeton University. For his doctoral research, Compton studied the angular distribution of X-rays reflected from crystals. Upon graduation, he married Betty McCloskey, an undergraduate classmate from Wooster College. The couple had two sons: Arthur Allen and John Joseph.

After spending a year as a physics instructor at the University of Minnesota, Compton worked for two years in Pittsburgh, Pennsylvania, as an engineering physicist with the Westinghouse Lamp Company. Then, in 1919, he received one of the first National Research Council fellowships awarded by the U.S. government. Compton used his fellowship to study gamma ray scattering phenomena at Baron Ernest Rutherford's (1871–1937) Cavendish Laboratory in England. While working with Rutherford, he verified the puzzling results obtained by other physicists—namely, that gamma rays experienced a variation in wavelength as a function of scattering angle.

The following year, Compton returned to the United States to accept a position as head of the department of physics at Washington University in Saint Louis, Missouri. There, working with X-rays, he resumed his investigation of the puzzling mystery of photon scattering and wavelength change. By 1922 his experiments revealed that there definitely was a measurable shift of X-ray (photon) wavelength with scattering angle—a phenomenon now called the "Compton effect." He applied special relativity and quantum mechanics to explain the results, presented in his famous paper "A Quantum Theory of the Scattering of X-rays by Light Elements," which appeared in the May 1923 issue of *The Physical Review*. In 1927 Compton shared the Nobel Prize in physics with Charles Wilson (1869–1959), for his pioneering work on the scattering of high-energy photons by electrons.

It was Wilson's cloud chamber that helped verify the behavior of Compton's X-ray scattered recoiling electrons. Telltale cloud tracks of recoiling electrons provided corroborating evidence of the particle-like behavior of electromagnetic radiation. Compton's precise experiments depicted the increase in wavelength of X-rays due to the scattering of the incident radiation by free electrons. Since his results implied that the scattered X-ray photons had less energy than the original X-ray photons, Compton became the first scientist to experimentally demonstrate the particle-like "quantum" nature of electromagnetic waves. His book *Secondary Radiations Produced by X-Rays* (1922) described much of this important research and his experimental procedures. The discovery of the Compton effect served as the technical catalyst for the acceptance and rapid development of quantum mechanics in the 1920s and 1930s.

The Compton effect is the physical principle behind many of the advanced X-ray and gamma ray detection techniques used in contemporary high-energy astrophysics. In recognition of his uniquely important contributions to modern astronomy, NASA named a large orbiting high-energy astrophysics observatory, the *Compton Gamma Ray Observatory* (CGRO), in his honor. NASA's space shuttle placed this important scientific spacecraft into orbit around Earth in April 1991. Its suite of gamma ray instruments operated successfully until June 2000 and provided scientists with unique astrophysical data in the gamma ray portion of the electromagnetic spectrum—an important spectral region not observable by instruments located on Earth's surface.

In 1923 Compton became a physics professor at the University of Chicago. Once settled in at the new campus, he resumed his world-changing research with X-rays. An excellent teacher and experimenter, he wrote the 1926 textbook *X-Rays and Electrons* to summarize and propagate his pioneering research experiences. From 1930 to 1940 Compton led a worldwide scientific study to measure the intensity of cosmic rays and to determine any geographic variation in their intensity. His precise measurements showed that cosmic ray intensity actually correlated with geomagnetic latitude rather than geographic latitude. Compton's results implied that cosmic rays were very energetic charged particles interacting with Earth's magnetic field. His pre–space age efforts became a major contribution to space physics and stimulated a great deal of scientific interest in understanding the Earth's magnetosphere—an interest that gave rise to many of the early satellite payloads, including JAMES ALFRED VAN ALLEN's instruments on the *Explorer I* satellite.

During World War II Compton played a major role in the development and use of the atomic bomb. He served as a senior scientific adviser and was also the director of the Manhattan Project's Metallurgical Laboratory at the University of Chicago. Under Compton's leadership in the "Met Lab" program, the brilliant Italian-American physicist Enrico Fermi (1901–54) was able to construct and operate the world's first nuclear reactor on December 2, 1942. This successful uranium-graphite reactor, called Chicago Pile One, became the technical ancestor for the large plutonium-production reactors built at Hanford, Washington. The Hanford reactors produced the plutonium used in the world's first atomic explosion, the Trinity device detonated in southern New Mexico on July 16, 1945, and also in the Fat Man atomic weapon dropped on Nagasaki, Japan, on August 9, 1945. Compton described his wartime role and experiences in the 1956 book *Atomic Quest—A Personal Narrative*.

Following World War II, he put aside physics research and followed his family's tradition of Christian service to education by accepting the position of chancellor at Washington University in St. Louis, Missouri. He served the university as its chancellor until 1953 and then continued there as a professor of natural philosophy,

until failing health forced him to retire in 1961. He died in Berkeley, California, on March 15, 1962.

Compton received numerous awards throughout his illustrious scientific career. In addition to the Nobel Prize in physics, he also received the American Academy of Arts and Sciences Rumford Gold Medal in 1927. The Radiological Society of North America presented him with its Gold Medal in 1928. The Royal Society in London bestowed its Hughes Medal on him in 1940. That same year, he also received the Franklin Medal from the Benjamin Franklin Institute in Philadelphia, Pennsylvania. Compton served as the president of the American Physical Society (1934) and the president of the American Association for the Advancement of Science (1942).

⊠ Copernicus, Nicolas (Mikolaj Kopernik, Niklas Koppernigk, Nicolaus Copernicus, Nicholas Copernicus)
(1473–1543)
Polish
Astronomer

Revolutionary ideas gave Nicolas Copernicus his permanent place of prestige in the field of astronomy. His proposal that the Sun, rather than the Earth, was the center of our universe forever changed how we view ourselves in the cosmos. But Copernicus was neither a revolutionary nor an astronomer by trade. And his ideas of the structure of our solar system, called a heliocentric system, were ones he discovered as a student in his late 20s at the University of Cracow, studying astronomy and ancient philosophy, among other subjects.

Yet it was his interpretation of what to do with these radical theories, and his original thoughts surrounding his discoveries, that lead to revolutionary ideas supported by the likes of GIORDANO BRUNO, JOHANNES KEPLER, and

The 16th-century Polish astronomer and church official Nicolas Copernicus launched the scientific revolution with the deathbed publication of his book, *On the Revolution of Heavenly Spheres* in 1543. By advocating heliocentric cosmology, Copernicus helped overthrow the Ptolemaic system and nearly two millennia of mistaken adherence to the Earth-centered cosmology of the ancient Greek philosophers. *(AIP Emilio Segrè Visual Archives, T.J.J. See Collection)*

GALILEO GALILEI, all of whom suffered at the hands of the Roman Catholic Church for supporting Copernicus's published work, which was ironically dedicated to, and gratefully received by, the pope.

Born in Thorn, Poland, in 1473, the younger of two brothers, Copernicus was taken into the care of his uncle at age 10 after his father's death. This man, Lucas Waczenrode, with whom Copernicus would ultimately spend most of his life, was dedicated to two things: serving the Catholic Church and providing his young nephews with an education.

Copernicus's university studies began at the University of Cracow, in the capital of Poland, where art, mathematics, astrology, and astronomy were part of his course of study. In 1496 he went to Italy, taking up studies in Bologna, Padua, and Ferrara, mostly in canon law and medicine. In 1501, when he began his studies in medicine at the University of Padua, Copernicus took courses in Greek and Latin. In addition to learning new languages, he was for the first time exposed to the writings of ancient philosophers. As with any medical student of the time, he also undertook more courses in astronomy and astrology. The exposure to philosophy and astronomy would ultimately have the biggest impact on his contribution to science.

He returned to Poland in 1503, rejoining his uncle, who had risen through the ranks of the Catholic Church to the position of bishop of Ermeland. Copernicus, now 30 years old, accepted a position in the Chapter of Frauenberg, and in addition to his duties as canon became his uncle's personal physician. His life was to be one of dedication to the church, even to the point of ending a relationship with his housekeeper when the church told him to do so. Astronomy was something he worked on in his spare time.

It was sometime around 1512 that Copernicus wrote his *Commentariolus*, a "short comment" outlining his thoughts on the order of the universe. This new world system was based on PTOLEMY's geocentric system, with a twist—instead of the Earth being the center of the universe, Copernicus theorized that everything revolved around the Sun. He also believed that the Earth rotated on its own axis, and he used the rotation of the Earth and its revolution around the Sun to explain the idea of planets spinning in retrograde.

Seventeen centuries earlier, according to the work of Archimedes (287–212 B.C.E.), the Pythagorean ARISTARCHUS OF SAMOS had first proposed that the Earth rotated on its axis and revolved around the Sun. His contemporaries thought his ideas were absurd, if not blasphemous, based on simple observation of everything that was "natural." Ptolemy's geocentric model was upheld as the one true system that "worked," even though it was extremely complicated and very flawed. Common practice throughout the centuries was to construct excuses to prove that Ptolemy's model was right. This was a normal part of science that became referred to as "saving the appearances" that the geocentric system ruled.

When Copernicus worked out the same ideas proposed by Aristarchus, he was the last to think that he was contradicting Ptolemy. He felt that his theories were an extension of Ptolemy's work, offering some simplicity and clarity. He worked on this process of clarification, making computations and creating new tables, for a few decades. He mostly kept his *Commentariolus* to himself, sharing it only with a few good friends, who also happened to be members of the church. Surprisingly, the Catholics he entrusted with his work applauded it, and repeatedly urged him to publish his ideas, which he declined to do for several years. But word of his findings traveled, eventually making its way to a young mathematician in Wittenberg named Rheticus (1514–74), who in 1539 traveled to meet the man who had done what had never been done before—combine a simpler theory with personal observation, and computation with new tables, to explain a new solar system. Copernicus gave Rheticus permission to tell others about this theory, and the mathematician wrote about Copernicus's work in his own publication *Narratio prima* (*First Communication*), crediting the originator as "The Reverend Father Dr. Nicholas of Torun, Canon of Ermland." The Catholics embraced the work of the good Reverend Father. The Lutherans, however, were another matter.

The Protestant leader Martin Luther (1483–1546) blasted Copernicus for his radical and heretical ideas, quoting the scripture "Joshua bade the sun and not the earth to stand still," as proof that Copernicus was doing the devil's work, and emphatically stating that he believed the "fool" Copernicus would "reverse the entire Ars Astronomiae," the science of all known astronomy, if he continued putting forth such ludicrous ideas.

But the persistent Rheticus would not let Copernicus hide his work any longer, and he convinced Copernicus to let him get the work printed in Nuremberg. In 1542 Rheticus took Copernicus's manuscript to Nuremberg, where it soon found its way into the publisher's house and the hands of a Lutheran priest, Andreas Oslander. Not wanting to stir things up with his fellow Lutherans, Oslander took it upon himself to write an anonymous preface, making it look as though Copernicus had written it himself, stating that the hypotheses contained in the book were not necessarily true, "nor even probable."

While many of Copernicus's ideas were flawed—for example, his belief, like Ptolemy's, that all orbits were perfectly circular—they provided the foundation for a major shift in thinking about the order of the universe. But not in Copernicus's lifetime. His book, *De revolutionibus orbium coelestium* (On the revolutions of the heavenly spheres) was finally published and a printed copy delivered, reportedly on May 24, 1543, to him in Frauenberg. A victim of stroke, he died that same day.

Oslander's disclaimer served to lead many to believe that Copernicus's theories were inconsequential to the author himself, although the newly calculated tables were considered useful. As for the Lutherans, the publication angered them even more. Seven years later, the Lutheran leader Melanchthon (1497–1560) was still staunchly denouncing Copernicus, quoting the Bible, and decreeing that with "these divine testimonies we will cling to the truth." Overall acceptance of Copernicus's theories did not happen. In addition to the religious flak the book was taking, the scholars of the time dismissed the heliocentric system completely. His book was not printed again until 1566, and then not again until 1617. But the work caught the spirit and minds of a few important figures who not only perpetuated the theories but also gave them new legs to stand on.

Giordano Bruno was one brave soul who believed in Copernicus's heliocentric system. But by the time he started to spread the word about his beliefs, the Catholic Church had decreed that the act of declaring the Sun as the center of the universe was heresy, and Bruno was burned alive at the stake for speaking his truths.

The biggest shift leading to the acceptance to Copernicus's new order happened when Johannes Kepler learned about the heliocentric system as a student at the university in Tubingen. Kepler communicated with Galileo Galilei, who confessed that for "many years," he had agreed with Copernicus. He told Kepler that he had even written arguments for and against the ideas, but had not told anyone about this "until now . . . deterred by the fate of Copernicus himself, our master, who although having won immortal fame with some few, to countless others appears . . . as an object of derision and contumely." He told Kepler that "Truly, I would venture to publish my views if more like you existed; since this is not so, I will abstain."

Kepler's passionate reply called upon Galileo to show his proofs to the world: "Be confident Galilei and proceed!" It was 1597. An undeniable change must have been written in the stars. Galileo surely saw it when he made his first telescope in 1609. Unfortunately, the Catholic Church changed its once-favorable stance on Copernicus, and the writings of all three men were declared heresy.

Centuries later, in 1835, after Galileo's important work had finally been back in circulation for nearly 100 years, the church and the scientific community were finally beginning to coexist. Acceptance of heliocentric thinking could freely take place, as the books of Galileo, Kepler, and the man who started it all, Copernicus, were removed from the church's list of forbidden books.

⊠ Curtis, Heber Doust
(1872–1942)
American
Astronomer

Heber Curtis gained national attention in 1920 when he engaged in the "Great Debate" over the size of the universe with fellow American astronomer HARLOW SHAPLEY. Curtis supported the daring hypothesis that spiral nebulas were actually "island universes"—that is, other galaxies, existing far beyond the Milky Way Galaxy. This radical position meant that the observable universe was much larger than anyone dared imagine. By 1924 EDWIN POWELL HUBBLE was able to support Curtis's hypothesis, when he demonstrated that the great spiral Andromeda "nebula" was actually a large galaxy similar to, but well beyond, the Milky Way.

Heber Curtis was born in Muskegon, Michigan, on June 27, 1872. As a child, he went to school in Detroit. He then studied the classic languages (Latin and Greek) at the University of Michigan for five years, receiving his bachelor of arts degree in 1892 and his master of arts degree in 1894. Following graduation, Curtis moved to California to accept a position at Napa College in Latin and Greek studies. However, while teaching the classics, he became interested in astronomy and volunteered as an amateur observer at the nearby Lick Observatory. In 1896 he became a professor of mathematics and astronomy at the College of the Pacific, following a merger of that institution with Napa College.

In 1900 Curtis served as a volunteer member of the Lick Observatory expedition that traveled to Thomaston, Georgia, to observe a

The American astronomer Heber Curtis stands on a ladder at the eyepiece of the 24-inch Crossley reflector telescope at the Lick Observatory. Curtis gained national attention in 1920 when he engaged in the "Great Debate" over the size of the universe with fellow American astronomer Harlow Shapley. Curtis supported the then-daring hypothesis that spiral nebulas were actually "island universes"—that is, other galaxies far beyond the boundaries of the Milky Way Galaxy. *(Courtesy AIP Emilio Segrè Visual Archives, Shapley Collection)*

solar eclipse. His enjoyment of and outstanding performance during the eclipse expedition converted Curtis's amateur interests in astronomy into a lifelong profession. After his first solar

eclipse expedition, he entered the University of Virginia on a fellowship and graduated in 1902 with his doctorate in astronomy. After graduation he joined the staff of the Lick Observatory as an astronomer. From 1902 to 1909 he made precise radial velocity measurements of the brighter stars in support of WILLIAM WALLACE CAMPBELL's observation program. Starting in 1910, Curtis became interested in photographing and analyzing spiral nebulas. He was soon convinced that these interesting celestial objects were actually isolated independent systems of stars, or "island universes," as IMMANUEL KANT had called them in the 18th century.

Many astronomers, including Harlow Shapley, opposed Curtis's hypothesis about the extragalactic nature of spiral nebulas. On April 26, 1920, Curtis and Shapley engaged in their famous "Great Debate" on the scale of the universe and the nature of the Milky Way Galaxy at the National Academy of Sciences, in Washington, D.C. At the time, the vast majority of astronomers considered the extent of the Milky Way Galaxy synonymous with the size of the universe—that is, they thought the universe was just one big galaxy. However, neither astronomer involved in this highly publicized debate was completely correct. Curtis argued that spiral nebulas were other galaxies similar to the Milky Way—a bold hypothesis later proven to be correct. But he also suggested that the Milky Way was small and that the Sun was near its center—both of these ideas were subsequently proven incorrect. Shapley, on the other hand, incorrectly opposed the hypothesis that spiral nebulas were other galaxies. He argued that the Milky Way Galaxy was very large (much larger than it actually is) and that the Sun was far from the galactic center.

In the mid–1920s the great American astronomer EDWIN POWELL HUBBLE helped to resolve one of the main points of controversy when he used the behavior of Cepheid variable stars to estimate the distance to the Andromeda Galaxy. Hubble showed that this distance was much greater than the size of the Milky Way Galaxy proposed by Shapley. So, as Curtis had suggested, the great spiral nebula in Andromeda and other spiral nebulas could not be part of the Milky Way and must be separate galaxies. In the 1930s astronomers proved that Shapley's comments were more accurate concerning the actual size of the Milky Way and the Sun's relative location within it. Therefore, when viewed from the perspective of science history, both eminent astronomers had argued their positions using partially faulty and fragmentary data. The Curtis-Shapley debate triggered a new wave of astronomical inquiry in the 1920s that allowed astronomers like Hubble to determine the true size of the Milky Way and to recognize that it is but one of many other galaxies in an incredibly vast, expanding universe.

In 1920 Curtis became the director of the Allegheny Observatory of the University of Pittsburgh. During the next decade he improved instrumentation at the observatory and participated in four astronomical expeditions to observe solar eclipses. In 1930 he returned to the University of Michigan as a professor of astronomy and the director of the university's astronomical observatories. He served in those positions until his death in Ann Arbor, Michigan, on January 9, 1942. His wife, Mary D. Rapier Curtis, his daughter, and three sons survived him. He was president of the Astronomical Society of the Pacific (1912), a fellow of the American Association for the Advancement of Science (1924), vice president of the American Astronomical Society (1926), a member of the American National Academy of Sciences, and foreign associate of the Royal Astronomical Society in London.

D

Dirac, Paul Adrien Maurice
(1902–1984)
British
Physicist, Mathematician

Speaking at a dedication of a memorial plaque being unveiled at London's Westminster Abbey, the famous physicist SIR STEPHEN WILLIAM HAWKING said, "Paul Dirac has done more than anyone this century, with the exception of ALBERT EINSTEIN, to advance physics and change our picture of the universe."

Born in Bishopston, Bristol, United Kingdom, to a Geneva-born Swiss-French father Charles Dirac, and a Cornwall-bred mother Florence Hannah Holton, Dirac, his older brother Reginald, and their younger sister Beatrice endured a harsh childhood under their father's strict discipline and generally spent their time in an unhappy household. Both boys would become alienated from their father by the time they left for college.

Dirac found solace in his studies, and by the time he graduated from Bishop Primary School his exceptional talent for mathematics had become apparent. In 1914, at age 12, he entered the secondary school at Merchant Venturers Technical College, where his father taught French. Dirac would later comment that as World War I broke out, the older students went off to military service, allowing the younger students better and more frequent access to the science laboratories and other facilities.

He graduated at the top of his class, and in 1918, he went to the University of Bristol to study electrical engineering before applying for and receiving a scholarship to Cambridge University in 1921. But the scholarship required additional money for full support, which he was unable to obtain, so he enrolled in Bristol University. There, he continued to excel, and graduated with a B.S. degree in mathematics with honors in 1923. At this time he earned a research grant at St. John's College at Cambridge.

During this period, classical physicists (Newtonian scholars) were having trouble reconciling SIR ISAAC NEWTON's physics with the behavior of electrons and atoms, and were laboring to develop a new theory to explain it. At Cambridge, Dirac studied under Ralph Fowler (1889–1944), the leading theoretician versed in the new theory of quantum physics, and became enamored of the algebraic commutators put forth by Werner Heisenberg (1901–76).

Dirac soon developed his own theory, a combination of quantum mechanics and special relativity, which included wave mechanics and matrix mechanics. By the time Dirac's doctoral thesis, "Quantum Mechanics," earned him his Ph.D. in 1926, he already had 11 papers in print. The thesis was widely acclaimed and he was elected a Fellow of St. John's College in 1927.

After receiving his degree, Dirac went to Copenhagen to work with Niels Bohr (1885–1962), then went to Göttingen where he worked with Robert Oppenheimer (1904–67), Max Born (1892–1970), James Franck (1882–1964), and the Russian Nobel laureate Igor Tamm (1895–1971). He made it a point to visit the Soviet Union several times, first in 1928 and then almost once a year throughout the 1930s.

In 1928 Dirac published his famous "spin-1/2 Dirac Equation," which is still widely used today, and which explained the mysterious magnetic and "spin" properties of the electron. He used this equation to predict the existence of a particle with the same mass as an electron, but with an opposite charge (that is, a positively charged "antimatter" electron). This antimatter was subsequently discovered and called a positron in 1932 by CARL DAVID ANDERSON. Today positrons are used every day in medicine, in PET (positron emission tomography) scanners that pinpoint places in the brain, such as where drugs are chemically active, which works by detecting the radiation as the positrons emitted from radioactive nuclei annihilate ordinary electrons nearby.

In 1930 Dirac published his epic *The Principles of Quantum Mechanics*, which is still in print, and in 1932 he was appointed Lucasian Professor of Mathematics at Cambridge, a post he held for 37 years, which was formerly held by Newton and since 1980 by Hawking. Also in 1930, he was made a Fellow of the Royal Society, and would be awarded that society's Royal Medal in 1939 and the prestigious Copley Medal in 1952.

The publication of his book took the physics community by storm, transformed the current image of the atomic universe, and resulted in his sharing the 1933 Nobel Prize in physics at 31 years of age with Austrian Erwin Schrödinger

(1887–1961). A shy and retiring fellow who abhorred the limelight, Dirac at first wanted to refuse the Nobel Prize because of the publicity it would generate, but changed his mind when he was told that refusing it would bring only more publicity than ever upon him.

The years 1934 and 1935 were equally dramatic, for both professional and personal reasons. Dirac accepted an invitation to visit Princeton University to meet the famous Hungarian physicist and future Nobel laureate (1963) Eugene Wigner (1902–95). While Dirac was there, Wigner's sister, Margit, came from Budapest to visit her brother, and Dirac fell in love. Married in London, the two eventually had two daughters. Margit had two sons by a previous marriage, and they took Dirac's name; one of them, Gabriel Andrew Dirac, became a famous pure mathematician and was a professor of pure mathematics at the University of Aarhus in Denmark.

During World War II Dirac worked on uranium separation and nuclear weapons with a Birmingham group investigating atomic energy, an assignment that led the British government to ban him from visiting the Soviet Union until 1957.

Dirac shunned all honorary degrees, of which dozens were showered on him, but he readily accepted honorary admission to respected academies and scientific societies. Of these there are also dozens, but primary among them were the USSR Academy of Sciences in 1931; Indian Academy of Science in 1939; Chinese Physical Society in 1943; the Royal Irish Academy in 1944; the Royal Society of Edinburgh in 1946; Institut de France in 1946; the National Institute of Sciences of India in 1947; the American Physical Society in 1948; the National Academy of Sciences in 1949; the National Academy of Arts and Sciences in 1950; Accademia delle Scienze di Torino in 1951; Academia das Ciencias de Lisboa in 1953; Pontifical Academy of Sciences, Vatican City, in 1958; Accademia Nazionale dei Lincei, Rome, in 1960; the Royal Danish Academy in 1962; and the Académie des Sciences de Paris in 1963. He was appointed to the Order of Merit in 1973.

After a glorious career reshaping the way physicists looked at the inner and outer universes, Dirac retired from Cambridge in 1969 and moved his family to Florida. He held visiting appointments at the University of Miami and at Florida State University (FSU). In 1971 he was appointed professor of physics at FSU, where he continued his research and continued to travel as well. In 1973 and 1975 he lectured at the Physical Engineering Institute in Leningrad.

"One of the most influential scientists of the twentieth century," as the editor of *Physics World* called him, this Bristol-born giant of the world of quantum physics died in Tallahassee in 1984.

⊠ Doppler, Christian Andreas
(1803–1853)
Austrian
Mathematician

Anyone who has ever gotten a speeding ticket can thank Christian Doppler. It was he who developed the famous "Doppler effect," which is the principle used in speed guns and traffic radar.

Doppler's family had been stonemasons in Salzburg, Austria, since 1674, and it was expected that he, too, would enter the family business. However, he was born to ill health and by the time he reached his teens he was too frail for the demanding work of stonemasonry.

After primary school in Salzburg, he attended secondary school in nearby Linz, but was only an average student. When his parents consulted a mathematics professor at the Salzburg Lyceum as to young Doppler's potential, the professor suggested that he might have a talent in mathematics, and recommended that Doppler

enter the Vienna Polytechnic Institute. Doppler enrolled in 1822.

Doppler finally began to excel and, indeed, mathematics turned out to be his forte. After graduating in 1825, he returned to Salzburg and studied philosophy at the Lyceum, then returned to Vienna, this time to study astronomy, mechanics, and higher mathematics at the university. In 1829 Doppler began a four-year tenure as an assistant to a professor of higher mathematics and mechanics. He published his first technical paper, "A Contribution to the Theory of Parallels," plus three others over the four-year period.

By the time Doppler was 30 years old, he had grown weary of being someone's assistant, and he began to think in terms of finding a permanent position in his own right. However, it was not easy at this time in Austria. Applicants for vacant professorships had to enter a public competition for the available jobs, in which the applicant had to take a day-long written exam and then give a lecture to an appointed panel of examiners. Politics played a key role in getting the final appointment.

Doppler applied for professorships at several schools in this manner, including the universities in Gorizia, Linz, Ljubljana, Salzburg, the Technical Secondary School in Prague, and Vienna Polytechnic Institute. Meanwhile, to earn a living he took a job as a bookkeeper at a cotton spinning factory. Through all these stressful efforts, his health, which was not good to begin with, began to deteriorate.

The public competition for all those jobs took almost two years to complete. At one point Doppler gave up and decided to move to America, but just when he was interviewing the American consul in Munich, he received the happy news in 1835 that he had won the job at the Technical Secondary School in Prague.

Once again, Doppler was unhappy with the situation. He considered teaching elementary mathematics in secondary school to be beneath him, which indeed it was, and he cast around for a better job. He failed again at getting a position at the Polytechnic Institute, but in 1836, he got a part-time job teaching higher mathematics four hours a week, and kept this job for two years.

In 1837 politics reared its ugly head again. The post of professor of geometry and mathematics at the Polytechnic became vacant, and Doppler filled the post. However, even as he was discharging his duties, the institute held a public competition for his job behind his back. When he found out about it, he expressed his dismay in strong terms, and was told he did not have to take part in the competition. It remains a mystery why it was even held at all, because Doppler was finally appointed formally to the position of full professorship in 1841.

It was during this time that Doppler presented his magnum opus, a paper entitled "On the Colored Light of the Double Stars and Certain Other Stars of the Heavens," which explained Doppler's idea that sound was composed of longitudinal waves that could be compressed or expanded, and that the frequency of the wave varied according to its velocity in relation to an observer.

As a practical proof of his theory, Doppler arranged for a trumpeter to sound a series of the same notes over and over as he rode on a flat car pulled by a speeding railroad train. On the ground he stationed another musician with the ability to identify musical tones by ear. As the train approached closer and then rushed past and moved quickly away, the second musician recorded that the notes from the trumpeter sounded different when the train was approaching than when it was moving away. Thus was born the famous "Doppler effect," which everyone has experienced when riding a swiftly moving vehicle past a ringing bell.

Doppler also tried to prove that his theory applied to light, but was unable to and it was left for Armand Fizeau (1819–96) to generalize and then prove Doppler's work as it applies to light. Doppler's work placed him in the pantheon of astronomy because it was the first major contribution to the discovery that the universe is expanding.

Even with his new fame, he was having a hard time of it at work. When students complained that his examinations were too difficult, he was reprimanded by the authorities and the stress further affected his health. At about this time he was also under great duress to participate in the examination of hundreds of students applying for competitive positions. For example, one report states that in January and February 1843, he had to examine 256 students in 17 days, including six hours of both oral and written exams in arithmetic and algebra. The same number of students sat for theoretical geometry exams in June and July, and in August Doppler had to examine 145 students in eight days. The strain was too much for him, and in 1844 he finally requested a sick leave, which lasted two years before he was well enough to return to work.

In 1846 Doppler changed venues, accepting a professorship of mathematics, physics, and mechanics at the Academy of Mines and Forests in Banska Stiavnica, mostly to escape the political unrest as the monarchy began to topple in Prague. By now he was becoming famous because of his widely celebrated Doppler effect thesis, and was appointed professor again at the Polytechnic Institute. In 1848 Doppler was elected to the Imperial Academy of Sciences in Vienna and received an honorary doctorate from the University of Prague. In 1850 he was appointed the first director of the newly established Institute of Physics at Vienna University.

But his health took a serious turn for the worse and by 1852 he decided to move to Venice for the warmer climate. He could not recover, though, and he died there in March 1853.

Although Doppler also published significant papers on electricity and magnetism, described the variation of magnetic declination with time, and published papers on optics and astronomy, his definition of the Doppler effect was his crowning glory. Today the Doppler effect is the principle behind technologies used across the spectrum of industry. In medicine, Doppler ultrasound is used widely in imaging instrumentation and vascular studies. Doppler radar is both an important military tool and useful in weather forecasting. Oceanographers use Doppler sonar to map ocean currents. Law enforcement uses speed guns and traffic radar. And in astronomy, his principle is still useful for measuring the speed of bodies and searching for new planets.

✂ Douglass, Andrew Ellicott
(1867–1962)
American
Astronomer, Environmental Scientist

Early in the 20th century, the American astronomer Andrew Douglass postulated that there might be a measurable relationship between sunspot activity and the terrestrial climate. To explore this hypothesis, he started the field of dendrochronology (tree-ring dating). Despite his years of investigation, he could only link tree ring data with local climate episodes, since this data could not provide an unambiguous link between past global climate conditions and sunspot activity. However, his pioneering efforts provided both archaeologists and climatologists with an important new research tool and also anticipated the efforts by modern scientists who employ sophisticated satellites in their contemporary investigation of Sun-Earth environmental relationships.

Douglass was born in Windsor, Vermont, on July 5, 1867. He came from a family with a long tradition of academic service; his father and grandfather were both college presidents. He graduated with honors from Trinity College in Hartford, Connecticut, in 1889, and upon graduation accepted a research assistantship at the

Harvard College Observatory. From 1891 to 1893 Douglass served as the chief assistant on the astronomical expedition that founded the Harvard Southern Hemisphere Observatory in Arequipa, Peru.

When Douglass returned from Peru, PERCIVAL LOWELL hired him to help locate the most suitable site in the "Arizona Territory" (statehood did not occur until 1912) for a new astronomical observatory dedicated primarily to support Lowell's obsession with Mars. Douglass recommended the city of Flagstaff and moved there after the founding of the Lowell Observatory in 1894. He was a good astronomer and soon became Lowell's chief assistant. Douglass had the primary task of collecting data about Mars, but the young astronomer from Vermont soon fell into sharp disagreement with Lowell on how certain data were being selectively applied to support Lowell's rather unscientific approach to prove Mars was inhabited by intelligent beings. Lowell was actually a skilled observational astronomer, but his results were often blurred by the preconceived notion that GIOVANNI VIRGINIO SCHIAPARELLI's *canali* (channels) were canals built by a race of intelligent beings. After numerous "scientific method" clashes, Lowell wanted no more internal dissidence, and he fired Douglass in 1901.

Unperturbed, Douglass remained in the Flagstaff area until 1906, when he joined the University of Arizona in Tucson as its first professor of astronomy. He also taught physics at the university and became the first astronomer to photograph the zodiacal light. In 1916 he secured funding from the Steward family to construct an astronomical observatory at the university. Two years later Douglass began his long and productive tenure as the director of the new Steward Observatory. Construction of the observatory was completed in 1923 and the new facility represented the first of several important milestones that would make the University of Arizona an internationally recognized center of astronomy. As the city of Tucson grew with statehood, Douglass found it necessary to move the observatory to Kitt Peak (south of the city) to avoid the problems of light pollution from the expanding urban environment. Douglass served as the director of the Steward Observatory until 1937, when he became the director of the university's Laboratory for Tree-Ring Research. Although a skilled astronomer, Douglass is now best remembered for his pioneering work in tree-ring dating.

While working in Flagstaff, Douglass had the great inspiration that led him to create the field of dendrochronology (tree-ring dating). At the time, he was searching for a measurable environmental relationship that would link the sunspot cycle and past climate episodes. He reasoned that vegetation changes, especially as portrayed by the growth of certain species of trees such as ponderosa pines, Douglas firs, and the great sequoias (giant redwoods), might provide the solution. Douglass hypothesized that the motion of the planets and the behavior of the Sun were responsible for weather and climate changes. Working in Arizona and portions of New Mexico, Douglass was soon able to analyze the rings of local trees, primarily pines and Douglas firs, and relate the size of each ring to previous annual levels of rainfall. A tree produces a ring each year and, depending on climate conditions, certain species of trees produce wide rings during wet years and narrow rings during dry years. In effect, the cross-sections of these trees recorded past rainfall and served as a natural clock.

Recognizing that tree-ring data in Arizona represented a record of past rainfall, Douglass further hypothesized that the tree-ring data might also represent a quantitative indication about the abundance of solar energy and the influence of the Sun on Earth's past climate. By 1911 he began studying the rings of the great sequoias, trees that are estimated to be between 1,800 and 2,700 years old. Between 1919 and 1936 he summarized his pioneering efforts in his three-volume treatise entitled *Climate Cycles and Tree Growth*. While his tree-ring data from Arizona and New Mexico indicated past periods of drought, to his disappointment these data generally did not correlate with similar climate episodes in other parts of the world. He and other scientists could only conclude that the past climate changes revealed by such tree-ring "clocks" were most likely only local environmental episodes, and not necessarily correlated with sunspot activity and climate stress on a planetary scale. Contemporary Earth system science models, using sophisticated spacecraft and computers, now help scientists investigate linkages between the Sun's long-term behavior and terrestrial weather and climate trends.

While never able to personally prove his original hypothesis and link sunspot activity with vegetation changes and terrestrial climate, Douglass established tree-ring dating as a valuable tool for archaeologists who needed to accurately date ancient structures and for environmental scientists who were searching for definitive records of past climate episodes. As a historic note, the American chemist Willard Libby (1908–80) invented radiocarbon dating in 1949. Libby's carbon-14 radioisotope technique provided archaeologists a universal clock with which to study the past. Nevertheless, dendrochronology still remains a useful tool in archaeology and environmental studies and often provides a useful calibration tool for the radiocarbon technique.

After an additional two decades of service to the University of Arizona, Douglass retired from his position as director of the Laboratory of Tree-Ring Research in 1958. He died in Tucson on March 20, 1962. The world-famous Steward Observatory and the field of dendrochronology are two important legacies of this remarkable scientist. His accomplishments earned him an appointment as research associate to the Carnegie Institution of Washington and a life membership to the National Geographic Society.

⊠ Drake, Frank Donald
(1930–)
American
Radio Astronomer

Frank Drake was born in 1930 to Richard and Winifred Drake and raised on Chicago's South Shore with his sister, Alma, and brother, Robert. Like millions of others of his generation, he spent hundreds of hours experimenting with motors, clocks, chemistry sets, radios, and gazing up at the stars on clear summer nights. Also like millions of kids who became hooked on science fiction, young Drake began to wonder about the existence of life elsewhere in the universe.

Drake won a Reserve Officers' Training Corps (ROTC) scholarship to Cornell University, majored in electronics, became fascinated by astronomy, and in 1952 received a B.A. degree in engineering physics. However, in his junior year he attended a lecture by the famous astrophysicist Otto Struve (1897–1963), during which Struve presented accumulating evidence that many stars in the Milky Way Galaxy had planetary systems around them. Further, Struve went on to speculate that with such a vast number of solar systems, the chances were good that one of them could sustain life. At last, Drake had found someone else who agreed with his growing contention that the odds were almost overwhelming that life did exist somewhere else.

Drake then spent three years in the U.S. Navy as an electronics officer on the U.S.S. *Albany*, accumulating experience in fixing and operating the most modern high-tech electronic equipment then available. When he finally left the navy, Drake decided to go to Harvard and study optical astronomy, but as fate would have it the only course available was in radio astronomy.

This turned out to be a perfect fit with his navy experience.

Now hooked on radio astronomy, Drake received his Ph.D. in astronomy from Harvard in 1958. He then took a position at the newly founded National Radio Astronomy Observatory (NRAO) in Green Bank, West Virginia, where he eventually became head of the Telescope Operations and Scientific Services Division. There Drake organized a formal search for extraterrestrial signals called Project Ozma, a two-month observation of the close stars Tau Ceti and Epsilon Eridani, using the 85-foot single channel antenna tuned to the 1420 megahertz frequency of hydrogen. No signals were detected.

In 1959 two physicists at Cornell, Giuseppi Cocconi and Philip Morrison, published a paper in which they described the possibility of using a microwave radio for interstellar communication, a potential apparatus that Drake had also been contemplating.

In 1961 Drake and J. Peter Pearman, of the Space Science Board of the National Academy of Sciences, organized the now famous Search for Extra-Terrestrial Intelligence (SETI) conference at the NRAO. It was a small meeting, with a dozen or so scientists attending to share their views and interest in searching for extraterrestrial life. In preparation for this meeting, Drake formulated what became known as the Drake equation to estimate the number of possible technologically advanced civilizations that could be sending radio signals in Earth's direction:

$$N = N * f_p n_e f_l f_i f_c L$$

in which

$N*$ represents the number of stars in the Milky Way Galaxy, about 200 billion;

f_p is the fraction of stars that have planets around them; thanks to the Hubble telescope new planets are being discovered every month, but Drake estimates this to be 20 percent;

n_e is the number of planets per star that are capable of sustaining life; based on this solar system, the quantity is given as 4—Earth, Venus, Mars, and possibly one of Jupiter's moons;

f_l is the fraction of planets in ne where life evolves; current estimates range from 100 percent to close to 1 percent, so 50 percent is used as a good arbitrary figure;

f_i is the fraction of fl where intelligent life evolves; again, guesses range from 100 percent to 1 percent, the logic being that intelligence is such a survival advantage that it will certainly evolve;

f_c is the fraction of fi that communicate; again impossible to guess, but experts give it 10 percent to 20 percent;

L is the length of time the communicating civilizations send detectable communications into space.

Using us as an example, the expected lifetime of the Sun and Earth as roughly 10 billion years, and thus far humans have been communicating with radio waves for less than 100 years. The figure for this variable depends solely on estimates of how long the Earth's civilization will survive.

All of these variables multiplied together results in N, the number of communicating civilizations in the galaxy.

After a brief tour with the Jet Propulsion Laboratory in Pasadena, California, in 1963, Drake joined the Center for Radiophysics and Space Research at Cornell, and in 1965 became director of Cornell's Arecibo Observatory in Puerto Rico. He returned to Cornell in 1968 as chairman of the astronomy department, and was Goldwin Smith Professor of Astronomy until 1984. He then joined the faculty of the University of California at Santa Cruz (UCSC) as dean of the natural sciences division, then professor of astronomy and astrophysics. He continues his tenure at UCSC today and is still chairman of the Board of Trustees of the SETI Institute.

Besides sharing in the discovery of the radiation belts of Jupiter, conducting the first SETI organized searches, and positing Drake's equation, Drake's methods were used in sending additional messages to outer space on the *Pioneer 10* and *Pioneer 11* (with plaques designed by Drake, CARL SAGAN, and Linda Sagan), and on board the *Voyager* spacecraft, conceived by Drake and compiled by several scientists.

Drake has received numerous honorary degrees and appointments, was chairman of the National Research Council Board of Physics and Astronomy, a member of three NAS/NRC Astronomy Survey committees, and president of the Astronomical Society of the Pacific. The Frank D. Drake Planetarium was dedicated in Norwood, Ohio, in 1983. He has also written more than 150 articles, and the book *Is Anyone Out There?* (1992), an autobiographical work coauthored with Dava Sobel.

Readers who wish to play with Drake's equation may enter their own variable values on the Internet at http://www.planetarysystems.org/drake_equation.html. Drake's own solution to his equation is N=10,000 communicative civilizations in the Milky Way Galaxy.

Draper, Henry
(1837–1882)
American
Astronomer, Photographer

Henry Draper took the first photograph of a distant astronomical object, the Orion Nebula, giving a push to celestial photography that opened modern astronomy to a new vision of the universe. All the men in the Draper family made pioneering strides in science, especially in the fields of meteorology and astronomy in the 19th century. Draper's father, John William Draper, who was primarily a chemist, took the first photograph, a daguerreotype, of the Moon in 1839, one of the first human photographs in 1840, and the first photograph of the diffraction spectrum. Henry's brother John was a noted physician and chemist, and his brother Daniel was a meteorologist who established the New York Meteorological Observatory in Central Park in 1868.

Draper was born in Virginia to John William Draper and Antonia Coetana de Paiva Pereira Gardner Draper, whose father was the personal physician to the emperor of Brazil. As a boy, Draper assisted his father in pursuing photographic techniques, and at age 13 helped him take pictures of microscope slides for a textbook. When he graduated from medical school in 1857, he used this experience to write his thesis on the spleen using photographs of microscope slides as illustrations. However, the law at the time dictated that he could not receive his medical degree until he was 21, so he took a year off and traveled through Europe.

During his travels, Draper visited Lord Rosse's observatory in Ireland, which at the time utilized the world's largest telescope, the 72-inch Leviathan reflector. The experience kindled within him an urgent desire to investigate the use of photography in astronomical observations when he returned home. To this goal, he built his own observatory on his father's estate at Hastings-on-Hudson, New York, while he was beginning his medical career. Draper joined the staff at Bellevue Hospital in New York City, and later became first a professor and then dean of the New York University School of Medicine.

In 1867 Draper married Anna Mary Palmer, a wealthy socialite who not only became an amiable and capable hostess, frequently entertaining the prominent scientists and celebrities of the day, but who also became a proficient and dedicated laboratory assistant to Draper.

Draper is best known for recording on photographic plates the Great Nebula of Orion on September 30, 1880. The images were not of the best quality, but the techniques were improved upon rapidly after his death in England. Draper took the first stellar spectrum photograph of Vega in August 1872, and, with Tebbutt's comet in 1881, made the first wide-angle photograph of a comet's tail and the first spectrum of a comet's head. Refining his father's work to a significant degree, he produced a great number of photographs of the Moon and made a benchmark spectrum of the Sun in 1873. He also took spectra of the Orion Nebula, the Moon, Mars, Venus, and several first magnitude stars.

From there Draper went on to invent the slit spectrograph, advancing the state-of-the-art in telescope clock drives and instrument optics. He also wrote a chemistry textbook and published many papers on his astronomical work and telescope design. He was one of the first to recommend building an observatory in the Andes Mountains to avoid atmospheric pollution.

Before his sudden death on November 20, 1882, of double pleurisy during a hunting trip to the Rocky Mountains, he had received honorary law degrees from New York University and the University of Wisconsin, a congressional medal for directing a U.S. expedition to photograph the transit of Venus in 1874, and was elected both to the National Academy of Sciences and Germany's Astronomische Gesellschaft. He was a member of the American Photographic Society, the American Philosophical Society, the American Academy of Arts and Sciences, and the American Association for the Advancement of Science.

After his death, Draper's wife established the Henry Draper Memorial Fund to support the advancement of photography in astronomical efforts. This fund supported a massive photographic stellar spectrum survey performed by ANNIE JUMP CANNON and other women at Pickering's observatory at Harvard University, which became the celebrated *Henry Draper Catalogue*, or the *HD Catalogue*, as it is still known and used today.

⊠ Dyson, Sir Frank Watson
(1868–1939)
British
Astronomer

Sir Frank Watson Dyson participated with SIR ARTHUR STANLEY EDDINGTON during the May 1919 solar eclipse expedition that measured the deflection of a beam of starlight as it passed close to the Sun. This effort provided the first experimental evidence in support of ALBERT EINSTEIN's theory of general relativity. A specialist in solar eclipses and in stellar motion studies, Dyson also served as Astronomer Royal of Scotland from 1905 to 1910 and then as England's Astronomer Royal from 1910 to 1933.

Dyson was born at Measham, near Ashby-de-la-Zouch, Leicestershire, England, on January 8, 1868. He was the son of a minister and won scholarships for his secondary and college education. Dyson graduated from Trinity College at Cambridge University in 1889 with a degree in mathematics. Following graduation, he became a Fellow at Trinity College in 1891 and then chief assistant to the Astronomer Royal at the Royal Greenwich Observatory in 1894. His primary astronomical studies involved the proper motions of stars and participating in solar eclipse expeditions on which he made special observations of the Sun's outer regions.

Dyson managed the Carte du Ciel project of photographing the entire sky—a project that involved an investigation of the proper motion of many stars and eventually led other astronomers to discover that the Milky Way Galaxy was rotating. In astronomy, the proper motion of a star is its apparent annual angular motion on the celestial sphere—a tiny effect that after thousands of years causes groups of stars with differences in the direction of their proper motions to change shape in some appreciable manner. In collaboration with William G. Thackeray, Dyson measured the position of 4,000 circumpolar stars and then compared these measurements with early 19th-century observations to determine proper motions. Dyson's efforts helped extend the "star streaming" work of JACOBUS CORNELIUS KAPTEYN to fainter stars. In 1904 Kapteyn first noticed that instead of an anticipated random distribution, the proper motions of stars in the Sun's neighborhood actually seemed to favor two opposite directions, called star streams. This was actually the earliest observational evidence that the galaxy was rotating, but astronomers did not immediately recognize the phenomenon.

Dyson enjoyed participating in solar eclipse expeditions. At the start of the 20th century, he was a member of eclipse expeditions to Portugal in 1900, to Sumatra in 1901, and to Tunisia, North Africa, in 1905. These activities made him a recognized expert on the Sun's corona and chromosphere. In 1901 he was elected as a fellow of the Royal Society, and he was knighted in 1915.

Dyson is best remembered for organizing and coordinating the two most significant solar eclipse expeditions in 1919, when he sent expeditions to the island of Principe, in the Gulf of Guinea off the coast of West Africa, and to Sobral in Brazil to carefully measure the positions of stars near the Sun's rim during the eclipse. Any deflection of starlight would provide evidence to support Einstein's general theory of relativity. While Dyson led the British expedition to Sobral, Brazil, Eddington led the British expedition to Principe. It was on Principe during the solar eclipse of May 29, 1919, that Eddington successfully measured the tiny deflection of a beam of starlight as the star's light just grazed the rim of the Sun—a subtle, gravitationally induced bending that provided scientists with their first experimental verification of general relativity. Dyson also participated on solar eclipse expeditions to Australia (1922), Sumatra (1926), and Malaya (1929).

Except for his five years in Edinburgh as Scotland's Astronomer Royal from 1905 to 1910, Dyson spent his entire career as an astronomer at the Royal Greenwich Observatory in southeast

London. In 1910 he was appointed as England's ninth Astronomer Royal and served in this prestigious position until his retirement in 1933. He made improvements in precision timekeeping at the observatory and was instrumental in the first radio ("wireless") signal broadcast of time in 1924. Upon retirement he and his wife, Lady Dyson, focused their attention on the welfare of their community. While on a voyage from Australia to the United Kingdom Dyson died at sea on May 25, 1939.

His numerous contributions to astronomy were honored in many ways. Dyson received the Bruce Gold Medal from the Astronomical Society of the Pacific in 1922 and the Royal Astronomical Society bestowed its Gold Medal on him in 1925. He became a Knight of the British Empire (KBE) in 1926. Dyson published his important book, *Eclipses of the Sun and Moon*, in 1937 in collaboration with the British astronomer Sir Richard van der Riet Woolley (1906–86).

E

Eddington, Sir Arthur Stanley
(1882–1944)
English
Astrophysicist

Arthur Eddington was born in Kendall, England, to strict Quaker parents, but his father died when he was two years old and he was raised in Somerset by his mother. When he was 16, Eddington was awarded a local scholarship and attended Owens College (now the University of Manchester) where he studied physics and mathematics. After graduating in 1902 he was awarded a scholarship to study at Cambridge, from which he graduated in 1905, and then won another scholarship to study at Trinity College, where his research work won him the coveted Smith Prize, an honor still awarded to graduate research students devoted to mathematics, applied mathematics, and theoretical physics. From 1906 to 1913 he was chief assistant director of the Royal Observatory in Greenwich, and in 1914 he was named Plumian Professor of Astronomy and became director of the observatory.

As a Quaker, his conscientious objector status kept him out of World War I and thereby allowed him to continue his studies at Cambridge from 1914 to 1918. ALBERT EINSTEIN had published his theory in 1915, which created a worldwide stir among astrophysicists, claiming that in a space-time continuum, light is composed of matter that will be affected by strong gravitational forces. Eddington grasped the significance of the theory, and when fellow astronomer SIR FRANK WATSON DYSON pointed out that an upcoming total solar eclipse would be an ideal phenomenon to test the theory, Eddington seized the opportunity. In 1919 he led an expedition to Brazil to observe a total eclipse of the Sun, and confirmed with photographic evidence the predictions of relativity that light rays are indeed bent when they pass stellar objects of great gravitational fields. Eddington thus became the first person to prove Einstein's theory to be correct, and he confirmed it again on another solar eclipse expedition to Príncipe Island in West Africa.

During this period at Cambridge he became recognized throughout the scientific community as the world's foremost authority on the theory of relativity aside from Einstein himself. In fact, Einstein called Eddington's *Mathematical Theory of Relativity*, published in 1923, "the finest presentation of the subject in any language."

Eddington went on to study the internal composition of stars. Among his discoveries was that a star's interior is under radiative equilibrium, involving the three forces of gravity, gas pressure, and radiation pressure. He demonstrated that energy could be transported by radiation as well as convection, and that the temperature of the center of a star is in the millions of degrees. He also put forth the theory that the chief source of star energy was subatomic, supplied mostly by hydrogen that was fused into helium. Most of this research was published in his 1926 book, *The Internal Constitution of Stars.*

In his later years Eddington came to believe that the basic constants of nature, such as the mass of the proton and the charge of the electron, were not coincidental but, as he put it, were part of a "natural and complete specification for constructing a Universe." He did not live to complete this thesis, but his book on the subject, *Fundamental Theory*, was published after his death.

Eddington spent most of his life reviewing and critiquing the works of his contemporaries, an endeavor that often leads to controversy and hurt feelings. Eddington's most famous battle was with SUBRAHMANYAN CHANDRASEKHAR, the celebrated Indian American astrophysicist who was an expert on white dwarfs, stars that have burned off their internal supply of hydrogen and have collapsed into themselves as intensely hot but very small stars. Chandra, as he was called, had postulated that there was an upper limit to the mass of a white dwarf, a concept that Eddington dismissed as impossible. Because Eddington was the most respected astrophysicist of his day, Chandra was crushed by the rejection, appealing to several leading physicists of the time, and it was eventually accepted that Chandra was right and Eddington wrong. The upper limit of 1.4 solar masses for white dwarfs is today known as Chandrasekhar's Limit.

This academic defeat notwithstanding, Eddington went on to establish himself as the most prominent and important astrophysicist of the interwar years, making several valuable contributions to the field of physics. During his respected career, Eddington was elected a fellow of the Royal Society in 1914, was president of the Royal Astronomical Society from 1921 to 1923,

was awarded the Bruce Medal in 1924, received the Royal Society Royal Medal in 1928, was knighted in 1930, and then was president of the Physical Society from 1930 to 1932. Eddington died in Cambridge on November 22, 1944, and has been remembered as "the father of modern astrophysics." A crater on the Moon bears the name Eddington in his honor.

⊠ **Ehricke, Krafft Arnold**
(1917–1984)
German/American
Rocket Engineer

This talented rocket engineer conceived advanced propulsion systems for use in the U.S. space program of the late 1950s and 1960s. One of Krafft Ehricke's most important technical achievements was the design and development of the Centaur upper-stage rocket vehicle—the first American rocket vehicle to use liquid hydrogen (LH_2) as its propellant. His Centaur vehicle made possible many important military and civilian space missions. As an inspirational space visionary, Ehricke's writings and lectures eloquently expounded upon the positive consequences of space technology. He anchored his far-reaching concept of an "extraterrestrial imperative" with the concept of a permanent human settlement of the Moon.

Ehricke was born in Berlin, Germany, on March 24, 1917, at a turbulent time when imperial Germany was locked in a devastating war with much of Europe and the United States. He grew up in the political and economic chaos of Germany's postwar Weimar Republic. Yet, despite the gloomy environment of a defeated Germany, Ehricke developed his lifelong positive vision that space technology would serve as the key to improving the human condition.

Following World War I, his parents, both dentists, attempted to find schooling of sufficient quality to challenge him. Unfortunately, Ehricke's frequent intellectual sparring contests with rigid Prussian schoolmasters earned him widely varying grades. By chance, at age 12, Ehricke saw Fritz Lang's 1929 motion picture, *Die Frau im Mond* (The woman in the moon). The Austrian filmmaker had hired the German rocket experts HERMANN JULIUS OBERTH and Willy Ley (1906–69) as technical advisers during the production of this film. Oberth and Ley gave the film an exceptionally prophetic two-stage rocket design that startled and delighted audiences with its impressive blast-off. Advanced in mathematics and physics for his age, he appreciated the great technical detail that Oberth had provided to make the film realistic, and he viewed Lang's film at least a dozen times. This motion picture served as Ehricke's introduction to the world of rockets and space travel, and he knew immediately what he wanted to do for the rest of his life. He soon discovered KONSTANTIN EDUARDOVICH TSIOLKOVSKY's theoretical concept of a very efficient chemical rocket that used hydrogen and oxygen as its liquid propellants. Then, as a teenager, he attempted to tackle Oberth's famous 1929 book *Roads to Space Travel*, but struggled with some of the more advanced mathematics.

In the early 1930s he was still too young to participate in the German Society for Space Travel (Verein für Raumschiffahrt, or VFR), so he experimented in a self-constructed laboratory at home. As Adolf Hitler (1889–1945) rose to power in 1933, Ehricke, like thousands of other young Germans, got swept up in the Nazi youth movement. His free-spirited thinking, however, soon earned him an unenviable position as a conscripted laborer for the Third Reich. Then, just before the outbreak of World War II, he was released from the labor draft so he could attend the Technical University of Berlin. There, he majored in aeronautics, the closest academic discipline to space technology. One of his professors was Hans Wilhelm Geiger (1882–1945), the noted German nuclear physicist. Geiger's lectures introduced Ehricke to the world of nuclear energy. Impressed, Ehricke would later recommend the use of nuclear power and propulsion in many of the space development scenarios he presented in the 1960s and 1970s.

Wartime conditions played havoc with Ehricke's attempt to earn his degree. While enrolled at the Technical University of Berlin, he was drafted for immediate service in the German army and sent to the western front. Wounded, he came back to Berlin to recover and resume his studies and in 1942 obtained a degree in aeronautical engineering from the university. But while taking postgraduate courses in orbital mechanics and nuclear physics, he was again drafted into the German army, promoted to the rank of lieutenant, and ordered to serve with a Panzer (tank) division on the eastern (Russian) front. But fortune played a hand, and in June 1942 the young engineer received new orders, this time reassigning him to rocket development work at Peenemünde. From 1942 to 1945 he worked on the German army's rocket program under the overall direction of WERNHER MAGNUS VON BRAUN.

As a young engineer, Ehricke found himself surrounded by many other skilled engineers and technicians whose goal was to produce the world's first modern liquid-propellant ballistic missile, the A-4 rocket. This rocket is better known as Hitler's Vengeance Weapon Two, or simply the V-2. After World War II the German V-2 rocket became the ancestor to many of the larger missiles developed by both the United States and the Soviet Union during the cold war, a period ranging roughly from 1946 to 1989.

Near the end of World War II, Ehricke joined the majority of the German rocket scientists at Peenemünde and fled to Bavaria to escape the advancing Soviet army. Swept up in Operation Paperclip along with other key German rocket personnel, Ehricke delayed accepting a contract to work on rockets in the United States by almost a year. He did this in order to locate his wife, Ingebord, who was then somewhere in war-torn Berlin. After a long search that culminated in a happy reunion, Ehricke, his wife, and their first child journeyed to the United States in December 1946 to begin a new life.

For the next five years, Ehricke supported the growing U.S. Army rocket program at White Sands, New Mexico, and Huntsville, Alabama. In the early 1950s he left his position with the U.S. Army and joined the newly formed Astronautics Division of General Dynamics (formerly called Convair). There he worked as a rocket concept and design specialist and participated in the development of the first U.S. intercontinental ballistic missile, the Atlas. He became a U.S. citizen in 1955.

Ehricke strongly advocated the use of liquid hydrogen as a rocket propellant. While at General Dynamics, he recommended the development of a liquid hydrogen/liquid oxygen propellant upper-stage rocket vehicle. His recommendation became the versatile and powerful Centaur upper-stage vehicle. In 1965 he completed his work at General Dynamics as the director of the Centaur program and joined the advanced studies group at North American Aviation in Anaheim, California. From 1965 to 1968 this new position allowed him to explore pathways of space technology development across a wide spectrum of military, scientific, and industrial applications. The excitement of examining possible space technologies and their impact on the human race remained with him for the rest of his life.

From 1968 to 1973 Ehricke worked as a senior scientist in the North American Rockwell Space Systems Division in Downey, California. In this capacity, he fully developed his far-ranging concepts concerning the use of space technology for the benefit of humankind. After departing Rockwell, he continued his visionary space advocacy efforts through his own consulting company, Space Global, located in La Jolla,

California. As the U.S. government wound down Project Apollo in the early 1970s, Ehricke continued to champion the use of the Moon and its resources. His extraterrestrial imperative was based upon the creation of a selenospheric (Moon-centered) human civilization in space. Until his death in late 1984, he spoke and wrote tirelessly about how space technology provides the human race with the ability to create an unbounded "open world civilization."

Ehricke was a dedicated space visionary who not only designed advanced rocket systems (such as the Atlas-Centaur configuration) that greatly supported the first "golden age" of space exploration, but also addressed the important, but frequently ignored, social and cultural impacts of space technology. He created original art to communicate many of his ideas. He coined the term *androsphere* to describe the synthesis of the terrestrial and extraterrestrial environments. The androsphere relates to human integration of Earth's biosphere—the portion of Earth that contains all the major terrestrial environmental regimes, such as the atmosphere, the hydrosphere, and the cryosphere—with the material and energy resources of the solar system. This is space age philosophy on a very grand scale.

In spring 1980 Ehricke was the prime lecturer during an innovative short course on space industrialization held at California State University in Northridge. Here, in a number of wonderful two-hour sessions, he had the time to properly introduce many of the complex ideas that supported his grand vision of the extraterrestrial imperative. He carefully pointed out to an enthusiastic audience of college students, faculty, and aerospace professionals how space technology was providing a crucial open-world pathway for human development. Ehricke's lectures included such visionary concepts as "astropolis" and the "androcell." Astropolis is his concept for a large urban extraterrestrial facility that orbits in Earth-Moon (cislunar) space and supports the long-term use of the space environment for basic and applied research, as well as for industrial development. Ehricke's androcell concept is bolder and involves a large human-made world in space, totally independent of the Earth-Moon system. These extraterrestrial city-states, with human populations of 100,000 or more, would offer their inhabitants the excitement of multigravity-level living at locations throughout heliocentric space. Just weeks before his death on December 11, 1984, Ehricke was a featured speaker at the national symposium on lunar bases and space activities for the 21st century held that October in Washington, D.C. Despite being terminally ill, Krafft traveled across the United States from his home in San Diego, California, to give a moving presentation on the importance of the Moon in creating a polyglobal civilization for the human race.

Ehricke authored many innovative technical papers on topics that ranged from hydrogen rocket propulsion systems to space industrialization and settlement. He received the first Gunter Löser Medal from the International Astronautical Federation in 1956.

⊠   Einstein, Albert
(1879–1955)
Swiss-American
Physicist, Philosopher

The most recognizable face in science belongs to the physicist Albert Einstein—a man who reportedly suffered greatly from being thrown into the celebrity spotlight in the press over his scientific epiphanies because, according to a friend, the mathematician Arnold Sommerfeld (1868–1951), "every form of vanity was foreign to him." Einstein credited philosophy as much as his scientific education with his ability to think in cosmically new ways, and he relied on physical relationships instead of mathematics to conceive the ideas that were considered truly genius.

Einstein was born on March 14, 1879, in Württemberg, Germany, to a nonobservant Jewish family. He reportedly did not speak his first words until he was three years old, going in an instant from complete silence to complete sentences. He began school at the Humanistiche Gymnasium in Munich, around age six. He also began playing the violin around this time, taking lessons for the next seven years and developing a skill and love for music that he kept for a lifetime. Never athletic, he acknowledged as an adult that he had no use for physical exertion, having a fondness only for the sport of sailing.

As a child, the precocious Einstein was indoctrinated with religion via the "traditional education-machine," as he called it in his later biographical writings, but discovered science around age 12 and "soon reached the conviction that much in the stories of the Bible could not be true. The consequence was a positively fanatic orgy of freethinking coupled with the impression that youth is intentionally being deceived by the state through lies." This was, he said, "a crushing impression." The result left Einstein suspicious of authority and skeptical of the generally accepted truths and norms of society—a skepticism that was according to his writings, "an attitude which has never again left me."

For the next four years, Einstein read voraciously. He was interested in mathematics, particularly differential and integral calculus, and stated that he had "the good fortune of hitting on books which were not too particular in their logical rigor but which made up for this by permitting the main thoughts to stand out clearly and synoptically." One mathematics book in particular that had an effect on the young Einstein was a book on Euclidean plane geometry. But even more important to Einstein than mathematics was the study of the natural sciences. He devoured books on science, particularly a multivolume work entitled *People's Books on Natural Science*, with, he confessed, "breathless attention."

At age 17 Einstein entered the Polytechnic Institute of Zurich, focusing on mathematics and physics, where one of his professors was Rudolf Minkowski (1895–1976). But Einstein wrote that he "neglected mathematics," because he felt that any one discipline in the field could "easily absorb the short lifetime granted to us," and besides, he was really more interested in science.

Einstein was critical, however, of an educational system that he felt did more to discourage freethinking than it did to promote it, even though he felt that he was better off in Zurich than in universities elsewhere. He referred to the system's practices as "coercion" to "cram all this stuff into one's mind" for the sole purpose of passing examinations. "This coercion had such a deterring effect," he wrote, that after passing his final exams he "found the consideration of any scientific problems distasteful to me for an entire year." Einstein believed that this type of forced indoctrination "smothers every truly scientific impulse," and added that "It is, in fact, nothing short of a miracle that the modern methods of instruction have not yet entirely strangled the holy curiosity of inquiry." Einstein graduated in 1900 with barely passing grades.

After that, Einstein wandered before settling into, and settling for, a "temporary" job at the Swiss Patent Office in Bern in 1902. This year marked two other big events in Einstein's life. One was his father's death. The other was the birth of his daughter. Einstein's classmate in theoretical physics from the university, Mileva Maric, with whom he had a long and passionate relationship, became pregnant in her final year in school and gave birth to their baby daughter, Leiserl, while staying in the care of her parents. When Mileva went to Bern to help Einstein get settled, she left the baby with her parents. Within the year Einstein and Mileva were married, but when they received news that their

daughter had scarlet fever, Mileva returned to tend to the child. Leiserl's fate is unclear—whether she succumbed to scarlet fever, or she survived and was adopted, or taken in by one of Mileva's closest friends, remains a mystery. But it is known that Mileva returned to Bern alone. The following year, Einstein's job was made permanent, and in 1904 his wife had their first son, Hans, who would remain their "only child" until Eduard came along six years later.

During this time, Einstein began to write. His ideas on physics were haunting him. "By and by I despaired of the possibility of discovering the true laws by means of constructive efforts based on known facts. How could such a universal principle be found?" he wrote. The known facts he referred to were, of course, Sir Isaac Newton's theories on physics, which Einstein felt were "the dogmatic rigidity [that] prevailed in matters of principles: In the beginning (if there was such a thing) God created Newton's laws of motion together with the necessary masses and forces."

The idea that this is all there is did not fit with his intuition, and yet he needed a sign that there was another way. The answer came to him in Maxwell's theory of electromagnetism. The electric industry was a new and exciting one, and Einstein credited James Maxwell (1831–79) as freeing up the Newtonian "dogmatic faith," calling Maxwell's work a profound influence on his ability to rethink the problems he strove to solve.

Within two years, Einstein began to publish his work. His ideas were revolutionary—the discovery of light quantum, which validated the field of quantum physics introduced by Max Karl Planck, and his explanations of Brownian motion, explaining how particles submerged in a liquid will "jitter" as a result of being bombarded with molecules, which he based on statistical thermodynamics. He earned a Ph.D. from the University of Zurich for his paper "On a New Determination of Molecular Dimensions."

In 1905 Einstein presented a theory that he entitled "On the Electrodynamics of Moving Bodies—The special theory of relativity." This idea resulted, he admitted, "from a paradox upon which I had already hit at the age of 16: If I pursue a beam of light with the velocity of light in a vacuum, I should observe such a beam of light as a spatially oscillatory electromagnetic field at rest. However, there appears to be no such thing."

The special theory of relativity shows that two observers who are moving with respect to one another will disagree on their observations of length, time, velocity, and mass. "Space and time cease to possess an absolute nature," he wrote. "In space-time, each observer carves out in his own fashion his space and his time." He went on to explain, "Each observer, as his time passes, discovers, so to speak, new slices of space-time, which appear to him as successive aspects of the material world, though in reality the ensemble of events constituting space-time exist prior to his knowledge of them."

This theory was based on two postulates. First, that there is no absolute way to measure absolute motions in space, but instead *relative* motion can be obtained on an object as compared to another object; and second, that the speed of light always travels at the same rate. Experiments began to prove the theory, and before long Einstein had the most famous equation of all time—one showing that mass and energy are equivalent:

$$E = mc^2.$$

Einstein credited his ability to come up with this theory to two sources. "The special theory of relativity owes its origin to Maxwell's equations of electromagnetic field," he said. But, "the type of critical reasoning which was required for the discovery . . . was decisively furthered in my case . . . by the reading of David Hume's and Ernest Mach's philosophical writings."

Einstein's special theory of relativity affected many disciplines—physics, philosophy, and, of course, mathematics. When Minkowski started using the theory in his work on world geometry, Einstein said, "Since the mathematicians have invaded the theory of relativity, I do not understand it myself any more."

Einstein's published work succeeded in grabbing the attention of the universities that had ignored him after his graduation. He left his job at the patent office and became an associate professor at the University of Zurich, then a professor of theoretical physics in Prague in 1911, then went back to Zurich in 1912 before accepting a research fellowship, in 1914, at the University of Berlin. Here, the absent-minded Professor Einstein, who was notorious for losing his lecture notes, was free to give up lecturing and focus entirely on research.

Between 1905 and 1914, while Einstein was trying his hand at teaching in various institutions, he continued pondering the ideas that led to the theory of special relativity. His memoirs reveal, "That the special theory of relativity is only the first step of a necessary development became completely clear to me only in my efforts to represent gravitation in the framework of this theory." He found the special theory "too narrow."

Back in 1907, while working at the patent office, Einstein had an image in his head of a person falling through space, and realized that the person would not feel his own weight. He called this the "happiest thought of my life." He needed to somehow explain this with the theory of relativity, but he could not figure out how to work in the gravity.

In Berlin Einstein finally had the creative freedom to pursue these thoughts that had seemed so difficult to him during the past seven years because, as he put it, "It was not so easy to free oneself from the idea that coordinates must have an immediate metrical meaning." But the four-dimensional work that Minkowski produced based on the special theory of relativity was now starting to make sense, and Einstein credited it as indispensable in the theory that would be the grandfather of them all: the general theory of relativity.

The generalization of the theory of relativity was presented in 1916, and it put forth new laws of motion and gravitation. The idea that space curves around a mass is explained today in high school science classes as the "bowling ball on the water bed" experiment: If a bowling ball (a heavy mass) is dropped onto a water bed, the mattress of the water bed curves around the bowling ball, and when an object of less mass, for example a marble, is dropped in the vicinity of the bowling ball, it falls toward the heavier mass. Thus, the object of smaller mass is attracted to the object of greater mass because it is traveling along the curves of space.

Additionally, Einstein imagined a person encapsulated in an elevator in space, noting that the person would experience the sensation of weightlessness until the elevator accelerated upward, at which point the person would hit the floor, experiencing the sensation of gravity; if the person dropped something during the acceleration, the floor would come up to meet it, but it would also appear that the object "fell" to the floor. The person would feel no different in the accelerating elevator in space than he or she would in the gravitational field on Earth. The concept, then, is that gravity and acceleration are the same. Einstein also proposed that light, when passing near the surface of the sun, is deflected, and that the wavelengths of light from extremely dense stars, such as white dwarfs, are lengthened.

Einstein was so certain about the simplicity of his theory that he felt no compunction to explain it to those who questioned its validity. He wrote to Sommerfeld in February 1916, "Of the general theory of relativity you will be convinced, once you have studied it. Therefore I am not going to defend it with a single word."

A solar eclipse expedition in 1919 organized by SIR FRANK WATSON DYSON and led by SIR ARTHUR STANLEY EDDINGTON proved that Einstein's ideas about deflected light were correct. Able to compare positions of stars during a blocked-out Sun to their positions when the Sun was in full view, scientists found that the light had been diverted. On November 7, 1919, the *Times* of London proclaimed his accomplishment with the headline "Revolution in science—New theory of the Universe—Newtonian ideas overthrown." According to Sommerfeld, "We owe the completion of his general theory of relativity to his leisure while in Berlin."

Einstein's wife left him and returned to Switzerland with their two children shortly after the family moved to Berlin. Einstein had rekindled a romantic relationship with his first cousin, Elsa, and when he began placing demands on his wife that did not suit that educated woman's nature, nor their relationship, she left the country and refused a divorce. In 1919 the Einsteins finally made their dissolution legal, and Albert married Elsa within a few months. The war was inciting anti-Semitic meetings, and Einstein considered leaving Berlin, but changed his mind when he attended a meeting at the Nauheim Congress of Natural Scientists in 1920 and Planck urged him to stay.

In 1921 Einstein received the Barnard Medal and then the Nobel Prize in physics, but surprisingly not for his theories on relativity—they were still too controversial—but rather for his discoveries from 1905 on quantum light. He sent his ex-wife the money from the prize, which he had promised (assuming he would some day win a Nobel Prize) as a condition of their divorce, going to great lengths to hide the transaction. In 1925 he received the Royal Society's Copley Medal, and in 1926, its Gold Medal. After extensive travel—to the United States, Japan, France, Palestine, and South America—the scientist eventually collapsed from exhaustion, and he took the better part of 1928 to recuperate.

In 1933 Einstein returned to the United States. Sommerfeld writes that "Einstein was driven out of Berlin and robbed of his possessions," in 1933, and that "a number of countries vied for his immigration." With an offer to teach at Princeton, Einstein, his wife, and his secretary (who ultimately worked for him for 50 years) stayed in the United States and never returned to Germany. He became a U.S. citizen in 1940, giving him dual citizenship in the United States and Switzerland.

In August 1948 Einstein's first wife, Mileva, died after a series of strokes three months after a nervous breakdown brought on by repeated confrontations with their son Eduard, who suffered from violent attacks of schizophrenia.

But Einstein's scientific brilliance at this point was behind him. His work at Princeton was dedicated to finding one system that explained the subatomic realm of the microcosm with the macrocosmic universe. His ideas were perceived as unachievable, and even he recognized that he was digging himself into a deeper and deeper hole, writing, "I have locked myself into quite hopeless scientific problems."

Einstein lived a personal life of campaigning for peace—his last effort was signing a manifesto that urged the elimination of all nuclear weapons. His professional life was one of imagination—visualizing problems as they physically exist rather than thinking about them purely from a mathematical perspective, and giving their solutions with unprecedented simplicity, allowing others to do the detailed work that developed their proof. In explaining his concepts on thinking, he said, "All our thinking is of this nature of a free play with concepts."

Einstein wrote his autobiographical notes at the age of 67, a task he likened to writing his own obituary. In his notes he gives a glimpse of his perception of his place in the world. "The essential in the being of a man of my type lies precisely in *what* he thinks and *how* he thinks, not in what he does or suffers."

Einstein's papers stayed in a basement file at Princeton until his secretary passed away, at which point they were willed to the Hebrew University in Jerusalem. Included in his documents were letters he had exchanged with Mileva, and the discovery that stunned the scientific world—it was clear that the two had corresponded about scientific theories and principles for years before they married, and Einstein himself wrote to her about "our theory," and "our work" on relativity. These letters have sparked heated debates in the scientific community, with various interpretations about Mileva's contributions to Einstein's great discoveries. It is well accepted that her background and education gave her the access and knowledge to be a collaborator, but whether she played a substantial role will most likely remain a source of contention among the scientific community and historians for years to come.

The crater Einstein on the surface of the Moon was named in his honor in 1964, and serves as a small reminder of one of the most brilliant thinkers in the history of humankind.

⊠ **Euler, Leonhard**
(1707–1783)
Swiss
Mathematician, Physicist

One of the most brilliant and productive mathematicians of all time, Leonhard Euler developed advanced analytical techniques to treat the complex orbital motion of the Moon and to efficiently determine the orbital paths of newly discovered comets. His improved methods in celestial mechanics advanced 18th-century observational astronomy and also supported important improvements in maritime navigation by making possible the preparation of more accurate lunar tables.

Leonhard Euler was born on April 15, 1707, in Basel, Switzerland. His father was the Reverend Paul Euler and his mother was Margaretha Brucker, a minister's daughter. While an undergraduate, Paul Euler became a classmate and friend of the great Swiss mathematician Johann Bernoulli (1667–1748). With a strong religious environment influencing his childhood, Leonhard Euler entered the University of Basel at the age of 14 to undertake general studies in preparation for his own career in the ministry. But Bernoulli, then a professor at the university, quickly recognized that the boy was a mathematical genius. In 1723 Euler completed his master's degree in philosophy and then sought his father's consent to abandon his ministerial study to pursue a new career in mathematics with the brilliant Bernoulli as his mentor.

The fact that Paul Euler and Bernoulli were undergraduates at the University of Basel helped secure paternal approval for Euler to make this major career change. Euler completed his formal studies in mathematics at the University of Basel in 1726 and soon began writing the first of his almost 900 scientific papers and books. As a recent graduate, he won the second-place prize in mathematics at the Paris Academy in 1727 for his innovative paper that addressed the mathematical arrangement of masts on sailing ships. Altogether, Euler would win 12 international academy prizes in mathematical competition.

Although reluctant to leave Switzerland, Euler departed Basel in April 1727 for St. Petersburg, Russia, to accept a position in the newly founded Academy of Sciences. The academy in St. Petersburg provided Euler an enriched working environment that encouraged him to explore all aspects of the mathematical sciences. Like many great mathematical figures of the 18th century, he applied his talents to theoretical astronomy, especially the treatment of planetary motions and the orbit of comets.

In January 1734 he married Katharina Gsell, the daughter of the Swiss painter George Gsell, who was then working in St. Petersburg. The couple had 13 children, although only five survived infancy. Normally the demands of such a large family might prove distracting, but Euler proudly claimed that some of his greatest mathematical inspirations and discoveries came while he was holding a baby in his arms and had other children playing around his feet. In 1735 he contracted a severe case of fever that almost proved fatal. Then he experienced eyesight problems in 1738 and by 1740 had completely lost vision in his right eye.

During his first St. Petersburg period, Euler's reputation as a mathematician grew. In 1736 he wrote *Mechanica* (Mechanics)—an important treatise that completely transformed Newtonian dynamics into a comprehensive and readily understandable form of mathematical analysis. He wrote award-winning papers for which he shared the Grand Prize of the Paris Academy in both 1738 and 1740.

To escape the harsh winters and the growing political turmoil in Russia, Euler accepted an invitation from the Prussian king Frederick the Great. Euler traveled to Berlin in 1741 on a mission to revitalize the Prussian Academy of Sciences. During the 25 years he spent in Berlin, Euler wrote numerous papers and important books on algebra, calculus, planetary orbits, and on the motion of the Moon. In 1744 Euler published *Theorium motuum planetarium et cometarium* (Theory of the motion of planets and comets)—an influential precursor to the more precise orbital mechanics work of COMTE JOSEPH LOUIS LAGRANGE. In 1753 Euler wrote *Theoria Motus Lunaris* (Theory of the lunar motion), his first attempt to analyze the exact motion of the Moon, a challenging problem that had baffled astronomers and mathematicians since the time of JOHANNES KEPLER.

By 1759 Euler had assumed the virtual leadership of the Berlin Academy of Sciences and his numerous contributions to theoretical and applied mathematics were internationally recognized. However, because of disagreements with King Frederick, Euler was never offered the position of academy president. So in 1766 Euler returned to St. Petersburg, accepting an invitation from the Empress Catherine II (Catherine the Great). Euler's departure from Berlin greatly angered the Prussian king, whose own bumbling had encouraged the brilliant Swiss mathematician to leave.

Back in St. Petersburg, Euler again experienced physical difficulties, but maintained an incredibly high level of productivity. In 1771 his home burned to the ground in the great St. Petersburg fire. Later that year, he went completely blind. Despite his blindness, he used his phenomenal memory to continue to publish papers and books at an enormous rate. In 1772 he published a second, more detailed book on the Moon's orbital behavior, *Theoria Motuum Lunae novo modo pertractata* (Theory of lunar motion, applying new rules)—a monumental analytical effort dealing with the difficult three-body problem in orbital mechanics.

Because of Euler's efforts, 18th-century astronomers were able to more quickly and precisely estimate the orbital path of newly discovered comets, using just a few precise observations. These analytical procedures greatly improved the business of "comet chasing" by astronomers such as CHARLES MESSIER. Euler's sophisticated treatment of the Moon's motion enhanced maritime navigation by providing detailed lunar and planetary tables to assist in longitude determination.

What is especially amazing about Euler during the later part of his life is the level of productivity he maintained even after he became totally blind and his first wife, Katharina, died. In 1776 Euler married her half sister, Abigail Gsell. He continued to use his phenomenal memory and some kindly assistance from two of his sons to remain history's most prolific mathematician until his death in St. Petersburg on September 18, 1783.

But even death did not stop his contributions to mathematics. He left behind a huge legacy of unpublished works that kept the printing presses in Russia busy for another three and a half decades. Euler's mathematical contributions to astronomy and orbital mechanics stretched well into the next century. His analytical methods prepared the way for the great achievements of 19th-century astronomy, when mathematicians working closely with telescopic astronomers theoretically predicted the position of, and then soon discovered previously unseen, solar system objects such as Neptune and the minor planets.

F

⊠ Fleming, Williamina Paton Stevens (Mina Fleming)
(1857–1911)
American
Astronomer

Williamina Paton Stevens was born in Dundee, Scotland, on May 15, 1857. Her father, a picture framer and early hobbyist in photography, died when she was only seven. She was educated in local public schools, and at age 14 gave up her education to take a job as a teacher, which she pursued until 1877, when at age 20 she married James Orr Fleming.

In 1878 the Flemings left Scotland for the United States, immigrating to Boston, Massachusetts. A year later, James Fleming abandoned Mina, as she was known, when she was pregnant with their first child. Mina Fleming was then forced to take a job as a housemaid, an event that actually proved to be to her great benefit, because her employer was none other than the famous professor EDWARD CHARLES PICKERING, director of the Harvard Observatory.

Pickering had been having problems with his personnel and had long advocated using women in support positions. In 1881 he hired Fleming to do clerical work in the observatory and to do mathematical calculations, and she became the first "computer," as they were called, at the Harvard Observatory, and the first woman in a group of famous female astronomers that would become known as "Pickering's harem." Despite her lack of formal higher education, Fleming soon proved adept at both mathematics and astronomical science.

Under Pickering's guidance, Fleming devised the first system of classifying stars according to their spectra, determined by holding a prism up to the objective lens of a telescope. Using this system, which was later named after her, Fleming cataloged more than 10,000 stars during the next nine years. In 1890 her work was published in a book entitled *The Draper Catalogue of Stellar Spectra*.

Fleming was eventually put in charge of dozens of more women hired to do computations at the observatory, among them ANNIE JUMP CANNON, who eventually succeeded Fleming. Fleming was also named editor of all publications issued by the observatory, and in 1898 the Harvard Corporation appointed her as the first woman curator of astronomical photographs. Among Fleming's other accomplishments were the discovery of 59 nebulae, 10 novae, and more than 300 variable stars. The eminent British astronomer Herbert Hall Turner (1861–1930) once said of Fleming, "Many astronomers are deservedly proud to have discovered one variable star, but [her discoveries are] an achievement bordering on the marvelous."

In 1906 Fleming became the first American woman to be elected to the Royal Astronomical Society. In 1910 she discovered "white dwarfs," stars in the late stages of their existence that are hot, dense, and appear white or bluish in color. In 1911 Fleming was named an honorary member of the Astronomical Society of Mexico and the Astronomical Society of France. That same year she also received the Guadalupa Almendaro Medal and, noting her lack of a graduate degree, Wellesley College named her honorary fellow on astronomy.

Fleming was the most famous female astronomer of her time, and one of the first women to have a lunar crater named after her. At the time of her death from pneumonia in Boston, on May 21, 1911, she had discovered 10 of the 28 known novae and 94 of the 107 known Wolf-Rayet stars—more than any single astronomer in history.

Fowler, William Alfred
(1911–1995)
American
Physicist

"Willy" Fowler, as he was known internationally by all his colleagues, friends, and associates, was born August 9, 1911, in Pittsburgh, Pennsylvania, the first child of John MacLeod Fowler and Jennie Summers Watson Fowler. His younger brother, Arthur, came next and was soon followed by a sister, Nelda. Fowler was the paternal grandson of an immigrant coal miner from Scotland and an immigrant grocer from Northern Ireland. When he was two years old, Fowler's family moved to Lima, Ohio, where his father had been transferred.

Fowler attended Horace Mann Grade School and Lima Central High School, where he was a popular student and an active athlete in baseball, football, track, and tennis, earning his varsity letter in football his senior year. His teachers encouraged his interest in engineering and science and, perhaps sensing a future for him in higher education, insisted that he study Latin for four years instead of the traditional French and German. During this time, when the family spent their annual vacation back in Pittsburgh, Fowler developed an interest in steam locomotives, and would spend hours at the Pennsylvania Railroad switch yards in Pittsburgh and at local railroad yards in Lima. This interest became one of his passionate avocations throughout his life, along with baseball and music, and in 1973 he would embark on a 1,500-mile trek from Khabarovsk to Moscow on the steam-powered Trans-Siberian Railroad.

In high school Fowler had written a prizewinning essay on how Portland cement was produced, and when he enrolled in Ohio State University he thought he would pursue ceramics engineering as his major. However, in his lower-division physics and mathematics courses, Fowler became interested in physics. Upon learning that a new degree field was being offered he switched his major to engineering physics. In his junior year he was elected to the engineering honorary society, Tau Beta Pi, and in his senior year he was elected president of the Ohio State chapter of that society. He received his B.S. degree in 1933 and was a classmate of the renowned theoretical physicist Leonard I. Schiff, who became a lifelong friend.

Fowler's path from Ohio led him straight to the California Institute of Technology (Caltech) in Pasadena, California, where he became a graduate student at the Kellogg Radiation Laboratory under the renowned Charles C. Lauritsen (1892–1968). Throughout his career Fowler referred to Lauritsen as "the greatest influence in my life," beginning with Lauritsen's supervision of Fowler's doctoral thesis, "Radioactive Elements of Low Atomic Number," which according to a brief autobiographical piece, Fowler described as a study "in which we discovered mirror nuclei and showed that the nuclear forces are charged symmetric—the same between two protons and between two neutrons when charged particle Coulomb forces are excluded." He received his Ph.D. in 1936.

Fowler then joined the Caltech faculty as a research fellow and in 1939 was appointed assistant professor. Prior to World War II he married Ardiane Foy Olmsted, with whom he had two daughters. During the war he spent time in the South Pacific as a civilian with simulated military rank, doing research and development work on rocket ordnance and proximity fuses. Back at Caltech, he was appointed as full professor in 1946 and named Institute Professor of Physics in 1970.

During 1954–55, Fowler spent a sabbatical year as a Fulbright scholar in Cambridge, England, where he worked closely with Fred Hoyle (born 1915), who in 1946 had established the concept of nucleosynthesis in stars. While there, he was joined by ELEANOR MARGARET PEACHEY BURBIDGE and her husband Geoffrey, and in 1956 the Burbidges and Hoyle all came to the Kellogg Laboratory at Caltech. The following year, Fowler coauthored one of the most famous papers in astrophysics history, "Synthesis of the Elements in the Stars," which showed conclusively that all of the elements from carbon to uranium could be produced by nuclear fission processes in stars, beginning with only the helium and hydrogen produced in the big bang, made famous by RALPH ASHER ALPHER. This paper was cowritten by the Burbidges and Hoyle and is commonly referred to by the authors' initials, B²FH.

During his 60-year career, Fowler carried out both theoretical studies to calculate fusion rates in elements, and experimental studies with particle accelerators to corroborate his theories. He was honored with an array of international awards for his lifetime achievements. He was awarded the medal of merit, presented by President Harry S. Truman in 1948; elected a member of the National Academy of Sciences in 1956; awarded the Barnard Medal for Meritorious Service to Science in 1965; designated Benjamin Franklin Fellow of the Royal Society of Arts in 1970; awarded the G. Unger Vetlesen Prize in 1973; awarded the National Medal of Science by President Gerald Ford in 1974; designated Associate of the Royal Astronomical Society in 1975; elected president of the American Physical Society in 1976; awarded the Eddington Medal of the Royal Astronomical Society in 1978; awarded the Bruce Gold Medal of the Astronomical Society of the Pacific in 1979; awarded the Legion d'Honneur from French president François Mitterrand in 1989, and received honorary degrees from a dozen prestigious universities worldwide. He reached the apogee of his career when he was awarded the Nobel Prize in physics in 1983 "for his theoretical and experimental studies of the nuclear reactions of importance in the formation of the chemical elements in the universe." He also collected several awards for his nonscientific interests in baseball, steam locomotives, art, music, and outdoor activities.

Fowler was one of the pioneers in the new field of astrophysics, and his work defining the creation of new elements inside stars contributed significant new knowledge to astronomy and to the increasingly popular Big Bang theory of the origin of the universe. A few years before his death he observed, "It is a remarkable fact that humans, on the basis of experiments and measurements carried out in the lab, are able to understand the universe in the early stages of its evolution, even during the first three minutes of its existence."

Fowler's wife Ardiane died in May 1988, and he remarried in December 1989 to Mary Dutcher, a former schoolteacher. He wrote in his Nobel autobiography that after retirement he spent his time attending weekly seminars at Caltech and "in general try to stay out of trouble."

On March 14, 1995, Fowler died of natural causes in his home in Pasadena, California.

⊠ Fraunhofer, Joseph von
(1787–1826)
German
Optician, Physicist

Early in the 19th century, Joseph von Fraunhofer pioneered the important field of astronomical spectroscopy. A skilled optician, he developed the prism spectrometer in about 1814. Then, while using his new instrument, he discovered more than 500 mysterious dark lines in the Sun's spectrum—lines that now carry his name. In 1823 he observed similar (but different) lines in the spectra of other stars, but left it for other scientists, among them ROBERT WILHELM BUNSEN and GUSTAV ROBERT KIRCHHOFF to solve the mystery of the "Fraunhofer lines."

Fraunhofer was born in Straubing, Bavaria, on March 6, 1787. His parents were involved in the German optical trade, primarily making decorative glass. When they died, leaving him an orphan at age 11, his guardians apprenticed him to an optician (mirror maker) in Munich. While completing this apprenticeship, Fraunhofer taught himself mathematics and physics. He joined the Untzschneider Optical Institute, located near Munich, as an optician in 1806. He learned quickly from the master glassmakers in this company and then applied his own talents to greatly improve the firm's fortunes. His lifelong contributions to the field of optics laid the foundation for German supremacy in the design and manufacture of quality optical instruments in the 19th century.

Fraunhofer's great contribution to the practice of astronomy occurred quite by accident, while he was pursuing the perfection of the achromatic lens, a lens capable of greatly reducing the undesirable phenomenon of chromatic aberration. As part of this quest, he devised a clever way to measure the refractive indices of optical glass by using the bright yellow emission line of the chemical element sodium as his reference, or standard. To support his calibration efforts, in 1814 he created the modern spectroscope by placing a prism in front of the object glass of a surveying instrument called a theodolite. Using this instrument, he discovered (or more correctly rediscovered) that the solar spectrum contained more than 500 dark lines, a pair of which corresponded to the closely spaced bright double emission lines of sodium found in the yellow portion of the visible spectrum.

In 1802 the British scientist William H. Wollaston (1766–1828) first observed seven dark lines in the solar spectrum. But he did not know what to make of them and speculated (incorrectly) that the mysterious dark lines were merely divisions between the colors manifested by sunlight as it passed through a prism. Working with a much better spectroscope 12 years later, Fraunhofer identified 576 dark lines in the solar spectrum and labeled the position of the most prominent lines with the letters A to K—a nomenclature physicists still use today in discussing the Fraunhofer lines. He also observed that a dark pair of his so-called D-lines in the solar spectrum corresponded to the position of a brilliant pair of yellow lines in the sodium flame he produced in his laboratory. However, the true significance of this discovery escaped him and, busy making improved lenses, he did not pursue the intriguing correlation any further. He also observed that the light from other stars when passed through his spectroscope produced spectra with dark lines that were similar but not identical to those found in sunlight. But once again, Fraunhofer left the great discovery of astronomical spectroscopy for other scientists.

Fraunhofer was content to note the dark line phenomenon in the solar spectrum and then continue in his quest to design an achromatic objective lens—a task he successfully accomplished in 1817. He did such high-quality optical work that his basic design for this type of lens is still used by opticians.

In 1821 he built the first diffraction grating and used the new device instead of a prism to form a spectrum and make more precise measurements of the wavelengths corresponding to the Fraunhofer lines. But again, he did not interpret these telltale dark lines as solar absorption lines. In 1859 Bunsen and Kirchhoff would finally put all the facts together and provide astronomers with the ability to assess the elemental composition of distant stars through spectroscopy.

Because Fraunhofer did not have a formal university education, the German academic community generally ignored his pioneering work in optical physics and astronomical spectroscopy. At the time, it was considered improper for a "technician"—no matter how skilled—to present a scientific paper to a gathering of learned academicians. Yet 19th-century German astronomers, such as FRIEDRICH WILHELM BESSEL, used optical instruments made by Fraunhofer to make important discoveries. In 1823 Fraunhofer was appointed as the director of the Physics Museum in Munich and given the honorary title of professor. He died of tuberculosis in Munich on June 7, 1826. On his tombstone is inscribed the following Latin epithet: *"Approximavit sidera"* (in contemporary English: "He reached for the stars.")

In his honor, scientists now refer to the dark (absorption) lines in the visible portion of the Sun's spectrum as Fraunhofer lines. These lines are the phenomenological basis of astronomical spectroscopy and give scientists the ability to learn what stars are really made of.

G

Galilei, Galileo
(1564–1642)
Italian
Astronomer, Mathematician, Physicist

As the incomparable "founding father of modern science," Galileo Galilei used his own version of the newly invented telescope to make detailed astronomical observations that ignited the scientific revolution of the 17th century. In 1610 he announced some of his early telescopic findings in the publication *Sidereus Nuncius* (The starry messenger)—including his discovery of the four major moons of Jupiter, now called the Galilean satellites in his honor. Their orbital behavior, like a miniature solar system, stimulated his enthusiastic support for the heliocentric cosmology of NICOLAS COPERNICUS. Unfortunately, this scientific position led to a direct clash with ecclesiastical authorities bent on retaining Ptolemaic cosmology for a number of political and social reasons. By 1632 this conflict earned the fiery Galileo an Inquisition trial at which he was found guilty of heresy for advocating Copernican cosmology. Because of his advancing age and blindness, he was confined to house arrest for the remainder of his life. This unjust incarceration could not prevent his brilliant scientific work in astronomy, physics, and mathematics from creating the revolution in thinking

now known as the scientific method—an entirely new way of looking at the universe and one of the greatest contributions of Western civilization to the human race.

The Italian astronomer, mathematician, and physicist Galileo Galilei was born in Pisa on February 15, 1564. Galileo is commonly known by his first name only. His father, Vincenzo Galilei (ca. 1520–91) was a scholar and musician. When Galileo entered the University of Pisa in 1581, his father encouraged him to study medicine. However, his inquisitive mind soon became more interested in physics and mathematics than medicine. While still a medical student, he attended church services one Sunday and noticed a chandelier swinging in the breeze. He began to time its swinging motion using his own pulse as a crude clock. When he returned home, he immediately set up an experiment that revealed the pendulum principle. After just two years of study, Galileo abandoned medicine and focused on mathematics and science. This change in career pathways also changed the history of modern science.

In 1585 Galileo left the university without receiving a degree to focus his activities on the physics of fluids and solid bodies. The following year, he published a small booklet that described the clever hydrostatic balance he invented to determine relative densities. This was Galileo's

A 1640 portrait of the incomparable "Founding Father of Modern Science," Galileo Galilei. This fiery Italian astronomer, physicist, and mathematician used his own version of the newly invented telescope to make detailed astronomical observations that flamed the scientific revolution in the 17th century. In 1610 his *Sidereus Nuncius* (The starry messenger), proclaimed the discovery of the four major moons of Jupiter—now called the Galilean satellites in his honor. Because they behaved like a miniature solar system, Galileo threw his enthusiastic support behind the controversial heliocentric cosmology of Nicolas Copernicus. Unfortunately, this scientific position ultimately earned him a conviction for heresy in 1632 and confinement to house arrest for the remainder of his life. *(Courtesy of the National Aeronautics and Space Administration [NASA])*

first splash into science and it brought him attention from the academic world.

In 1589 Galileo became a mathematics professor at the University of Pisa. He was a bril-

liant lecturer and students came from all over Europe to attend his classrooms. Galileo often used his tenacity, sharp wit, and biting sarcasm to win philosophical arguments against his university colleagues. His fiery, unyielding personality earned him the nickname "the Wrangler." While his personal characteristics drove Galileo to many great discoveries, they also left a trail of bitter opponents who waited patiently for him to stumble so they could exact their revenge. By 1592 he had sufficiently offended his academic colleagues at the University of Pisa so that university officials chose not to renew his teaching contract.

In the late 16th century, European professors usually taught physics (then called "natural philosophy") as an extension of Aristotelian philosophy and not as a science, based on observation and experimentation, as it is practiced today. Through his skillful use of mathematics and innovative experiments, Galileo singlehandedly changed that approach. The motion of falling objects and projectiles intrigued him, and his scientific experiments constantly challenged the 2,000-year tradition of ancient Greek learning.

ARISTOTLE stated that heavy objects would fall faster than lighter objects. Galileo disagreed and declared that, except for air resistance, the two objects would fall at the same rate regardless of their mass. It is not certain whether he actually performed the legendary musket ball–cannonball drop experiment from the Leaning Tower of Pisa to prove this point. However, he did conduct a sufficient number of experiments with objects on inclined planes to upset Aristotelian physics and establish the science of mechanics.

Throughout his life, Galileo was limited in his motion experiments by an inability to accurately measure small increments of time. Despite this impediment, he conducted many important experiments that produced remarkable insights into the physics of free fall and projectile motion. A century later, SIR ISAAC NEWTON would build upon Galileo's pioneering work to create his universal law of gravitation and three laws of motion.

As previously mentioned, by 1592 Galileo's anti-Aristotelian research and abrasive behavior had sufficiently offended his colleagues at the University of Pisa to the point that university officials not so politely invited him to teach elsewhere. So Galileo moved to the University of Padua—the main institution of higher learning for the powerful Republic of Venice. This university had a more lenient policy of academic freedom, encouraged in part by the progressive Venetian government. Galileo eagerly accepted an appointment as a professor of mathematics. He even wrote a special treatise on mechanics to accompany his lectures. In Padua he taught courses on Euclidian geometry and on astronomy. The late 16th-century astronomy courses at Padua were primarily intended for medical students who needed to learn enough Ptolemaic astronomy to practice medical astrology. Italian physicians of the High Renaissance were expected to "consult with the stars" as part of treating patients. Such academic duties gave Galileo his first formal contact with astronomy, and the practice of science would never be the same.

In 1597 the German astronomer JOHANNES KEPLER, gave Galileo a copy of Copernicus's book—even though the book was then officially banned in Italy. Up to this point in his life Galileo had not been seriously interested in astronomy, but in this "forbidden" book he discovered heliocentric cosmology and immediately embraced it. Galileo and Kepler continued to correspond until about 1610.

Almost coinciding with his interest in Copernican theory, Galileo experienced several important changes in his personal life while living in Padua. The first of his three out-of-wedlock children by Marina Gamba of Venice was born in Padua on August 21, 1600. She was named Virginia Galilei and would later assume the name Sister Maria Celeste. His second

daughter, Livia Galilei, was born in Padua in 1601, and his son, Vincenzio Galilei, was also born in Padua in 1604. Galileo did not marry his children's mother because in 17th-century Italy, the relationship was considered inappropriate—Gamba was much younger than Galileo and from a socially inferior family. Yet Galileo always remained financially responsible to her and his children—although he did so always under extreme financial stress. When his father died at a young age in 1591, Galileo followed contemporary social custom and assumed familial financial responsibility by providing a substantial dowry for his sister, Virginia Galilei. This constant drain on his meager salary as a mathematics professor made him resort to tutoring and casting horoscopes to make financial ends meet.

Between 1604 and 1605, Galileo performed his first public work involving astronomy. He observed the supernova of 1604 in the constellation Ophiuchus and used the phenomenon to refute the cherished Aristotelian belief that the heavens were immutable (that is, unchangeable). He boldly delivered this challenge on Aristotle's doctrine in a series of public lectures. Unfortunately, these well-attended lectures brought him into direct conflict with the university's pro-Aristotelian philosophy professors.

In 1609, Galileo learned that a new optical instrument (called a magnifying tube) had just been invented in Holland. Within six months, he devised his own version of the instrument, then in 1610 he turned this improved telescope to the heavens and started the age of telescopic astronomy. With his crude instrument, he made a series of astounding discoveries, including the existance of mountains on the Moon, many new stars, and four moons orbiting Jupiter. His published discoveries in The starry messenger stimulated both enthusiasm and anger. The moons orbiting Jupiter proved that not all heavenly bodies revolve around Earth; direct observational evidence for the Copernican model.

Personally unable to continue teaching the Aristotelian doctrine at the university because of the overwhelming evidence against it, Galileo left Padua in 1610 and went to Florence where he accepted an appointment as chief mathematician and philosopher to the grand duke of Tuscany, Cosimo II. By now Galileo's fame had spread throughout Italy and the rest of Europe. His telescopes were in demand and he obligingly provided them to selected astronomers throughout Europe, including Kepler. In 1611 he proudly took his telescope to Rome and let church officials personally observe some of his discoveries. While in Rome, he also became a member of the prestigious Accademia dei Lincei (Lyncean Academy), the first true scientific society, founded in 1603.

In 1613 Galileo published his "Letters on Sunspots" through the academy. He used the existence and motion of sunspots to demonstrate that the Sun itself changes, again attacking Aristotle's doctrine of the immutability of the heavens. In so doing, he also openly endorsed the Copernican model. This started Galileo's long and bitter fight with church authorities.

Above all, Galileo believed in the freedom of scientific inquiry. Late in 1615 Galileo went to Rome and publicly argued for the Copernican model. Angered by Galileo's open defiance, Pope Paul V (1605–21) formed a special commission to review the theory of Earth's motion. Dutifully, the unscientific commission concluded that the Copernican theory was contrary to biblical teachings and possibly even represented a form of heresy. In late February 1616 Cardinal Robert Bellarmine officially warned Galileo to abandon his support of the Copernican hypothesis or face harsh punishment.

The cardinal's stern message toned down Galileo's behavior—at least for a few years. In 1623 Galileo published *Il saggiatore* (The assayer), a book in which he discussed the principles for scientific research, but carefully avoided

support for Copernican theory. He even dedicated the book to his lifelong friend, the new pope, Urban VIII (1623–44).

However, in 1632 Galileo pushed his friendship with the pope to the limit by publishing *Dialogue on the Two Chief World Systems*. In this masterful satirical work, Galileo uses two main characters to present scientific arguments, concerning the Ptolemaic and Copernican worldviews, to an intelligent third person. Galileo's fictional Copernican, Salviati, cleverly wins these lengthy arguments. In contrast, Galileo presents Aristotelian thinking and the Ptolemaic system through an ineffective character he calls Simplicio. The pope regarded Galileo's fictional Simplicio as an insulting personal caricature, and within months the Inquisition summoned Galileo to Rome. Under threat of execution, the aging Italian scientist publicly retracted his support for the Copernican model on June 22, 1633. Tradition suggests that a feeble, but defiant, Galileo whispered these words as he rose from his knees after renouncing Copernican cosmology: *"Eppur si muove."* ("And yet it moves.")

Instead of torture and death, the Inquisition sentenced him to life in prison—a more lenient term that he was allowed to serve under house arrest at his villa in Arceti near Florence. The ecclesiastical authorities also banned his book *Dialogue*. However, Galileo's loyal supporters smuggled copies out of Italy, and soon its pro-Copernican message was spreading across Europe.

While under house arrest, Galileo spent his final years working on less controversial physics. In 1638 he published *Discourses and Mathematical Demonstrations Relating to Two New Sciences*. Fearing for his life, he carefully avoided astronomy in this seminal work and concentrated his creative energies on summarizing the science of mechanics, including uniform acceleration, free fall, and projectile motion.

Through Galileo's pioneering work and personal sacrifice, the Scientific Revolution ultimately prevailed over misguided adherence to centuries of Aristotelian philosophy. He never really opposed the church nor its religious teachings. He did, however, come out strongly in favor of the freedom of scientific inquiry and he would not tolerate supposedly learned academicians or ecclesiastics who remained mentally blind to demonstrable scientific facts. Physical blindness struck the brilliant scientist in 1638, and he died at home on December 25, 1642, the same day Newton was born.

Some three and a half centuries later, on October 31, 1992, Pope John Paul II formally retracted the sentence of heresy passed on Galileo by the Inquisition. As the pope was announcing this official retraction, a NASA spacecraft named *Galileo* was speeding through interplanetary space on course for its successful rendezvous mission with Jupiter and the giant planet's family of interesting moons, including the Galilean moons that helped establish the Scientific Revolution.

⊠ Galle, Johann Gottfried
(1812–1910)
German
Astronomer

By good fortune and some careful telescopic viewing, Johann Gottfried Galle became the first person to observe and properly identify the planet Neptune. Acting upon a recommendation from URBAIN-JEAN-JOSEPH LEVERRIER, Galle began a special telescopic observation at the Berlin Observatory on September 23, 1846. His search quickly revealed the new planet precisely in the region of the night sky predicted by Leverrier's calculations. Improved mathematical techniques helped Galle and Leverrier convert subtle perturbations in the orbit of Uranus into one of the major discoveries of 19th-century astronomy.

Galle was born in Pabsthaus, Germany, on June 9, 1812. He received his early education at the Wittenberg Gymnasium (secondary school) and then studied mathematics and physics at the Berlin University. After graduating from the university in 1833, Galle taught mathematics in the gymnasiums at Guben and Berlin. Then, in 1835, his former professor Johann Franz Encke (1791–1865) invited Galle to become his assistant at the newly founded Berlin Observatory.

Galle proved to be a skilled telescopic astronomer. In 1838 he discovered the C or "crêpe" ring of Saturn. He was also an avid comet hunter, and during winter 1839–40 Galle discovered three new comets. In 1846 good fortune smiled on Galle and allowed him to play a major role in one of the most important moments in the history of planetary astronomy. Independent of the British astronomer JOHN COUCH ADAMS, the French astronomer and mathematician Urbain-Jean-Joseph Leverrier, studied the perturbations in the orbit of Uranus and made his own mathematical predictions concerning the location of another planet beyond Uranus. Leverrier corresponded with Galle on September 18, 1846, and asked the German astronomer to investigate a section of the night sky to confirm his mathematical prediction that there was a planet beyond Uranus.

Upon receipt of Leverrier's letter, Galle and his assistant, Heinrich Ludwig d'Arrest (1822–75), immediately began searching the portion of the sky recommended by the French mathematical astronomer. D'Arrest also suggested to Galle that they use Encke's most recent star chart to assist in the search for the elusive trans-Uranian planet. In less than an hour Galle detected a "star" that was not on Encke's latest chart. A good scientist, Galle waited 24 hours to confirm that this celestial object was indeed the planet Neptune. On the following night (September 24), Galle again observed the object and carefully compared its motion relative to that of the so-called fixed stars. There was no longer any question: The change in its position clearly indicated that this "wandering light" was another planet.

Galle discovered Neptune pretty much in the position predicted by Leverrier's calculations. He immediately wrote to Leverrier, informing him of the discovery and thanking him for the suggestion on where to search. The telescopic discovery of Neptune by Johann Galle on September 23, 1846, is perhaps the greatest example of the marriage of mathematics and astronomy in the 19th century. Subtle perturbations in the calculated orbit of Uranus suggested to mathematicians and astronomers that one or more planets could lie beyond. Yet this discovery became one of the biggest controversies in the scientific community. Today, credit for the combined predictive and observational discovery of Neptune is awarded jointly to Adams, Galle, and Leverrier.

In 1851 Galle accepted a position as the director of the Breslau Observatory and remained there until his retirement in 1897. He made additional contributions to astronomy by focusing his attention on determining the mean distance from Earth to the Sun—an important reference distance called the astronomical unit (AU). Astronomers relied on the transits of Venus across the face of the Sun to make an estimate of the astronomical unit. At the time, this was a rather difficult measurement, since astronomers had to determine precisely the moment of first contact.

However, in 1872 Galle suggested a more accurate and reliable way of measuring the scale of the solar system. He recommended that astronomers use the asteroids whose orbits come very close to Earth to measure solar parallax. Galle reasoned that the minor planets offered more precise, pointlike images and their large number provided frequent favorable oppositions. In an effort to demonstrate and refine this technique, Galle observed the asteroid Flora in

1873. Astronomers SIR DAVID GILL and SIR HAROLD SPENCER JONES greatly improved measurement of the Earth-Sun distance using Galle's suggestion.

Of the three codiscoverers of Neptune, only Galle survived until 1896 to receive congratulations from the international astronomical community as it celebrated the 50th anniversary of their great achievement. He died on July 10, 1910, in Potsdam, Germany.

⊠ **Gamow, George** (Georgi Antonovich)
(1904–1968)
Ukrainian/American
Physicist, Cosmologist

A successful nuclear physicist, cosmologist, and astrophysicist, George Gamow was a pioneering proponent of the big bang theory, which boldly speculated that the universe began with a huge ancient explosion. As part of this hypothesis, he also participated in the prediction of the existence of a telltale cosmic microwave background, observational evidence for which was discovered by ARNO ALLEN PENZIAS and ROBERT WOODROW WILSON in the early 1960s.

George Gamow was born on March 4, 1904, in the city of Odessa in the Ukraine. His grandfather was a general in the Russian army and his father was a teacher. On his 13th birthday, Gamow received a telescope—a life-changing gift that introduced him to astronomy and helped him decide on a career in physics. Having survived the Russian Revolution, he entered Novorossysky University in 1922 at 18 years of age. He soon transferred to the University of Leningrad (now called the University of St. Petersburg), where he studied physics, mathematics, and cosmology. Gamow completed his Ph.D. in 1928, the same year he proposed that alpha decay could be explained by the phenomenon of the quantum mechanical tunneling of alpha particles through the nuclear potential

A successful nuclear physicist, cosmologist, and astrophysicist, George Gamow pioneered the big bang hypothesis in the late 1940s. Building up the preliminary concepts of others, he boldly concluded that the universe began with a huge explosion—sarcastically called the "big bang" by his detractors. Gamow further suggested that the modern remnants of this ancient explosion would be a curtainlike microwave ("cold light") signal at the very edge of the observable universe. Almost two decades before the event, Gamow had brilliantly anticipated the detection of this telltale cosmic microwave background by the Nobel Prize–winning radio astronomers Arno Penzias and Robert W. Wilson in 1964. *(AIP Emilio Segrè Visual Archives)*

barrier in the atomic nucleus. This innovative work represented the first successful extension of quantum mechanics to the atomic nucleus.

As a bright young nuclear physicist, Gamow performed postdoctoral research throughout Europe, visiting the University of Göttingen in Germany, working with Niels Bohr (1885–1962) in Copenhagen, Denmark, and studying beside Ernest Rutherford (1871–1937) at the Cavendish Laboratory in Cambridge, England. In 1931 he was summoned back to the Soviet Union to serve as a professor of physics at Leningrad University. However, after living in western Europe for several years, he found life in Stalinist Russia not to his liking. A highly creative individual, he obtained permission to attend the 1933 International Solvay Congress in Brussels, Belgium, and then never returned to Stalin's Russia. He emigrated from Europe to the United States in 1934, accepting a faculty position at George Washington University, in Washington, D.C., and becoming a U.S. citizen. In 1936 he collaborated with the Hungarian-American physicist Edward Teller (1908–2003) in developing a theory to explain beta decay, the process whereby the nucleus emits an electron.

At this point in his career, Gamow began exploring relationships between cosmology and nuclear processes. He used his knowledge of nuclear physics to interpret the processes taking place in various types of stars. In 1942, again in collaboration with Teller, Gamow developed a theory about the thermonuclear reactions and internal structures within red giant stars. These efforts also led Gamow to conclude that the Sun's energy results from thermonuclear reactions. He was also an ardent supporter of EDWIN POWELL HUBBLE'S expanding universe theory, an astrophysical hypothesis that made him an early and strong advocate of the Big Bang theory.

RALPH ASHER ALPHER, a student under Gamow, introduced the Big Bang cosmology which Gamow had published in the 1948 paper "The Origin of the Chemical Elements." He based some of this work on cosmology concepts previously suggested by ABBE-GEORGES-ÉDOUARD LEMAÎTRE. Starting with the explosion of Lemaître's "cosmic egg," Gamow proposed a method of thermonuclear reactions by which the various elements could emerge from a primordial mixture of nuclear particles that he called the *ylem*. Although later nucleosynthesis models introduced by WILLIAM ALFRED FOWLER and other astrophysicists would supersede this work, Gamow's pioneering investigations provided an important starting point. Perhaps the most significant postulation to emerge from these efforts was the prediction that the Big Bang would produce a uniform cosmic microwave background— a lingering remnant at the edge of the observable universe. The discovery of this cosmic microwave background in 1964 by Penzias and Wilson renewed interest in the Big Bang cosmology and provided scientific evidence that strongly supported an ancient explosion hypothesis. In his 1952 book, *Creation of the Universe*, Gamow presented a detailed discussion of these concepts in cosmology.

Just after James Watson (born 1928) and Francis Crick (born 1916) proposed their DNA model in 1953, Gamow turned his scientific interests from the physics taking place at start of the universe to the biophysical nature of life. He published several papers on the storage and transfer of information in living cells. While his precise details were not terribly accurate by today's level of understanding in the field of genetics, he did introduce the important concept of "genetic code"—a fundamental idea confirmed by subsequent experimental investigations.

In 1956 Gamow became a faculty member at the University of Colorado in Boulder and remained at that institution for the rest of his life. He was a prolific writer who produced highly technical books, such as the *Structure of Atomic Nuclei and Nuclear Transformations* (1937), and also very popular books about science for general audiences. Several of his most popular books were *Mr. Tompkins in Wonderland* (1936), *One, Two, and Three . . . Infinity* (1947), and *A Star Called the Sun* (1964).

George Gamow was elected to the Royal Danish Academy of Sciences in 1950 and the U.S. National Academy of Sciences in 1953. The United Nations Educational, Scientific, and Cultural Organization (UNESCO) bestowed its Kalinga Prize on Gamow in 1956 for his literary efforts that popularized modern science around the world. He died in Boulder on August 19, 1968.

⊠ Giacconi, Riccardo
(1931–)
Italian/American
Astrophysicist

Using special instruments carried into space on sounding rockets and satellites, Riccardo Giacconi helped establish the exciting new field of X-ray astronomy. This great astrophysical adventure began quite by accident in June 1962, when Giacconi's team of scientists placed a specially designed instrument package on a sounding rocket. As the rocket climbed above Earth's sensible atmosphere over White Sands, New Mexico, the instruments unexpectedly detected the first cosmic X-ray source—that is, a source of X-rays from beyond the solar system. Subsequently named Scorpius X-1, this "X-ray star" is the brightest of all nontransient X-ray sources ever discovered. He shared the 2002 Nobel Prize in physics for his pioneering contributions to astrophysics.

Riccardo Giacconi was born in Milan, Italy, on October 6, 1931. He earned his Ph.D. degree in cosmic ray physics at the University of Milan in 1956 and then immigrated to the United States to accept a position as a research associate at Indiana University in Bloomington. In 1959 he moved to the Boston area to join American Science and Engineering, a small scientific research firm established by scientists from the Massachusetts Institute of Technology, where he began his pioneering activities in X-ray astronomy.

Collaborating with Professor BRUNO BENEDETTO ROSSI at the Massachusetts Institute of Technology and other researchers in the early 1960s, Giacconi designed novel X-ray detection instruments for use on sounding rockets and eventually spacecraft. His June 1962 experiment

The Italian-American astrophysicist Riccardo Giacconi as he appeared in the mid-1960s, during that exciting research period when he helped establish the foundations for modern X-ray astronomy. His first major achievement occurred in 1962 when he collaborated with Bruno Rossi and placed a new type of X-ray detection instrument on a sounding rocket that fortuitously discovered the first cosmic X-ray source called Scorpius X-1. In the 1970s he supervised the development of several X-ray astronomy spacecraft for NASA. For these pioneering contributions to astrophysics and to X-ray astronomy, he shared the 2002 Nobel Prize in physics. *(AIP Emilio Segrè Visual Archives, Physics Today Collection)*

proved especially significant when his team, in search of possible solar-induced X-ray emissions from the lunar surface, accidentally detected the first known X-ray source outside the solar system from Scorpius X-1. Previous rocket-borne experiments by personnel from the U.S. Naval Research Laboratory that started in 1949 had detected X-rays from the Sun, but the new X-ray star was the first source of X-rays known to exist beyond the solar system. This fortuitous discovery is often regarded as the beginning of X-ray astronomy, an important field within high-energy astrophysics.

Because Earth's atmosphere absorbs X-rays, instruments to detect and observe X-rays produced by astronomical phenomena must be placed in space high above the sensible atmosphere. Sounding rockets provide a brief (typically just a few minutes) way of searching for cosmic X-ray sources, while specially instrumented orbiting observatories provide scientists with a much longer period of time to conduct all-sky surveys and detailed investigations of interesting X-ray sources. Astrophysicists investigate X-ray emissions because they provide a unique insight into some of the most violent and energetic processes taking place in the universe.

In the mid-1960s, as an executive vice president and senior scientist at American Science and Engineering, Giacconi headed the scientific team that constructed the first imaging X-ray telescope. In this novel instrument, incoming X-rays graze or ricochet off precisely designed and shaped surfaces and then collect at a specific place called the focal point. The first such device flew into space on a sounding rocket in 1965 and made captured images of X-ray hot spots in the upper atmosphere of the Sun. The National Aeronautics and Space Administration (NASA) used greatly improved versions of this type of instrument on the *Chandra X-ray Observatory*.

The following analogy illustrates the significance of Giacconi's efforts in X-ray astronomy and also serves as a graphic testament to the rapid rate of progress in astronomy brought on by the space age. By historic coincidence, Giacconi's first, relatively crude X-ray telescope was approximately the same length and diameter as the astronomical telescope used by GALILEO GALILEI in 1610. Over a period of about four centuries, optical telescopes improved in sensitivity by 100 million times, as their technology matured from Galileo's first telescope to the capability of

NASA's *Hubble Space Telescope*. About 40 years after Giacconi tested the first X-ray telescope on a sounding rocket, NASA's magnificent *Chandra X-ray Observatory* provided scientists with a leap in measurement sensitivity of about 100 million. Much as Galileo's optical telescope revolutionized observational astronomy in the 17th century, Giacconi's X-ray telescope triggered a modern revolution in high-energy astrophysics.

Giacconi functioned well both as a skilled astrophysicist and as a successful scientific manager capable of overseeing the execution of large, complex scientific projects. He envisioned and then supervised development of two important orbiting X-ray observatories: the *Uhuru* satellite and the *Einstein X-Ray Observatory*.

In the late 1960s his group at American Science and Engineering supported NASA in the design, construction, and operation of the *Uhuru* satellite—the first orbiting observatory dedicated to making surveys of the X-ray sky. *Uhuru*, also known as *Small Astronomical Satellite 1*, was launched from the San Marco platform off the coast of Kenya on December 12, 1970. By coincidence, that date marked the seventh anniversary of Kenyan independence, so NASA called this satellite *Uhuru*, which is the Swahili word for "freedom." The well-designed spacecraft operated for more than three years and detected 339 X-ray sources, many of which turned out to be due to matter from companion stars being pulled at extremely high velocity into suspected black holes or superdense neutron stars. With this successful spacecraft, Giacconi's team firmly established X-ray astronomy as an exciting new astronomical field.

In 1973 Giacconi moved his X-ray astronomy team to the Harvard-Smithsonian Center for Astrophysics, headquartered in Cambridge, Massachusetts. There, he supervised the development of a new satellite for NASA called *High Energy Astronomical Observatory 2* (*HEAO-2*). Following a successful launch in November 1978, NASA renamed the spacecraft the *Einstein*

X-Ray Observatory. It was the first orbiting observatory to use the grazing incidence X-ray telescope (a concept initially proposed by Giacconi in 1960) to produce detailed images of cosmic X-ray sources. The *Einstein X-Ray Observatory* operated until April 1981 and produced more than 7,000 images of various extended X-ray sources, such as clusters of galaxies and supernova remnants.

Giacconi's display of superior scientific management skills during the Uhuru and Einstein observatory projects secured his 1981 appointment as the first director of the Space Telescope Science Institute—the astronomical research center on the Homewood campus of Johns Hopkins University, in Maryland, responsible for operating NASA's *Hubble Space Telescope*. From 1993 to 1999 Giacconi served as the director general of the European Southern Observatory (ESO) in Garching, Germany. This was his first scientific leadership role involving ground-based astronomical facilities. Prior to leaving this European intergovernmental organization, he successfully saw the "first light" for ESO's new *Very Large Telescope* in the Southern Hemisphere on Cerro Paranal in Chile.

In July 1999 Giacconi returned to the United States to become president and chief executive officer of Associated Universities, Inc.—the Washington, D.C.–headquartered not-for-profit scientific management corporation that operates the National Radio Astronomy Observatory in cooperation with the National Science Foundation. In October 2002 the Nobel Committee awarded Giacconi one-half of that year's Nobel Prize in physics for his pioneering contributions to astrophysics.

A renowned scientist and manager, Giacconi has received numerous awards, including the Bruce Gold Medal from the Astronomical Society of the Pacific (1981) and the Gold Medal of the Royal Astronomical Society in London (1982). He is the author of more than 200 scientific publications that deal with the

X-ray universe, ranging from black hole candidates to distant clusters of galaxies.

⊠ Gill, Sir David
(1843–1914)
British
Astronomer

Sir David Gill was the Scottish watchmaker turned astronomer who passionately pursued the measurement of solar parallax. His careful observations, often involving travel with his wife, Lady Isabel Gill, to remote regions of Earth, established a value of the astronomical unit that served as the international reference until 1968. For example, in 1877 the couple lived for six months on tiny Ascension Island, in the middle of the South Atlantic, just so Gill could estimate a value for solar parallax by observing Mars during its closest approach. The couple also resided in South Africa for 28 years so he could serve as the British royal astronomer in charge of the Cape of Good Hope Observatory.

Gill was born in Aberdeen, Scotland, on June 12, 1843. From 1858 to 1860 he studied at the Marischal College of the University of Aberdeen, but left without a degree because of family circumstances: As the eldest surviving son, he abandoned college and trained to be a watchmaker, so he could take over his aging father's business. However, while at Marischal College, Gill had the brilliant Scottish physicist James Clerk Maxwell (1831–79) as his professor of natural philosophy. Maxwell's lectures greatly inspired Gill, and the young watchmaker would ultimately pursue a career in science.

By good fortune, his activities in precision timekeeping led him directly into a career in astronomy. While operating his watch- and clock-making business, Gill developed a great facility with precision instruments. As a hobby, he restored an old, abandoned transit telescope in order to provide the city of Aberdeen with accurate time. He also purchased a 12-inch reflecting telescope and modified the device to take photographs. Gill's excellent photograph of the Moon taken on May 18, 1869, soon caught the attention of the young Lord Lindsay, who was persuading his father, the earl of Crawford, to build him a private astronomical observatory. After brief negotiations, Gill abandoned the family watchmaking business in 1872 to become the director of Lindsay's observatory near Aberdeen. As a competent instrument maker and self-educated observational astronomer, Gill set about the task of designing, equipping, and operating Lord Lindsay's observatory at Dun Echt, Scotland.

His years as a privately employed astronomer at Dun Echt brought Gill in contact with the most prominent astronomers in Europe. The work also prepared him well for the next step in his career as a professional astronomer, especially his experience with precision measurements using the heliometer. In the 19th century astronomers used this (now obsolete) instrument to measure the parallax of stars and the angular diameter of planets. David Gill became a world-recognized expert in the use of the heliometer—an instrument he applied relentlessly in pursuit of his lifelong quest to precisely measure solar parallax.

Astronomers define solar parallax as the angle subtended by Earth's equatorial radius when Earth is observed from the center of the Sun—that is, at a distance of one astronomical unit. The astronomical unit (149,600,000 kilometers) is the basic reference distance in solar system astronomy. By international agreement, the solar parallax is approximately 8.794 arc seconds.

Sponsored by Lord Lindsay, Gill participated in an astronomical expedition to Mauritius in 1874 to observe the transit of Venus. He took 50 precision chronometers on a long sea voyage around the southern tip of Africa to Mauritius, a small island in the Indian Ocean east of Madagascar. Gill focused on providing precise timing for the other astronomers.

A planetary transit takes place when one celestial object moves across the face of another celestial object of larger diameter—such as Mercury or Venus moving across the face of the Sun as viewed by an observer from Earth. The 1874 transit of Venus provided Gill with an opportunity to contribute to the elusive value of the astronomical unit. However, he soon discovered firsthand the difficulty of using the transit of Venus to determine solar parallax. Upon magnification, the image of the planetary disk became insufficiently sharp, so it was very difficult for astronomers to measure the precise moment of first contact, even with Gill's precision chronometers. In 1876 Gill departed on friendly terms from his position as director of Lord Lindsay's observatory.

Along with his wife, Gill then conducted his own privately organized expedition to Ascension Island in 1877 to measure solar parallax. The couple made their arduous journey to this tiny island in the middle of the South Atlantic Ocean just to precisely measure Mars during its closest approach to Earth. Gill used the distance from the Royal Greenwich Observatory to Ascension Island as a baseline. After six months on Ascension, he was able to obtain reasonable results, but he was still not satisfied. He also recognized and began to follow JOHANN GOTTFRIED GALLE'S suggestion to use asteroids with their pointlike masses in making measurements of solar parallax. He knew this was the right approach, because he had already made a preliminary (but not successful) effort, by observing the asteroid Juno during the Mauritius expedition.

In 1889 Gill organized an international effort that successfully made precise measurements of the solar parallax by observing the motion of three asteroids (Iris, Victoria, and Sappho). From the observations, Gill calculated a value of the astronomical unit that remained the international standard for almost a century. His value of 8.80 arc seconds for solar parallax was replaced in 1968 with a value of 8.794 arc sec-

onds obtained by radar observations of solar system distances. Then, between 1930 and 1931, SIR HAROLD SPENCER JONES followed Gill's work and used observations of the asteroid Eros to make refined estimates of solar parallax. After a decade of careful calculation, Jones obtained a value of 8.790 arc seconds.

The astronomical expeditions of 1874 and 1877 brought Gill recognition within the British astronomical community, and in 1879 he was appointed as Her Majesty's Royal Astronomer at the Cape of Good Hope Observatory in South Africa. He served with great distinction as royal astronomer at this facility until his retirement in 1907. Early in this appointment, he accidentally became a proponent for astrophotography, when he successfully photographed the great comet of 1882 and then noticed the clarity of the stars in the background of his comet photographs. He upgraded and improved the observatory so it could play a major role in photographing the Southern Hemisphere sky in support of such major international efforts as the Carte du Ciel project. He collaborated with JACOBUS CORNELIUS KAPTEYN in the creation of the *Cape Durchmusterung*—an enormous catalog published in 1904 that contained more than 400,000 Southern Hemisphere stars whose positions and magnitudes were determined from photographs taken at the Cape of Good Hope Observatory. Gill was knighted in 1900.

The Gills returned to Great Britain from South Africa in 1907 and retired in London. There, he busied himself by completing a book, *History and Description of the Cape Observatory,* published in 1913. He died in London on January 24, 1914. He was internationally recognized as one of the great observational astronomers of the period. He was elected as a fellow of the Royal Astronomical Society in 1867 and served as the society's president from 1909 to 1911. He received the society's Gold Medal twice: in 1882 and again in 1908. His numerous other awards included the 1900 Bruce Gold Medal from the Astronomical

Society of the Pacific and the 1899 Watson Medal from the U.S. National Academy of Sciences.

⊠   Goddard, Robert Hutchings
(1882–1945)
American
Physicist, Rocket Engineer

Robert Goddard was the brilliant, but reclusive, American physicist who promoted rocket science and cofounded the field of astronautics in the early part of the 20th century—independent of Konstantin Eduardovich Tsiolkovsky and Hermann Julius Oberth. However, unlike Tsiolkovsky and Oberth, who were primarily theorists, Goddard performed many hands-on pioneering rocket experiments. Included in his long list of accomplishments is the successful launch of the world's first liquid propellant rocket on March 16, 1926. Goddard's lifelong efforts resulted in more than 214 U.S. patents, almost all rocket-related. Today, Goddard is widely recognized as the father of modern rocketry.

Goddard was born in Worcester, Massachusetts, on October 5, 1882, to Nahum Danford Goddard and Fannie Louise Hoyt Goddard. When Robert was very young, the family moved to Boston, but returned to Worcester in 1898. His father was a machine shop superintendent, and visits to his father's shop helped nurture his fascination with gears, levers, and all types of mechanical gadgets. As a child he was very sickly and suffered from pulmonary tuberculosis. Forced to avoid strenuous youthful games and sports, Goddard retreated to more cerebral pursuits like reading, daydreaming, and keeping a detailed personal diary.

In the 1890s he became very interested in space travel. He was especially influenced by Jules Verne's classic science fiction novel *From the Earth to the Moon*, and by H. G. Wells's, *The War of the Worlds*, which appeared as a daily serial feature in a Boston newspaper. Many years later, Robert Goddard sent a personal letter to H. G. Wells, explaining what a career stimulus and source of inspiration that particular story was to him in his youth. The English author penned a letter back to Goddard, acknowledging the kind remarks.

Starting in childhood, Goddard began to keep a diary in which he meticulously recorded his thoughts and his experiences—the successes as well as the failures. For example, one day during his early teens, an explosive mixture of hydrogen and oxygen shattered the windows of his improvised home laboratory. Without much emotion he dutifully noted in his diary: ". . . such a mixture must be handled with great care." Later on, his wife would help him carefully document the results of his rocket experiments.

Goddard's youthful diary describes a very special event in his life that happened shortly after his 17th birthday. On October 19, 1899, he climbed an old cherry tree in his backyard with the initial intention of pruning it. But because it was such a pleasant autumn day, he remained up in the tree for hours, thinking about his future and whether he wanted to devote his life to rocketry. His thoughts even wandered to developing a device that could reach Mars. October 19 thus became Goddard's personal holiday or special "Anniversary Day"—the day on which he committed his life to rocketry and space flight. Of that life-changing day, Goddard later wrote: "I was a different boy when I descended the tree from when I ascended, for existence at last seemed very purposive."

Because of his frequent absence from school due to illness, he did not graduate from high school until 1904, when he was in his early 20s. At high school graduation, he concluded his oration with the following words: "It is difficult to say what is impossible, for the dream of yesterday, is the hope of today, and the reality of tomorrow." This is Goddard's most frequently remembered quotation.

Robert H. Goddard and the world's first liquid propellant rocket in its launch frame. On March 16, 1926, Goddard successfully flight-tested this historic liquid oxygen-gasoline rocket in a field at Auburn, Massachusetts. From 1930 to 1941 he made substantial progress in the development of larger liquid-propellant rockets. With his pioneering experiments he developed and demonstrated the fundamental principles of modern rocket technology. Along with visionary theoreticians Konstantin Tsiolkovsky and Hermann Oberth, Goddard is regarded as one of the founding fathers of astronautics. *(Courtesy of NASA)*

When he became an undergraduate student, most people, including many supposed scientific experts who should have known better, mistakenly believed that a rocket could not operate in the vacuum of space because "it needed an atmosphere to push on." Goddard knew better. He understood the theory of reaction engines, and would later use novel experimental techniques to demonstrate the fallacy of this common and very incorrect hypothesis about a rocket's inability to function in outer space. By the time he died, the large liquid-propellant rocket was not only a technical reality, but also was scientifically recognized as the technical means for achieving another 20th-century "impossible" dream—interplanetary travel.

Goddard enrolled at Worcester Polytechnic Institute in 1904. In his freshman year, he wrote a very interesting essay in response to one professor's assignment about transportation systems people might use far in the future—the professor's target year was 1950. Goddard's paper described a high-speed vacuum tube railway system that would take a specially designed commuter train from New York to Boston in only 10 minutes! His proposed train used electromagnetic levitation and accelerated continuously for the first half of the trip and then decelerated continuously at the same rate during the second half of the trip.

In 1907 he achieved his first public notoriety as a "rocket man" when he ignited a black powder (gunpowder) rocket in the basement of the physics building. As clouds of gray smoke filled the building, school officials became immediately "interested" in Goddard's work. Much to their credit, the tolerant academic officials at Worcester Polytechnic Institute did not expel him for the incident. Instead, he went on to graduate with his bachelor of science degree in 1908 and then remained at the school as an instructor of physics.

He began graduate studies in physics at Clark University in 1908. His enrollment began

a long relationship with the institution, first as a student and later as a faculty member. In 1910 he earned his master's degree and followed it the next year with his Ph.D. in physics. After receiving his doctorate, he accepted a research fellowship at Princeton University from 1912 to 1913. However, he suffered a near-fatal relapse of pulmonary tuberculosis in 1913 that incapacitated him for many months. Recovered, Goddard joined the faculty at Clark University the following year as an instructor of physics. By 1920 he was a full professor of physics and also the director of the physical laboratories at Clark. He retained these positions until he retired in August 1943.

Clark University provided Goddard with the small laboratory, supportive environment, and occasional, very modest amount of institutional funding necessary to begin serious experimentation with rockets. Following his return to Clark University, Goddard met his future wife, Esther Christine Kisk. Beginning in 1918, she started helping him carefully compile the notes and reports of his many experiments. Over time their professional relationship turned personal and they were married on June 21, 1924. They had no children.

Esther Kisk Goddard created a home environment that provided her husband with a continuous flow of encouragement and support. She cheered with him when an experiment worked; she was also there to cushion the blow of experimental failure or to buffer him from the harsh remarks of uninformed detractors. She was the publicly transparent, but essential, partner in "Goddard, Inc."—the chronically underfunded, federally ignored, but nevertheless incredibly productive husband-wife rocket research team that achieved many marvelous engineering milestones in liquid propulsion technology in the 1920s and 1930s.

As a professor at Clark University, Goddard responded to his childhood fascination with rockets in a methodical, scientifically rigorous way. Goddard recognized the rocket as the enabling technology for space travel and then, unlike Tsiolkovsky and Oberth, he went well beyond a theoretical investigation of these interesting reaction devices. He enthusiastically designed new rockets, invented special equipment to analyze their performance, and then flight-tested his experimental vehicles to transform theory into real world action. In 1914 Goddard obtained the first two of his numerous rocketry-related U.S. patents. One was for a rocket that used liquid fuel; the other for a two-stage, solid fuel rocket.

In 1915 Goddard performed a series of cleverly designed experiments in his laboratory at Clark University that clearly demonstrated a rocket engine generates thrust in a vacuum. Goddard's work provided indisputable experimental evidence showing that space travel was indeed possible. Yet his innovative research was simply too far ahead of its time. His academic colleagues generally did not understand or appreciate the significance of the rocket experiments, and Goddard continued to encounter great difficulty in gathering any financial support to continue. Since no one else thought rocket physics was promising, he became disheartened and even thought very seriously about abandoning his lifelong quest.

But somehow Goddard persevered and in September 1916 sent a description of his experiments along with a modest request for funding to the Smithsonian Institution. Upon reviewing Goddard's proposal, officials at the Smithsonian found his rocket experiments to be "sound and ingenious." One key point in Goddard's favor was his well-expressed desire to develop the liquid-propellant rocket as a means of taking instruments into the high-altitude regions of Earth's atmosphere—regions that were well beyond the reach of weather balloons. This early marriage between rocketry and meteorology proved very important, because this relationship provided officials in private funding institutions a tangible reason for supporting Goddard's proposed rocket work.

On January 5, 1917, the Smithsonian Institution informed Goddard that he would receive a $5,000 grant from the institution's Hodgkins Fund for atmospheric research. Not only did the Smithsonian provide additional funding to Goddard over the years, but equally important, it published Goddard's two classic monographs on rocket propulsion: the first in 1919 and the second in 1936.

In 1919 Goddard summarized his early rocket theory work in the classic report "A Method of Reaching Extreme Altitudes." It appeared in volume 71, number two of the *Smithsonian Miscellaneous Collections* in late 1919 and then as "Smithsonian Monograph Report Number 2540" in January 1920. This was the first American scientific work that carefully discussed all the fundamental principles of rocket propulsion. Goddard described the results of his experiments with solid-propellant rockets and even included a final chapter on how the rocket might be used to get a modest payload to the Moon. He suggested carrying a payload of explosive powder that would flash upon impact and signal observers on Earth. Unfortunately, nontechnical newspaper reporters missed the true significance of this great treatise on rocket propulsion and decided instead to sensationalize his suggestion about reaching the Moon with a rocket. The press gave him such unflattering nicknames as "Moony" and the "Moon man." The headline of the January 12, 1920, edition of the *New York Times* boldly proclaimed "Believes Rocket Can Reach Moon," and the accompanying article proceeded to soundly criticize Goddard for not realizing that a rocket cannot operate in a vacuum. Goddard became a newspaper sensation, but his newfound fame was all derogatory.

Offended by such uninformed adverse publicity, Goddard chose to work in seclusion for the rest of his life. He avoided further controversy by publishing as little as possible and refusing to grant interviews to the press. These actions caused much of his brilliant work in rocketry to go unrecognized during his lifetime. The German rocket scientists from Peenemünde, who emigrated to the United States after World War II, expressed amazement at the extent of Goddard's inventions, many of which they had painfully duplicated. They also wondered aloud why the U.S. government had chosen to totally ignore this brilliant man.

Independent of Tsiolkovsky and Oberth, Goddard recognized that the liquid-propellant rocket was the key to interplanetary travel. He then devoted his professional life to inventing and improving liquid-propellant rocket technology. Some of his most important contributions to rocket science are briefly summarized below.

In 1912 Goddard first explored on a theoretical and mathematical basis the practicality of using the rocket to reach high altitudes and even attain a sufficiently high velocity (called the escape velocity) to leave Earth and get to the Moon. He obtained patents in 1914 for the liquid-propellant rocket and for the concept of a two-stage rocket.

In 1915 he used a clever static test experiment in a vacuum chamber to show that the rocket would indeed function in outer space. As World War I drew to a close, Goddard demonstrated several rockets to U.S. military officials, including the prototype for the famous bazooka rocket used in World War II. However, with World War I almost over, the U.S. government simply chose to ignore Goddard and the value of his rocket research.

Undiscouraged by a lack of government interest, he turned his creative energy on the development of the liquid-propellant rocket and performed a series of important experiments at Clark University between 1921 and 1926. One test is particularly noteworthy. On December 6, 1925, Goddard performed a successful captive (that is, a static, or nonflying) test of a liquid-fueled rocket in the annex of the physics building. During this test, Goddard's rocket generated enough thrust to lift its own weight. Encouraged by the static test, Goddard proceeded to construct and flight-test the world's first liquid-fueled rocket.

On March 26, 1926, Goddard made space technology history by successfully launching the world's first liquid-propellant rocket. Even though this rocket was quite primitive, it is nonetheless the technical ancestor of all modern liquid-propellant rockets. The rocket engine itself was just 1.2 meters tall and 15.2 centimeters in diameter. Gasoline and liquid oxygen served as its propellants. Only four people witnessed this great historic event. The publicity-shy physicist took just his wife and two loyal staff members from Clark University (P. M. Roope and Henry Sachs) to a snow-covered field at Effie Ward's farm in Auburn, Massachusetts. "Auntie Effie," as he referred to her in his diary, was a distant relative.

Goddard assembled a metal frame structure that resembled a child's jungle gym to support the rocket prior to launch. His wife served as the team's official photographer. The launch procedure was quite simple, since there was no countdown: Goddard carefully opened the fuel valves, then, with the aid of a long pole, Sachs applied a blowtorch to ignite the engine. With a great roar, the tiny rocket vehicle jumped from the metal support structure, flew in an arc, and then landed unceremoniously some 56 meters away in a frozen cabbage patch. In all, this historic rocket rose to a height of about 12 meters and its engine burned for only two and one-half seconds. Despite its great technical significance, the world would not learn about Goddard's great achievement for some time.

With modest funding from the Smithsonian, Goddard continued his work. In July 1929 he launched a larger and louder rocket near Worcester that successfully carried an instrument payload (a barometer and a camera)—the first rocket to do so. However, it also created a major disturbance. As a consequence the local authorities ordered him to cease his rocket flight experiments in Massachusetts.

Fortunately, the aviation pioneer Charles Lindbergh (1902–74) became interested in Goddard's work and helped him secure additional funding from the philanthropist Daniel Guggenheim. A timely grant of $50,000 allowed Goddard to set up a rocket test station in a remote part of the New Mexico desert, near Roswell. There, undisturbed and well out of the public view, Goddard conducted a series of important liquid-propellant rocket experiments in the 1930s. These experiments led to Goddard's second important Smithsonian monograph, "Liquid-Propellant Rocket Development," which was published in 1936. On March 26, 1937, Goddard launched a liquid-propellant rocket nicknamed "Nell" that flew for 22.3 seconds and reached a maximum altitude of about 2.7 kilometers. This was the highest altitude ever obtained by one of Goddard's rockets.

Goddard retained only a small technical staff to assist him in New Mexico and avoided contact with the rest of the scientific community. Yet, his secretive, reclusive nature did not impair the innovative quality of his work. In 1932 he pioneered the use of vanes in the rocket engine's exhaust with a gyroscopic control device to help guide a rocket during flight. In 1935, he became the first person to launch a liquid-propellant rocket that attained a velocity greater than the speed of sound. Finally, in 1937, Goddard successfully launched a liquid-propellant rocket that had its motor pivoted on gimbals so it could respond to the guidance signals from its onboard gyroscopic control mechanism. He also pioneered the use of the converging-diverging (de Laval) nozzle in rocketry, tested the first pulse jet engine, constructed the first turbopumps for use

in a liquid-propellant rocket, and developed the first liquid-propellant rocket cluster.

In his lifetime he registered 214 patents on various rockets and their components. Unfortunately, since his pioneering rocket work went essentially ignored, he finally closed his rocket facility near Roswell and moved to Annapolis, Maryland. There he worked with the U.S. Navy to provide his rocket technology expertise in the use of solid-propellant rockets to assist seaplane takeoff. As a lifelong solitary researcher, Goddard found the social demands of team-developing jet-assisted takeoff (JATO) for the government not particularly to his liking.

Just before his death, he was able to inspect the twisted remains of a German V-2 rocket that had fallen into U.S. hands and been taken back to the United States for analysis. As he performed a technical autopsy on the rocket, a military intelligence analyst inquired: "Dr. Goddard, isn't this just like your rockets?" Without showing any emotion, Goddard calmly replied: "Seems to be."

America's visionary "rocket man" died of throat cancer on August 10, 1945, in Annapolis, Maryland, and did not see the dawn of the space age. In 1951 Mrs. Esther Goddard and the Guggenheim Foundation, which helped fund much of Goddard's rocket research, filed a joint lawsuit against the U.S. government for infringement on his patents. In 1960 the government settled this lawsuit and in the process officially acknowledged Goddard's great contributions to modern rocketry. That June, the National Aeronautics and Space Administration (NASA) and the Department of Defense jointly awarded his estate the sum of $1 million for the use of his patents. NASA's Goddard Space Flight Center in Greenbelt, Maryland, now honors his memory by carrying his name.

H

Hale, George Ellery
(1868–1938)
American
Astrophysicist

George Ellery Hale was born on June 29, 1868, in Chicago, Illinois. Because his mother, Mary, had lost two previous children in infancy, he was coddled as a child by ever-attentive nurses and teachers, as were two subsequent Hale children, Martha and William. Hale's father, William E. Hale, was a young struggling engineer-salesman for a company in nearby Beloit, Wisconsin, who later amassed a fortune in the emerging elevator business, which thrived as the city was rebuilt after the Great Chicago Fire of 1871. This family fortune was to have a crucial influence on the course of Hale's scientific life.

Early on, Hale became fascinated with the microscope. As he examined everything around him in great detail, taking copious notes, his father bought more and more powerful instruments. In his early teens he came upon a description of the spectrum, as seen as a rainbow of light through a prism, and became fascinated by the notion of determining physical properties of the known universe by corresponding spectroscopic analysis.

Hale's biographical notebooks indicate that he built his first telescope at the age of 13, but it was small and the images unclear. When he learned of the availability of a four-inch Clark refractor, his father bought it for him. Hale installed the telescope on the roof of the house, and after he observed Saturn, Jupiter, and the Moon for the first time, was inspired to pursue research in the field of astronomy. Soon after, he attached a plateholder to the telescope and observed a partial eclipse of the Sun, marking the beginning of a lifelong passion of observing sunspots.

With the continuing encouragement of his mother to pursue intellectual endeavors and satisfy his inquisitive nature, Hale became educated in the arts, literature, and music. Perhaps more important, Hale had his father's wholehearted financial support, which allowed him to accumulate gratings, lenses, telescopes, and other instrumentation as he built a sophisticated planetarium in the family backyard. By the time he was 17, Hale had met prominent astronomers in the Chicago area and had made observations at the Dearborn Observatory in nearby Douglas Park.

Such was Hale's scientific precocity that when he ordered a small diffraction grating from the famous instrument maker J. A. Brashear, Brashear was astounded to find Hale was only 17 years old when he finally met him after corresponding from his manufacturing facility in

Pittsburgh. Brashear had surmised during their correspondence that Hale was a mature and experienced astronomer.

Hale's college years were eventful indeed. He entered Massachusetts Institute of Technology (MIT) in 1886 and pursued a degree in physics. Since MIT at the time did not offer courses in astronomy, Hale wrote to the noted professor EDWARD CHARLES PICKERING, director of the Harvard College Observatory, offering his services as a volunteer assistant. Pickering, who was breaking new ground in astronomical photography and stellar spectral classification, was so impressed with the young Hale's knowledge of spectroscopy as it could be applied to astronomy that he hired Hale and kept him employed as an assistant for almost three years.

While a junior in college, and initially against his family's wishes, Hale became engaged to a Brooklyn girl, Evelina Conklin, whom he had met on a family summer vacation when he was 13. This event was followed by the publication of his first article, an essay describing the evolution of the spectroscope and its applications to astronomy. Also while a junior, he invented the spectroheliograph, a device that made it possible to observe the Sun's prominences and sunspots in daylight. He tested the instrument with Pickering at the Harvard Observatory, and the invention made him world famous by the time he graduated from MIT in 1890 with his degree in physics.

On their honeymoon in 1890, the Hales visited the Lick Observatory on Mount Hamilton, near San Jose, California. There, Hale had a chance to observe through Lick's 36-inch refractor, then the largest in the world, and was offered the opportunity to work on the large telescope with his spectroheliograph. But by now Hale was receiving job offers from all over the world, and he declined the Lick position after being warned by colleagues that daytime observations on Mount Hamilton were difficult due

to air turbulence caused by warming of the local terrain.

The Hales returned to Chicago, where Hale continued to build his own observatory with a 12-inch refractor his father bought him. He called it the Kenwood Physical Observatory, and as he worked there the huge 36-inch Lick telescope was always on his mind. During this time, when he was only 22, Hale was elected a Fellow of the British Royal Astronomical Society. Soon after, he was offered a position at the prestigious University of Chicago, but when he discovered that the president of the university really coveted his Kenwood telescope and instruments—and possibly was interested in a generous endowment by his father—Hale decided to travel to Europe instead.

In Europe, Hale met virtually every famous astronomer and astrophysicist and visited all the scientific institutions and laboratories he could. Back in the United States, Hale proposed and developed a scientific journal that would eventually become the esteemed *Astrophysical Journal*. In 1892 he again was offered a position in the University of Chicago's new physics department, which he accepted due to the fact that he could then work with the eminent professor ALBERT ABRAHAM MICHELSON (who would become the university's first Nobel laureate in 1908). Hale was named associate professor of astrophysics, the first astronomer in the world to hold such a title. Now came the work that would define Hale's position forever in the field of astronomy: his successive role in the creation of the world's three greatest observatories.

While at the University of Chicago, Hale became aware of the availability of two 42-inch glass discs and decided to devote his energies into creating a world-class observatory at the university. However, he needed even more money than his father had, so Hale persuaded a local millionaire, Charles Tyson Yerkes, to help fund the effort. When all was said and done, Hale found himself director of the Yerkes

Observatory at Williams Bay, Wisconsin, boasting a 40-inch refractor telescope, the largest astrophysical laboratory in the world. Hale was 24 years old.

Political strife and ego-driven financial arguments between Yerkes Observatory directors and the benefactors convinced Hale to move his family to California in 1904. There, on Mount Wilson, near Pasadena, Hale set up a small laboratory to continue his sunspot observations, and received a grant from the Carnegie Institution to form the Mount Wilson Solar Observatory, which would eventually house a 100-inch telescope, the largest in the world. The Mount Wilson Observatory under Hale's direction transformed the field of astrophysics. Here, galaxies were explained, EDWIN POWELL HUBBLE discovered the expanding universe, and the phenomena of sunspots and solar magnetism were analyzed. Also while at Mount Wilson, Hale played a major role in transforming the Throop Polytechnic Institute into the world-acclaimed California Institute of Technology (Caltech).

Although he never lived to see his final project realized, Hale also set in motion the efforts to persuade the Rockefeller Foundation to fund a 200-inch telescope, the largest ever built, at Mount Palomar, affiliated with Caltech. Completed in 1948, 10 years after his death, it was named the Hale Telescope.

Hale wrote three books: *Depths of the Universe* (1924), *Beyond the Milky Way* (1026), and *Signals from the Stars* (1931). He received every major scientific medal and award and was an esteemed member of every major scientific society of his time. He coined the word *astrophysics*, and his invention of the spectroheliograph ushered in the age of solar prominence and sunspot observation. He created the three most famous observatories of the 20th century, and he left the scientific community with what is to this day the world's leading research journal in the field of astrophysics.

Hale died of cardiac failure due to arterial sclerosis on February 21, 1938, in Pasadena, a few months short of his 70th birthday.

⊠ Hall, Asaph
(1829–1907)
American
Astronomer

In 1877, while he was a staff member at the U.S. Naval Observatory in Washington, D.C., Asaph Hall discovered the two small moons of Mars. He called the larger innermost moon Phobos ("fear"), and the smaller outermost moon Deimos ("terrified flight")—after the attendants of Mars, the god of war in Roman mythology (known as Ares in Greek mythology).

The American astronomer Hall was born in Goshen, Connecticut, on October 15, 1829. His father died when he was just 13 years old, so he had to leave school to support his family as a carpenter's apprentice. In the early 1850s, Hall and his wife, Angelina, were working as schoolteachers in Ohio, when he decided to become a professional astronomer. Since the difficult days of his childhood, he had always entertained a deep interest in astronomy and taught himself the subject as best he could. So in 1857, with only a little formal training (approximately one year) at the University of Michigan and a good deal of encouragement from his wife, he approached William Bond (1789–1859), the director of the Harvard College Observatory, for a position as an assistant astronomer. Recalling his own struggle to become an astronomer, Bond admired Hall's spunk and hired him as an assistant researcher to support his son, George Phillips Bond (1825–65), who also worked at the observatory. Hall's starting salary was just three dollars a week, but that did not deter him from pursuing his dream.

His experience at the Harvard College Observatory polished Hall's skills as a professional astronomer. By 1863 he had joined the U.S. Naval Observatory in Washington, D.C., and he eventually took charge of its new 26-inch refractor telescope. During this period, the observatory and its instruments were under the direction of SIMON NEWCOMB, who was actually more interested in the mathematical aspects of astronomy than in performing personal observations of the heavens.

In December 1876 Hall was observing the moons of Saturn when he noticed a white spot on the generally featureless, butterscotch-colored globe of the ringed planet. He carefully observed this spot and used it to estimate the rate of Saturn's rotation to considerable accuracy. Hall worked out the period of rotation to be about 10.23 hours—a value comparable to the value reported by SIR WILLIAM HERSCHEL in 1794. The contemporary value for the period of rotation of Saturn's equatorial region is approximately 10.66 hours.

In summer 1877 Mars was in opposition and approached within 56 million kilometers of Earth. Hall pondered whether the Red Planet had any natural satellites. Other astronomers had previously searched for Martian moons and were unsuccessful. However, curiously, the writer Jonathan Swift mentioned two Martian satellites in his famous 1726 novel, *Gulliver's Travels*. On the night of August 10, 1877, Hall made his first attempt with the powerful Naval Observatory telescope, but viewing conditions along the Potomac River were horrible and Mars produced a glare in the telescope that made searching for any satellites very difficult. He returned home very frustrated, but his wife encouraged him to keep trying. On the following evening, Hall detected a suspicious object near the planet just before fog rolled in from the Potomac River and enveloped the Naval Observatory. It was not until the evening of August 16 that Hall again observed a faint starlike object near Mars, which proved to be its tiny outermost moon, Deimos. He showed

the interesting object to an assistant and instructed him to keep quiet about the discovery. Hall wanted to confirm his observations before other astronomers could take credit for his finding. On the evening of August 17, as Hall waited for Deimos to reappear in the telescope, he suddenly observed the larger, inner moon of Mars, which he called Phobos. By August 18, Hall's excitement and log notes had informally leaked news of his discovery to the other astronomers at the Naval Observatory. That evening the observatory was packed with eager individuals, including Newcomb, each trying to snatch a piece of the "astronomical glory" resulting from Hall's great discovery. For example, Newcomb improperly implied in a subsequent newspaper story that his calculations helped Hall realize that the two new objects were satellites orbiting Mars. This attempt at "sharing" Hall's discovery caused a great deal of personal friction between the two astronomers that lasted for decades. Imitating Hall's work, astronomers at other observatories soon claimed they had discovered a third and fourth moon of Mars—hasty "discoveries" that not only proved incorrect but were based on proclaimed orbital motion data violating Kepler's laws.

When the dust settled, astronomers around the world confirmed that Mars possessed just two tiny moons, and Hall received full and sole credit for the discovery. The Royal Astronomical Society in London awarded him its prestigious Gold Medal in 1877.

Hall remained at the Naval Observatory until his retirement in 1891. He focused his attentions on planetary astronomy, determining stellar parallax, and investigating the orbital mechanics of binary star systems, such as the visual binary 61 Cygnus. Following retirement from the Naval Observatory, Hall became a professor of astronomy at Harvard from 1896 to 1901. He died in Annapolis, Maryland, on November 22, 1907.

Halley, Sir Edmund
(ca. 1656–1742)
English
Astronomer, Mathematician, Physicist

Edmund Halley was born in Haggerston, England, around October 29, 1656, a date that is approximate, although somewhat based on the year Halley claims he was born, due to a change in the calendar system around that time. His father, also named Edmund Halley, was a wealthy soap maker and landlord, which afforded his son many privileges in his early youth. When Halley was 10 years old, the Great London Fire of 1666 ravaged the family's business, but not completely. The shrewd businessman managed his finances well enough to provide private tutoring for his son and then enroll him in the prestigious St. Paul's School, where young Halley excelled in mathematics, began his fascination with instrumentation, and started on his lifelong path of studying the night sky.

When Halley entered Queen's College at Oxford in 1673, at 17 years old, he was already an accomplished astronomer. With an extensive set of instruments given to him as a gift from his father, and the knowledge to use them, Halley easily caught the attention of John Flamsteed (1646–1719), the Astronomer Royal, who was working both at Oxford and at the Greenwich Observatory. By 1675 Halley had become his assistant. One of Halley's observations at Oxford was an occultation, the concealing of one heavenly body by another, as for example of a star by a planet. He noted the occultation of Mars by the Moon on August 21, 1676, and featured this in his first published article in the journal *Philosophical Transactions of the Royal Society.*

The next significant event in Halley's life seems to be his travel in 1676, when he left Oxford to go on a voyage to the island of St. Helena, off the African coast, the southernmost territory under British rule. His dream, literally, was to construct an observatory to chart the southern skies, and he convinced the powers who could make this happen that it would complement the Greenwich Observatory's work in the northern skies. This venture was financially supported by his father and by influential Royal Society members. King Charles II also got involved, personally requesting the East India Company to transport Halley and a colleague. During the trip, Halley redesigned a sextant and made notes of oceanic and atmospheric phenomena.

Halley was on St. Helena for 18 months, and he made the first complete observation of a transit of Mercury, catalogued 341 Southern Hemisphere stars in his *Catalogus stellarum australium,* and discovered the star cluster in Centaurus—all of which he published upon his return to the Royal Society in 1678. This work made him a reputable astronomer, but Halley officially had no degree. Since he had left Oxford without graduating, which was very common for wealthy students, in 1678 he was granted an honorary degree from Oxford by edict of King Charles II. That same year he was made a member of the Royal Society, at the age of 22 one of its youngest Fellows.

In 1680 Halley decided to travel through Europe with a friend from school. En route, near Calais, he observed a comet. With his interest sparked, he continued his trip to Paris, where he met up with fellow astronomer GIOVANNI DOMENICO CASSINI and the two observed the comet together.

In 1681 he married Mary Tooke. The following year he decided to visit SIR ISAAC NEWTON in Cambridge for some professional camaraderie. Halley had been working on a proof of planets having elliptical orbits, and much to his amazement, Newton had already proved the thesis. Even more astonishing, he had no intention of publishing his proof. Halley then urged Newton to compose his epic *Principia Mathematica,* and even financed its publication, eventually earning back his investment when the work sold extremely well. Halley edited the work and volunteered to write an introduction, "Ode to Newton," which heaped great praise on Newton and his scientific brilliance. Famous discoveries resulted for both men out of this venture. Newton discovered gravity, and Halley discovered Newton.

In 1684 Halley's father, who had remarried in 1681 after being a widower for 10 years, disappeared. After a five-week absence, his body was found, the victim of a murder. In addition to the great emotional loss, it was a sudden end to Halley's lifelong funding from his father, who was a dedicated patron of his son's work at the rate of a £300 annual allowance. By 1685 Halley was working for the Royal Society as, among other things, editor of their publication *Philosophical Transactions*, a position he held until 1696.

Halley's vast interest in a variety of fields led him to make contributions in other areas of science over the next few years. By 1686 he had completed a new map of the world, in which he showed the prevailing winds over the oceans. This new work on trade winds and the tides was a great contribution to science, and ultimately earned him credit as the founder of geophysics.

By 1691, however, Halley felt ready to get back into academia, and he decided to pursue appointment to an astronomy chair at Oxford. Halley could have reasonably expected some support from his old friend and mentor Flamsteed, from his days at Oxford and the Greenwich Observatory. But Flamsteed disliked Newton, who owed his great success to Halley, and while it is fair to say that many people personally disliked Newton, it seemed not enough reason to abandon Halley in his quest for a job. So Flamsteed created other reasons. From a scientific perspective, Flamsteed was upset for what he perceived as a slap at the observatory by Newton, thinking that the Royal Observatory should have received more credit regarding Newton's theories. But his most convincing

diatribe, it turned out, was that Flamsteed considered Halley a bad Christian because Halley allegedly did not believe solely in the biblical rendition of creation. The common theme in astronomy of science versus religion reared its ugly head. Flamsteed told the university that he thought Halley would "corrupt" the students, and Halley did not get the job.

Unperturbed, Halley continued his research and work for the Royal Society. In 1693 he published another first, this time a study of mortality rates related to age that was researched from documents in Breslau, Germany. This ultimately became the basis for future actuarial tables used by insurance companies, and made Halley a pioneer in yet another new field of science, social statistics.

Newton had become warden of the Royal Mint in 1696, and quickly appointed Halley as deputy controller of the mint in Chester, England. Halley left the position in 1698 when King William III awarded him the command of a warship, the *Paramore Pink*, built specifically for scientific expeditions. Halley had been working on the determination of longitude using a specially designed compass, and the purpose of the voyage of the *Paramore Pink* was to test his new method and discover "new lands" south of the equator.

As a naval captain, Halley started the three-year stint a little shaky. His first voyage in 1698 was aborted when he reached Barbados, and he returned to Portsmouth. A second voyage, in 1699, charted the North American coast of the Atlantic, after which he published the first charts of equal lines of declination. Several other voyages along the coasts of England followed. In 1702 he spent a year inspecting a naval base and other ports in the Austrian Empire at the request of Queen Anne.

In 1704, waving goodbye forever to his oceanic duties, Halley finally made his goal of teaching at Oxford a reality. He was appointed to the chair of geometry at Oxford as Savilian professor, a seat he held for the rest of his life. Flamsteed was not happy about this new appointment, apparently feeling that Halley's time at sea had turned him into an even more wretched creature than ever, complaining that Halley now "talks, swears and drinks brandy like a sea captain." Halley gave an impressive inaugural speech, including "an account of the most celebrated of the ancient and modern geometricians." And whether for the sake of his friend Newton, or just for the fun of irritating Flamsteed, Halley confidently "gave his greatest encomiums [to] . . . Mr. Newton."

It was about this time that Halley started to focus his interest on comets. Newton had proposed that comets had parabolic orbits, but Halley suspected they might in fact be elliptical. Using his own theory, he calculated that the orbit of the Comet of 1682 was in fact periodic, and was the same comet that had appeared in 1531 and 1607. In 1705 Halley predicted that the comet would reappear in a pattern of 76 years. Halley would not live to see the comet return in December 1758, but when it did it was immediately dubbed "Halley's comet" and, of course, reappeared in 1836, 1910, and 1986; every 76 years since, precisely as predicted.

In 1710 Halley proposed that stars had their own motions and proved his theory by calculating the motions of three specific stars. The theory of proper stellar motion, although new to the Western world thanks to Halley's research and calculations, had been originally proposed by the Buddhist monk and astronomer Zhang Sui (683–727). Halley probably had no knowledge whatsoever of Sui's theory. This same year, most likely for his work on this theory, Halley was again awarded an honorary degree, an M.A., from Oxford.

Halley continued his work in astronomy in conjunction with his position in geometry at Oxford. He also served as an arbiter of several disputes. Most famously, he "resolved" Newton's dispute with Leibniz over who invented the calculus: Newton "won," although the dispute was never really settled during their lifetime.

Over the years Halley was also in and out of disputes and controversies with his old mentor-turned-critic, Flamsteed. Halley had the last laugh when he was appointed Royal Astronomer at Greenwich in 1720, succeeding Flamsteed after the latter's death. He held this post for 21 years, until age 85, and during this time he invented the use of a transit to make lunar observations to determine longitude at sea and completed an 18-year observation of the Moon.

In 1729 Queen Caroline wanted to make certain that England's beloved scientist Halley was financially secure. She decreed that Halley should be given an additional salary equal to half-pay of a naval captain—an amount he continued to receive until his death.

Halley's scientific and astronomical brilliance, as well as his aptitude for mathematical logic, gave the world knowledge that has affected even today the important fields of navigation, geophysics, meteorology, astronomy, cartography, ballistics, and the development of measurement instruments such as the thermometer, the backstaff, and the diving bell.

This preeminent scientist was a member of the Royal Society from 1678 to 1743, and of the Académie des Sciences at Paris from 1729 until his death. His contributions have earned him both a lunar and a Martian crater named in his honor. Halley died in Cambridge on January 14, 1742.

⊠ Hawking, Sir Stephen William
(1942–)
English
Cosmologist, Physicist

There are few disciplines in which the word *genius* is used to describe several of its members—physics is one of them. The most renowned genius of the late 20th and early 21st centuries is the Lucasian Professor of Mathematics at Cambridge, Stephen Hawking—the man who brought the mind-boggling principles of cosmology into the everyday realm of the average person, and who helped popularize and mainstream such science-fiction ideas as black holes, worm holes, time travel, and life elsewhere in the universe. Hawking is known as the scientist who explains to the masses the laws that govern the universe.

Born on January 8, 1942, in Oxford, England, Hawking is fond of saying that his birth date was the 300th anniversary of GALILEO GALILEI's death. This takes a bit of explaining. One of the legends in astronomy since the late 18th century has been that Galileo and SIR ISAAC NEWTON shared a date in their life—Newton was born the same day that Galileo died, and that day happened to be Christmas Day, 1642. More than 100 years later, in 1752, the calendar system was changed in England, and the Gregorian calendar was adopted. Going back in time, after the implementation of the Gregorian calendar, and updating the dates of the events that happened previous to the installation of this new system has the effect of changing the date that Newton was born and that Galileo died. The date, while they were alive, was December 25, 1642—although even this is in question! Some historians site Galileo's death in December 1641, so the new date would then be January 1642; and some say that Newton was born in December 1642, so the corrected year would be January 1643. In any event, if the dates are changed to the Gregorian system and corrected for today's time, December 25 becomes January 8, as it relates to our calendar. So, in essence, Hawking was born the same day Galileo died and the same day Newton was born. Perhaps all of this confusion is simply forewarning from the universe that if the time of his birth is confusing, it is nothing compared to the philosophies and theories he proposes. Yet this brilliant man has managed to make even the most complex subjects of space accessible to the world at large.

As a young boy, Hawking knew by age 14 that mathematics was his life's calling. He attended St. Albans School starting at age 11; then, at age 17, he went to University College in Oxford, his father's alma mater, where he took up physics because mathematics was not offered. He was notorious for missing classes and labs, a trait he shared with another great physicist, ALBERT EINSTEIN, whose work would play an important role in Hawking's future career. But unlike Einstein, who barely passed his exams, Hawking graduated at the top of his class without much effort. With his undergraduate degree in natural science behind him, Hawking left Oxford and went to Cambridge for his Ph.D. in cosmology—a new and relatively unexplored field.

But something was not quite right with Hawking—minor clumsiness escalated into falling over for no reason, and pretty soon he was having trouble tying his shoes. When Hawking went home for Christmas vacation, his father, a doctor, noticed the symptoms and took him to a specialist, where multiple sclerosis was ruled out. But within a few weeks, Hawking was diagnosed with amyotrophic lateral sclerosis (ALS—also called Lou Gehrig's disease), a degenerative motor neuron disease that causes the body's muscles to atrophy. Hawking admits that the shock of finding out he had an incurable disease was compounded by the fact that it was probably going to kill him before he finished his Ph.D., but when he witnessed the agonizing death of a boy with leukemia in the hospital bed across from him, he gained new insight, realizing that "Clearly, there were people who were worse off than me."

The young boy's death, and the ensuing dreams Hawking had, helped change Hawking's perspective, and he realized that he needed to do something good with the life he had left. He stated to enjoy "life in the present," and he felt a surge of hope and purpose. Shortly after his diagnosis, he resumed his research at Cambridge, earned a research fellowship at Caius College in Cambridge, and married an undergraduate at Westfield College in London named Jane Wilde.

Hawking's chosen field, theoretical physics, proved to be a blessing as his disease continued to take its toll on his body. He found that the worse his condition became, the more he could focus his time on research, rather than lecturing. He continued to research and live a relatively normal life until 1974, when he lost the ability to feed himself or manage the usual rituals associated with going to bed. By this time he and his wife had three children, and it was clear that she could not continue to take care of him alone, so they came up with the brilliant solution of giving a research student free board and personal access to Hawking in exchange for help. This lasted for six years, until 1980, when they brought in professional help and Hawking eventually needed round-the-clock medical assistance.

In the early 1980s Hawking's speech became slurred, and his lectures required the assistance of a close friend who could still understand him to "translate" for the audience. Then, in 1985, Hawking came down with pneumonia. After a tracheotomy to save his life, his speaking capacity was destroyed. The only way he could communicate was by "raising [his] eyebrows when someone pointed to the right letter on a spelling card," which made it nearly impossible to get the ideas out of his head and onto paper.

A California computer whiz named Walt Woltosz came to Hawking's rescue with a program he wrote in which Hawking could chose words from a computer screen with a click of a switch. Soon, Hawking had a computer on his wheelchair, and a voice synthesizer. The hardware and software combination suddenly gave him the freedom to write at a rate of up to 15 words a minute, save his compositions to the computer and print them out, or even have the computer speak for him. It is by this method of writing, 15 words a minute, that Hawking has created extensive papers on complicated theories, written two books that have sold in the mainstream market with international appeal, and lectured around the world. The only problem, Hawking says, is that he now speaks with an American accent.

The beginning of Hawking's development of his popular theories started at Cambridge when he met the brilliant mathematician Roger Penrose, whose work on what happens at the end of a star's life inspired Hawking to devote himself to the subject of collapsed stars. By 1970 the mathematician-and-cosmologist team informed the world that they had discovered that the universe had a definite beginning of time, about 15 billion years ago in the big bang—confirming the idea originally proposed by RALPH ASHER ALPHER.

The idea of a beginning of time was a concept that had been argued on the basis of religion or philosophy for centuries. Hawking wrote, "SIR ARTHUR EDDINGTON once said, 'Don't worry if your theory doesn't agree with the observations, because they are probably wrong.' But if your theory disagrees with the Second Law of Thermodynamics (in which disorder increases with time), it is in bad trouble."

Hawking and Penrose proved that Einstein's general theory of relativity supported their theories—that when a star collapsed it became a black hole, a place where "time came to an end," and that "the expansion of the universe is like the time reverse of the collapse of a star." They cited the uncertainty principle of quantum mechanics that states an object cannot have both a well-defined position and a well-defined speed, and used the debunked steady state theory (as galaxies move apart, new galaxies are formed in between them from new matter that is constantly being formed) for further proof.

By combining the uncertainty principle with the general theory of relativity, Hawking created a new theory, the quantum theory, introducing the concept of "imaginary time," which when combined with three-dimensional space forms "a Euclidean space-time," where space and time are "like the surface of the Earth"—with no boundaries. Using the laws of physics, the state of the universe can be calculated in imaginary time, and therefore can also be calculated in real time, and so the beginning of time can be pinpointed.

In 1992, when the satellite *COBE (Cosmic Background Explorer)* found "irregularities in the microwave background radiation," Hawking says, "we saw back to the origin of the universe." The work Hawking and Penrose started more than two decades prior suddenly had some observational teeth, although even Hawking admits that more data is required to make a definitive stance. Their "no boundary" theory explains that the universe is finite, self-contained in space and imaginary time. It also implies that what had a beginning must also have an end.

Famous for his great sense of humor, Hawking recounts the story of giving a presentation on the "beginning of time" in Japan, where his sponsors requested that he "not mention the possible re-collapse of the universe, because it might affect the stock market." Hawking's reply was to reassure "anyone who is nervous about their investment that it is a bit early to sell," because the eventual end will not happen for about 20 billion years.

Hawking also made headlines for his work on black holes, which are formed when stars become so dense that nothing can escape their gravitational fields, including light. This idea of black holes was originally proposed in 1783 by the English scientist John Michell (1724–93) at Cambridge, then independently shortly thereafter by PIERRE-SIMON, MARQUIS DE LAPLACE in his publication *The System of the World*. Neither of them, nor anyone else in the scientific community, took the ideas beyond their initial pub-

lications. Einstein's theory of general relativity was the next theory to address gravity and its effect on light. Then, in 1928, SUBRAHMANYAN CHANDRASEKHAR worked out his theory of how stars maintain themselves with their own gravity, and determined that once a star reaches a certain level of density, known as the Chandra limit, it can no longer support itself. Eddington rebuked this idea, but the idea of gravitational collapse was revived again in 1939 by Robert Oppenheimer (1904–67), then dropped during World War II, and rediscovered in the 1960s.

The resulting work was supported by general relativity, and suggested that light emitted from a dense enough star would turn back in on itself. This light, and everything else within its proximity, would be pulled into the field. An American scientist named John Wheeler labeled this phenomenon a black hole in 1969, and the boundary surrounding the black hole is called its event horizon. Hawking first proclaimed in his research that by its nature, a black hole emits nothing—it only devours, it does not spit out. But he later discovered, with the prodding of an American student at Princeton and two Russian scientists, that black holes actually emit particles and radiation, and in this emission of energy, there is a loss of mass, meaning the black hole will actually get smaller and smaller and eventually just evaporate, in a period of about 10^{67} years.

In addition to creating theories on the workings of the universe, Hawking gives presentations on such topics as time travel, predicting the future of the cosmos, and life elsewhere in the universe. Time travel involves the general theory of relativity, quantum theory, the uncertainty principle, and discussion of cosmic strings, and wormholes, along with such practical observations as the fact that we have not yet been flooded by time-traveling tourists from the future. He similarly discusses Einstein's famous quote "God does not play dice," and gives his theories on this postulate, based, of course, on scientific principles, to determine whether this is true and/or if it is possible to predict the future of the universe. The second law of thermodynamics, a definition of life, and a principle called the anthropic principle, along with information on the Big Bang, the makeup of DNA, and some comments on evolution, stoke the fire of his discussion of the question of life outside our solar system. This proprietary information can be found on his Web site, http://www.hawking.org.uk.

Hawking's famous book *A Brief History of Time* was published in 1988, and became an international best seller, translated into more than 40 languages, and selling more than 9 million copies. He updated the book with a new version entitled *The Illustrated A Brief History of Time*, in 1996, and also published in 2001 a book entitled *The Universe in a Nutshell*. He has also published numerous technical papers, all of which require a professional foundation in physics in order to understand. *The Theory of Everything: The Origin and Fate of the Universe* is a new compilation of his work, much of which is in print in *A Brief History of Time*, and is a book in which he emphatically states he did not authorize and does not endorse.

A father of three, Hawking left his wife in 1990 and remarried in 1995. When asked how he feels about his disease, his standard answer is "Not a lot. I try to lead as normal a life as possible and not think about my condition, or regret the things it prevents me from doing, which are not that many." Remembering the early days of his diagnosis and the boy who shared his room, Hawking has a strong reminder of how fortunate he is. "Whenever I feel inclined to be sorry for myself, I remember that boy."

Hawking continues to travel around the world, giving public lectures and attending scientific conferences and symposiums. He has been awarded 12 honorary degrees, and is a member of the Royal Society and the U.S. National Academy of Sciences.

⊠ Herschel, Caroline Lucretia
(1750–1848)
German
Astronomer

Caroline Lucretia Herschel has often been mentioned only in reference to "assisting" her famous astronomer brother, SIR WILLIAM HERSCHEL, yet it was her calculations, her energy, and her dedication to astronomy that led to her important contributions to the field, including the compilation of two catalogs and the discovery of the first comet ever found by a woman, winning her the praise of kings.

Born in 1750 in Hanover, Germany, Herschel was brought up in the somewhat traditional way used in rearing female children at the time. Her mother, Anna Ilse Moritzen Herschel, believed that girls should be raised to learn only household tasks, and was opposed to any education of her daughter. Her father, Isaac Herschel, was not necessarily in agreement, but given his wife's disposition and the large number of children to which the family had to attend, reported as six to 10 depending on the source, he gave as much attention to providing some knowledge to his daughter as time and his wife would allow. Most important, he introduced her to the stars, "to make me acquainted with several of the beautiful constellations, after we had been gazing at a comet," she wrote in her diary. But her father left the family's home to fight the French when she was only seven years old, and returned three years later much worse for the wear. The young girl was obliged to take on the role her mother prescribed, strictly relegated to household chores and tending to her ill father for the next 10 years, until he died in 1767.

The Herschel family was steeped in music. Isaac had been in the Hanoverian Foot Guard, first as oboist and then eventually as bandmaster, and Herschel's older brother William, who played a profound role in her life, became a member of the band as well before leaving home and becoming a music teacher in Bath, England.

In 1772, at age 22, Caroline Herschel decided to leave behind a life of knitting, dressmaking, and studies to become a nanny. Her brother William invited her to come and live with him in Bath. Over the next few years, in exchange for her help running William's household, a task at which by this time she excelled, the organist, composer, and music teacher gave his sister singing lessons, and began to indoctrinate her in the higher education that she was so suited to absorb, including English, algebra, geometry, spherical trigonometry, and astronomy. So much for her mother's ideas.

It is said that William's interest in astronomy was spurred shortly after his sister's arrival, in 1773, when at age 35 he bought a book by James Ferguson entitled *Astronomy Explained upon Sir Isaac Newton's Principles, and Made Easy to Those Who Have Not Studied Mathematics.* This book was also the inspiration for Robert Baily Thomas to create the *Old Farmer's Almanack* 20 years later, in 1793, which is still in publication today and has the distinction of being the longest-running almanac in history.

Apparently it took William next to no time to get hooked on astronomy, because before the year was over, he and his sister, along with their brother Alexander, who was also living with them at the time, began working on building their own telescopes. Herschel claims that "nearly every room in the house [was] turned into a workshop," for some aspect of production. And while the reality of seeing the house in a shambles "was to my sorrow," she managed to work around it and became quite good at grinding telescope mirrors. The resulting telescopes, most notably a seven-foot model, turned a brother's interest in peering at the night sky as often as the British weather would permit into what would become a lifelong family affair. Over the years, they continued to make and sell telescopes, but studying the stars was their main venture.

With William manning the eyepiece, the family's goal was to do a systematic survey of the sky. Herschel wrote down all of the measurements William called out or signaled to her, and then did the daily calculations and logged all of their findings. Her daily duties still included running the household, and by now she had also become an accomplished singer, giving performances as often as five times a week with her brother. But the most important thing seemed to be taking care of their work with the stars. Herschel noted that "every leisure moment was eagerly snatched at for resuming some work which was in progress."

In March 1781 Herschel was recording data from her brother as he observed a comet in the night sky. But calculations proved that this was not a comet at all. The pair had discovered a new planet, as far beyond Saturn as Saturn is from the Earth. "William's" discovery of the planet, eventually named Uranus the father of Saturn, Saturn being the father of Jupiter, catapulted the Herschel duo from local amateur astronomers to national treasures. King George III gave William an annual salary of £200, and as a new member of the Royal Society William decided to leave the music business for good, going from professional musician to professional astronomer overnight. While Herschel loathed giving up her singing career, she acquiesced, although she enjoyed singing privately throughout the rest of her life, and sometimes for very noble guests.

The following year, Herschel's brother moved them into a bigger house in Windsor, and she became the owner of a new telescope of her own, a gift from William so that she could start making her own observations. As she had learned with her brother, the best approach was systematic, so she began to methodically track the night sky in search of comets. But her work with her brother continued to occupy most of her time. He became more immersed in his own observations, and she was always present, recording data during night-time observations and calculating the work during the day. For the next several years, she spent most of her time recording and working on computations with her brother as he did work that led to discoveries of new binary stars, new moons, and eventually the doubling of the size of the known solar system.

Her own discoveries occurred, however, after a move in April 1786 to another house, a home they called Observatory House. Here, on August 1, 1786, Caroline Herschel discovered her first comet. She became instantly famous. Her comet was dubbed "the first lady's comet," because Herschel was in fact the first "lady" to ever discover a comet.

This began an entirely new life for Herschel. In 1787 the king gave her a £50 annual salary, and the Herschels became frequent guests at the castle. Princess Augusta would often peer into a telescope at the castle, and would invite guests to look and see "Miss Herschel's comet."

That same year, Herschel's brother married and, understandably, it was time for Herschel to leave her brother's house and make a home of her own. She was reportedly furious and bitter with her brother's wife for insisting that she leave, but she moved within walking distance and came to their house every day to work on his observations. Eventually, she softened toward her sister-in-law, and things smoothed out between them, especially after the birth of William's son John in 1792. As her nephew grew, Herschel was able to start teaching him mathematics and astronomy—a task that resulted in the formation of another renowned scientist in the family, SIR JOHN FREDERICK WILLIAM HERSCHEL.

Herschel continued both her work with her brother and her search for comets, discovering a total of eight comets during the next 11 years. She also embarked on another huge project, working on updating Royal Astronomer John Flamsteed's *Historia Coelestis Britannica* star catalog. In 1798 she presented her first publication, *Index to Flamsteed's Observations of the Fixed Stars,* in which she updated his catalog with 560 stars that were not previously included. This would be her sole publication for the next 25 years.

In 1822 William Herschel passed away, and Caroline, now 72, left England to live the rest of her years in Hanover. By 1828 her next publication was ready to be presented to the world. A catalog that she put together in part to help her nephew with his own astronomical contributions, the work was a compilation of 2,500 nebulae. This was a great achievement that was recognized with a gold medal in 1828 from the Royal Astronomical Society. In 1832 William's widow died, and their son John went to Hanover to visit his aunt, now aged 83, marveling that the petite woman "runs about the town with me, and skips up her two flights of stairs," sings, dances, and is full of life "quite fresh and funny at ten p.m."

The effervescent Herschel remained alert and active for many years to come. She was honored in 1835 when she and MARY FAIRFAX SOMERVILLE jointly became the first two women elected to the Royal Society. Leading scientists continued to visit her throughout her life, and she remained a celebrity for her amazing work in astronomy. In 1838 the Royal Irish Academy elected her a member, and her 96th birthday was marked with a gift from the king of Prussia, a gold medal for her scientific achievements.

At her 98th birthday celebration, the crown prince and princess of Prussia visited Herschel at her home, where she said, "Let's sing a catch," and proceeded to perform a song written much earlier in her lifetime by her brother William. She remained active and alert until her death at home on January 9, 1848, where she drifted away peacefully. The first home that she shared with her brother William in Bath, located at 19 King Street, is now the site of the William Herschel

Museum. In addition to her many awards, a minor planet was named Lucretia in 1889 in her honor, and a crater on the Moon bears her name.

Herschel, Sir John Frederick William
(1792–1871)
British
Mathematician, Astronomer

Sir John Herschel had such a wide range of interests in his life that he could easily be considered a Renaissance man of the Victorian age. He influenced many of the great minds of his time, including Charles Babbage (1791–1871), Charles Darwin (1809–82), and Michael Faraday (1791–1867). But as diverse as his interests were, many of them melded seamlessly together allowing him to make some of the greatest contributions in his life to astronomy—both inheriting and passing on his family's legacy of extraordinary scientific achievement.

John Frederick William Herschel was born on March 7, 1792, to Mary Pitt Herschel and the famous "King's Astronomer" SIR WILLIAM HERSCHEL. He was Mary's second child, from her second marriage. Mary's first husband and child died, leaving her a young widow until she married William in 1788. William's only child, the boy was raised in the home they called Observatory House, where William had built his 40-foot telescope with a £4,000 grant from the king.

Education began early in Herschel's life. CAROLINE LUCRETIA HERSCHEL, at that time one of the most famous women in the world, doted on her nephew and took every opportunity to teach him science and mathematics during her daily visits to Observatory House. Herschel's official schooling started at Dr. Gretton's School, in the town of Hitcham, where he went until the age of eight, when he changed to Eton College. But as the new boy in school, Herschel soon found himself a target for bullies, and his

mother pulled him out within a few months, turning to private tutoring for her son's education. At the age of 17, in 1809, Herschel went to St. John's College at Cambridge.

Mathematics was the field of choice for Herschel at Cambridge, and he teamed up with two other brilliant students to force changes in the way mathematics was being taught in England. In 1812 he and classmates Babbage and George Peacock (1791–1858) formed an organization they called the Analytical Society, hoping to create enough of a critical mass to convert the British educational system over to modern mathematics, specifically the teachings of Sylvestre Lacroix (1765–1843). Herschel translated from French Lacroix's treatise on calculus, and added his own examples to the work. The society was short-lived, but the translation was successful, and the work was adopted as a textbook, giving Herschel his first of many impressive successes.

While working on changing the educational system, the 20-year-old undergraduate student published his first paper—a mathematics piece entitled, "On a Remarkable Application of Cotes's Theorem"—in the *Transactions of the Royal Society*. The following year, after graduating from Cambridge in 1813, Herschel was elected a Fellow of the Royal Society based on the merits of his paper.

After graduation, Herschel took a detour from science and decided to become a lawyer. He studied for about a year and a half, then abandoned the law and returned to Cambridge. In 1816 he graduated from Cambridge again, this time with an M.A. in mathematics, and went back home to spend the summer with his 78-year-old father. This visit was a turning point in Herschel's professional life. He wrote to Babbage, saying that he was going back to Cambridge just long enough to pack up and take care of his personal affairs, and then "I am going under my father's directions . . . and continuing his scrutiny of the heavens."

But while Herschel was committed to following in his father's footsteps, he was not quite ready to give up his other interests—and he had many. One of them was photography. In 1819 Herschel began experimenting with "photography," a term that he made up. Interested in chemistry, Herschel had done many experiments with chemicals as they related to photography. He published his experiments, and they would prove to have a profound impact on the development of photography in about 20 years.

In 1820 Herschel began to make some serious inroads in astronomy. He joined his father at the first meeting of the Astronomical Society of London—the precursor of the Royal Astronomical Society—and his father became the first president of the society, with Herschel elected as vice president. That same year, he published a book coauthored with Babbage, *A Collection of Examples of the Applications of the Differential and Integral Calculus*. In 1821 the Royal Society awarded him his first Copley Medal for his publications.

But the collaborations with his famous family came to a sudden halt when, in 1822, his father died and his aunt Caroline moved back to Germany. Left to make his name in astronomy on his own, Herschel published his first astronomical paper, in which he described a new way to calculate lunar eclipses. He also spent some time studying light at a time when physicists were beginning to look at the spectra of nonsolar light. His experiments dealt with the spectra of metallic salts, in which he could determine that there were small quantities of salt in flames. This work marked the beginning of a new branch of physics called spectroscopy.

By 1824 Herschel was fully engaged in astronomy. He was elected secretary of the Royal Society, a position he kept for three years. He published his first major book, a catalog of double stars in the *Transactions of the Royal Society*, for which he received the Paris Academy's

Lalande Prize the following year and the Astronomical Society's gold medal. Then, in 1826, he began to focus his work on binary stars, which had been a major component of his father's work. He took time out only to write, and to marry a woman named Margaret Brodie Steward in 1829.

Herschel began to make many important contributions to encyclopedias, writing about scientific subjects to help educate the masses. His first such piece was a 245-page work on light, written in 1828, for the *Encyclopaedia Metropolitana*. Next came similar articles on mathematics for the 1830 edition of the *Edinburgh Encyclopaedia*. That same year, Herschel published one of his most famous works, *Preliminary Discourse on the Study of Natural Philosophy*, a book that influenced Darwin and prompted Faraday to write to Herschel telling him that the book "made me, if I may be permitted to say so, a better philosopher." The following year, Herschel was knighted. Then, in 1832, his mother died.

Herschel continued concentrating on the work his father had started with binary stars, and he devised a way to determine their orbits and discovered that they orbited a common center of gravity. He was awarded the Royal Medal from the Royal Society in 1833, for his paper "On the Investigation of the Orbits of Revolving Double Stars." He also published *A Treatise on Astronomy* as an encyclopedia entry.

Heavily engrossed in carrying on his father's work, Herschel packed his family, consisting of his wife and four children, and the 20-foot refractor telescope that he and his father had built, and sailed for the Cape of Good Hope in South Africa, where the Royal Observatory had built a new facility in 1828, with a mission of cataloging the heavenly objects that could not be seen in the Northern Hemisphere. He arrived in January 1834, and set up observations in Feldhausen, near Cape Town.

The next few years were busy for Herschel. In South Africa he was able to observe and chart the kinds of "exotic objects" that his father found so fascinating, nebulae and star clusters. An exceptional artist, he drew many of the observations he saw in the Southern Hemisphere sky. He had the extreme good fortune of being able to observe the return of Halley's comet in 1835, and he noticed that there seemed to be something other than gravity that affected its orbit, deciding that it was somehow being repelled by the Sun. In truth, the repulsion he observed was caused by solar wind, leading many to believe that Herschel was the first to discover this unseen but documented force. He also discovered that the comet was expelling evaporated gases.

In 1836 Herschel worked on establishing the relative brightness of nearly 200 stars. He used the Moon as a focal point and compared stars and their apparent brightness as they related to the distance of the Moon and its apparent brightness. In 1833 a paper he published on nebulae and star clusters in the *Philosophical Transactions* earned him another Royal Medal from the Royal Society. During his last two years in South Africa, Herschel also worked on measuring solar energy, and in 1838, loaded with data from his years of Southern Hemisphere observations, he headed home to England.

The next several years were full of big advancements in photography, and Herschel put himself right in the middle of it. A good friend of his, William Henry Fox Talbot (1800–77), who is credited as the inventor of this field, consulted with Herschel many times, as the two discussed the new inventions Talbot was making. Herschel is credited with discovering that hyposulfite of soda could be used as a photographic fixative, and for devising the terms *negative* and *positive* for paper photos. Seeing the obvious benefits of the new realm of photography, Herschel soon became a very vocal advocate for its use in astronomical observations. He wrote a paper in 1839, "On the Chemical Action of the Rays of the Solar Spectrum on Preparations of Silver,

and Other Substances, Both Metallic and Non-Metallic, and on Some Photographic Processes," printed in the *Philosophical Transactions*, which, in 1840, earned him another Royal Society Royal Medal.

In 1842 he left photography in the capable hands of Talbot and others and got back to the business of compiling his data from South Africa. During this time, he became head of the Marischal College in Aberdeen, and then, three years later, in 1845, president of the British Association at Cambridge. In August 1846 he heard about a new planet that fellow astronomer JOHN COUCH ADAMS had determined existed beyond Uranus, which was the planet Herschel's father had discovered. Herschel was elated by the news, and did everything he could to help Adams get support in his search for the planet, including setting up a presentation for Adams at a meeting of the British Association, for which Adams arrived the day after the meeting. When the London *Times* ran the headline "Le Verrier's Planet Found" on October 1, 1846, Herschel notified the press that this new planet, ultimately named Neptune, was really Adams's planet, which started an international incident.

When he was not involved in the Adams scandal, Herschel continued working on his observations, computations, and compilations, and in 1847, he published *Results of Astronomical Observations Made during the Years 1834, 1835, 1836, 1837, and 1838, at the Cape of Good Hope, Being the Completion of a Telescopic Survey of the Whole Surface of the Visible Heavens, Commenced in 1825*. This work resulted in his second Copley Medal from the Royal Society. In 1847, he also improved a device called a calorimeter, used to measure the heat from a chemical reaction, which had been originally constructed in 1783 by Antoine Lavoisier (1743–94) and PIERRE-SIMON, MARQUIS DE LAPLACE.

In 1849 Herschel expanded his work into the *Outlines of Astronomy*, originally written as an encyclopedia entry in an in-depth textbook, and it was an instant success. The book had 12 editions in England, was translated into many languages, and was used as a textbook around the world for decades to come.

As strange as Herschel's detour into law seemed in the beginning of his career, he made perhaps an even more bizarre stray from his scientific endeavors in 1850, when he decided to take on the role of Master of the Mint. For the next 13 years, Herschel overworked himself in this position, and turned to writing poetry, much of which was published in 1857; and some scientific work, published in 1861, on meteorology, geography, and telescopes. In 1863 he finally retired from his duties at the mint. His biggest publication was still to come.

In 1864 Herschel published his *General Catalogue of Nebulae and Clusters*, listing 5,097 star clusters and nebulae, of which 4,630 were discovered by Herschel and his father throughout their careers. This was revised in 1888 by L. E. Dreyer into the *New General Catalogue* (NGC) with 7,840 nebulae and clusters. The NGC numbers are still used to identify nonstellar objects. Herschel followed this work in 1866 with a general publication entitled *Familiar Lectures on Scientific Subjects*.

Herschel's life was full of accomplishment, fueled in part by the diversity of interests that kept his mind active, including astronomy, biology, chemistry, education, language, mathematics, music, poetry, philanthropy, physics, and public service. As a respected scientist, he corresponded with some of the most brilliant minds of the time.

Nearly all of Herschel's documents have survived, including a collection of 10,000 letters (of which 75 percent are from the world's leading scientific minds of the time), his astronomical observations, diaries, books, laboratory notes, and a collection of his photographic work. He won multiple awards for his contributions to science in both mathematics and astronomy, and

was involved in many organizations to help further the sciences.

Craters on the Moon and on Mars are named for this extraordinary scientist, who died on May 11, 1871, in Kent, England, and was buried in Westminster Abbey. Herschel had 12 children. His son, Alexander, became a professor of physics and carried on the Herschel tradition of working with the Astronomer Royal, advocating with George Biddell Airy (1801–92) the use of photography in astronomy.

⊠ **Herschel, Sir William (Friedrich Wilhelm Herschel)**
(1738–1822)
German
Astronomer

Sir William Herschel is the epitome of an amateur astronomer who turned a passion for his hobby into his life's work. Thanks to a book that caught his interest at age 35, and the help of his sister, Caroline, Herschel went from being a musician and amateur astronomer to being the King's Astronomer in just nine years. He also became the last person to discover a planet purely on observation rather than on mathematical computation.

Friedrich Wilhelm Herschel was born on November 15, 1738, in Hanover, Germany, to Anna Ilse Moritzen Herschel, the strict matriarch of her family, and Isaac Herschel, an oboist who eventually became the bandleader of the Hanoverian Foot Guard. In 1750, when Herschel was 12 years old, his sister Caroline was born, and in a little more than 20 years she would begin to play a profound role in his success as an astronomer.

The young Herschel was a dedicated musician, and as a teenager joined the Hanoverian Guard to be in the band with his father. But military bands are first and foremost military, and when the Seven Years' War broke out, Herschel and his father were called to duty in 1757 to fight at the battle of Hastenbeck.

Fighting in a war took its toll on both father and son. The elder Herschel did everything he could to help his son get out of the war, and in 1759, Herschel left Germany for England, where he would spend the rest of his life. He earned a living as a traveling musician, stopping in Halifax in 1765 and working as an organist for a year, then settling in Bath, England, where he became the organist at the Octagon Chapel and eventually acquired a job as city public concerts director.

Meanwhile, Isaac Herschel continued fighting, and when he finally returned home in 1760, he was in need of constant care. Caroline took care of their father until 1767, when he passed away.

Working successfully in Bath, Herschel became a music teacher, composer, and conductor. By 1772 he was a well-respected member of his community. Back home in Germany, Herschel's sister Caroline, who had so diligently tended their father, was now serving her overbearing mother, taking care of the household chores, and studying to become a nanny.

Herschel needed help running his household, so in 1772 he invited his 22-year-old sister to live with him in England. She jumped at the chance, and moved to England to join Herschel and one of their brothers, Alexander, in Bath. Herschel was soon teaching his sister music, and she began singing in concerts the two would give throughout the week.

But Herschel did more for Caroline than teach her music. He soon introduced her to the kind of intellectual stimulation that she had longed for as a child but was denied because their mother did not believe that girls should be educated. To their mutual benefit, he taught his sister algebra, geometry, trigonometry, and astronomy.

On May 10, 1773, Herschel purchased a book that would forever change his life, his sister's life, and the size of the known universe. *Astronomy Explained upon Sir Isaac Newton's Principles, and Made Easy to Those Who Have Not Studied Mathematics*, written by James Ferguson, put a burning desire in Herschel to study the stars. But in order to study them, Herschel needed a telescope, so he collected used parts and built one. It was small. It was also inadequate.

Encouraged, he enlisted the help of his siblings to build a bigger telescope, and Herschel, Caroline, and Alexander proceeded to turn their home into their own private telescope manufacturing facility. Caroline became adept at grinding mirrors, while Alexander helped with the overall construction. The family became so good at their avocation that they started selling their telescopes to other amateur astronomers and even to the Royal Observatory.

By 1774 they had constructed a seven-foot telescope, and Herschel started studying the heavens. Meticulous and methodical, Herschel devised a plan to study a strip of sky at a time. Standing on a ladder, peering into the telescope, he told his sister everything he saw, and she recorded it all in great detail. For the next several years, the brother-and-sister team observed the night sky as often as they could, concerts and local weather permitting, and Caroline recorded by night and computed by day in addition to managing the house.

This was an interesting time in the evolution of astronomy—a time when the observations that had been made up to this point were being compiled into useable data for navigators at sea. Certainly, Herschel's work could eventually be useful for navigators, but while that was the sole purpose of the work being done at the Royal Observatory, it was not the foremost thought on Herschel's mind. He wanted to make a systematic survey of the stars, and he set about observing and documenting everything he saw, with his sister's diligent help.

On the evening of March 13, 1781, Herschel was peering though his seven-foot telescope, equipped with a 6.2-inch reflector, at the constellation Gemini, when he saw a disk-shaped light shining back at him. He observed the object as it moved slowly across the sky, and was convinced that he had discovered his first comet. His sister wrote down the observations and performed the calculations as the object continued to move in the night sky.

But the "comet" was beyond Saturn, and if it really were a comet, it would have been too small to see. Herschel's suspicions were not correct. The Royal Observatory's fifth Astronomer Royal, Nevil Maskelyne, discovered that Herschel's "comet" was, instead, a planet that was farther beyond Saturn than Saturn is from the Sun. And since the galaxy and the universe were considered to be the same at this point in history, Herschel's discovery doubled the known size of the universe. Herschel commented, "It was this night its turn to be discovered."

Astronomers from around the world were ready to pounce on the chance to name the new planet, and the English were afraid that the French would steal the opportunity if Herschel did not come up with something fast. So Herschel suggested the most politically correct name he could think of: Georgium Sidas, the Georgian Planet, to honor the reign of King George III. Unfortunately, King George was not as beloved outside of England, so the French gave it their own name—Herschel. Other suggested names included Minerva, for the Roman goddess of wisdom, and Hypercronius, which means "above Saturn." But the Germans, who had nothing to do with the discovery, seemed to be very adept at naming things and JOHANN ELERT BODE took the logical approach and named the planet Uranus, who in mythology was the son of Gaia (the Earth), and the father of Saturn. The various countries continued calling the planet by the names they liked best until 1850, when JOHN COUCH ADAMS, who discovered Neptune, proposed that it should have the official name of Uranus.

But what's in a name? It was the discovery that counted. Saturn had been the farthest known planet, at a distance of 886,200,000 miles from the Sun. Uranus, a little less than half the size of Saturn, about four times as big as the Earth, was 1,782,000,000 miles from the Sun. This discovery was worthy of some recognition, even if an amateur made it. Within a few short months, on December 7, 1781, the Royal Society elected Herschel into its ranks as a Fellow, and Herschel was appointed the "King's Astronomer," and was given a lifetime salary of £200 per year by King George. The following year, in May 1782, Herschel and his sister quit the music business forever, and in August they moved into a bigger house in the town of Datchet, where they lived for nearly three years.

During their time in Datchet, Herschel focused his attention on new heavenly bodies. A friend named William Watson had given Herschel a copy of CHARLES MESSIER's catalog of star clusters and nebulae as a congratulatory gift upon his election into the Royal Society. Herschel was now interested in observing these "exotic objects," as he called them. In 1782 he and Caroline began their work on surveying the sky for these nebulae, and as a gift he gave his sister a telescope of her own. For the next few years, the duo worked diligently on recording and classifying their observations, being careful not to duplicate the work of Messier.

In 1785 Herschel and Caroline moved again, this time to Windsor to be near the king. Herschel was working on his theories about the shape of our galaxy. The small telescopes he had used up to this time were adequate for the work he had done, but his needs were now outgrowing his technology. So the king offered Herschel his patronage again, this time in the form of a £2,000 grant to build a telescope that would become the world's largest—a 40-foot reflector.

In April 1786 the Herschel siblings moved for the last time together, into a home they called Observatory House in Slough. The move was a good one for both of them. Herschel started the process of building his new impressive telescope, and in August Caroline made her first discovery—a comet. She was the first woman ever to make such a discovery. It was dubbed "the first lady's comet," and she was awarded her own salary of £50 per year from the king. From this point on, the two were frequent guests at the palace.

Herschel's next two big discoveries occurred in 1787, when he found the first two satellites of his planet, Titania and Oberon. The next two, Ariel and Umbriel, were discovered in 1787 by another British astronomer, and these four moons were eventually named by Herschel's son. The fifth moon, Miranda, was discovered in 1948 by GERARD PETER KUIPER. To date, 22 moons have been discovered in its orbit.

The following year marked perhaps the biggest change of all for the brother and sister astronomers. Herschel married a young widow named Mary Pitt, and Caroline was forced to move out of their home at Mary's insistence. This was the first time that Caroline had ever lived on her own, and she was extremely upset. Eventually she reconciled with her sister-in-law and regretted her anger. Caroline continued to work with her brother, and walked to his house daily to record their observations.

Three years and £4,000 after construction began, Herschel's 40-foot telescope was complete. His first discoveries with this new masterpiece were Mimas and Enceladus, the sixth and seventh moons of Saturn. He was also now able to make out some individual stars in some of the globular clusters he had observed with his older equipment.

In 1792, at age 55, Herschel became a father. His son John was born at Observatory House, and grew up with the 40-foot telescope as part of his daily life. The boy was influenced by the music and science around him, especially by his Aunt Caroline, and eventually joined his

father in astronomical work. In 1864 SIR JOHN FREDERICK WILLIAM HERSCHEL published *The General Catalogue of Nebulae*, a collection of more than 5,000 nebulae, of which 4,630 were discovered by John and his father.

Over the years, Herschel continued his observations of nebulae, star clusters, the Sun, and binary stars. He found that the Sun was gaseous in nature, observed sunspots, and discovered infrared light emanating from the Sun. He discovered nearly 1,000 binary stars, and proposed that these stars have a common center of gravity around which they rotate. His observations of nebulae led him to believe that they were clusters of stars he called "island nebulae," and he believed that these were "island universes," galaxies, outside the Milky Way. He was, in fact, the first person to accurately describe the Milky Way Galaxy, a feat he accomplished using the statistics he had accumulated during his observations with Caroline.

On January 12, 1820, Herschel attended a dinner at the Freemason's Tavern in London, where a group of 14 men talked about forming a society with the simple goal of promoting astronomy. On March 10 the Astronomical Society of London, later the Royal Astronomical Society, had its first official meeting, and Herschel became its first president.

Herschel was known for his tremendous enthusiasm for astronomy and for his exacting, methodical style of observation of the night sky. He was knighted in 1816 for his contributions to the field of astronomy. On August 25, 1822, Herschel died in Slough, England, and was laid to rest at St. Laurence's Church in Upton. His work was his life's gift, and his contributions earned him the respect and admiration of far more learned men than he. The house Herschel shared with his sister in Bath, the site of the discovery of Herschel's planet, Uranus, was opened in 1981 as the William Herschel Museum. The remains of his 40-foot reflector telescope from Observation House are now on display at the Royal Observatory in Greenwich. In honor of his work, a crater on the Moon was named Herschel for this amateur astronomer, as was one for his sister and another for his son.

⊠ Hertz, Heinrich Rudolf
(1857–1894)
German
Physicist

In 1888 Heinrich Hertz produced and detected radio waves for the first time. He also demonstrated that this form of electromagnetic radiation, like light, propagates at the speed of light. His discoveries form the basis of both the global telecommunications industry (including communications satellites) and radio astronomy. The hertz (Hz) is the Système Internationale (SI) unit of frequency, named in his honor.

Hertz was born on February 22, 1857, in Hamburg, Germany, into a prosperous and cultured family. Following a year of military service (1876–77), he entered the University of Munich to study engineering. However, after just one year, he found engineering not to his liking and began to pursue a life of scientific investigation as a physicist in academia. Consequently, in 1878 he transferred to the University of Berlin and started studying physics, with the famous German scientist Herman von Helmholtz (1821–94), as his mentor. Hertz graduated magna cum laude with his Ph.D. degree in physics in 1880. After graduation, he continued working at the University of Berlin as an assistant to Helmholtz for the next three years.

He left Berlin in 1883 to work as a physicist at the University of Kiel. There, following suggestions from his mentor, Hertz began investigating the validity of the electromagnetic theory recently proposed by a Scottish physicist, James Clerk Maxwell (1831–79). As a professor of physics at the Karlsruhe Polytechnic from 1885

In 1888 the German physicist Heinrich Rudolf Hertz became the first person to produce and detect radio waves. He also demonstrated that radio waves, like other forms of electromagnetic radiation, propagate at the speed of light. His scientific discoveries laid the foundation for communications satellites as part of the global telecommunications industry and for radio astronomy. *(Deutsches Museum, courtesy AIP Emilio Segrè Visual Archives, Physics Today Collection)*

to 1889, Hertz finally gained access to the equipment he needed to perform the famous experiments that demonstrated the existence of electromagnetic waves, and he verified Maxwell's equations. During this period, Hertz not only produced electromagnetic (radio frequency) waves in the laboratory but also measured their wavelength and velocity. Of great importance to modern physics and the fields of telecommunications and radio astronomy, Hertz showed that his newly identified radio waves propagated at the speed of light, as predicted by Maxwell's theory of electromagnetism. He also discovered that radio waves were simply another form of electromagnetic radiation, similar to visible light and infrared radiation, save for their longer wavelengths and shorter frequencies. Hertz's experiments verified Maxwell's electromagnetic theory and set the stage for others, such as Guglielmo Marconi (1874–1937) to use the newly discovered radio waves to transform the world of communications in the 20th century.

In 1887, while experimenting with ultraviolet radiation, Hertz observed that incidental ultraviolet radiation was releasing electrons from the surface of a metal. Unfortunately, he did not recognize the significance of this phenomenon, nor did he pursue further investigation of the photoelectric effect. In 1905 ALBERT EINSTEIN wrote a famous paper describing this effect, linking it to MAX KARL PLANCK's idea of photons as quantum packets of electromagnetic energy. Einstein earned the 1921 Nobel Prize in physics for his work on the photoelectric effect.

Hertz performed his most famous experiment in 1888, with an electric circuit in which he oscillated the flow of current between two metal balls separated by an air gap. He observed that each time the electric potential reached a peak in one direction or the other, a spark would jump across the gap. Hertz applied Maxwell's electromagnetic theory to the situation and determined that the oscillating spark should generate a very long electromagnetic wave that traveled at the speed of light. He also used a simple loop of wire, with a small air gap at one end, to detect the presence of electromagnetic waves produced by his oscillating spark circuit. With this pioneering experiment, Hertz produced and detected "Hertzian waves"—later called radiotelegraphy waves by Marconi and then, simply, radio waves. By establishing that Hertzian waves were electromagnetic in nature, the young German physicist extended human knowledge about the electromagnetic spectrum, validated Maxwell's electromagnetic theory, and

identified the fundamental principles for wireless communications.

In 1889 Hertz accepted a professorship at the University of Bonn. There, he used cathode-ray tubes to investigate the physics of electric discharges in rarefied gases, again just missing another important discovery—X-rays—accomplished by the German physicist Wilhelm Roentgen (1845–1923) at Würzburg in 1895.

Hertz was an excellent physicist whose pioneering research with electromagnetic waves gave physics a solid foundation upon which others could build. His major publications included *Electric Waves* (1890) and *Principles of Mechanics* (1894). He suffered from lingering ill health due to blood poisoning and died as a young man, in his late 30s on January 1, 1894, in Bonn. In his honor, the international scientific community named the basic unit of frequency the hertz (symbol Hz). One hertz corresponds to a frequency of one cycle per second.

⊠ **Hertzsprung, Ejnar**
(1873–1967)
Danish
Astronomer

At the start of the 20th century, Ejnar Hertzsprung made one of the most important contributions to modern astronomy, when he showed in 1905 how the luminosity of a star is related to its color, or spectrum. His work, independent of HENRY NORRIS RUSSELL, contributed to one of the great observational syntheses in astrophysics— the famous Hertzsprung-Russell (HR) diagram, an essential tool for anyone seeking to understand stellar evolution.

The Danish astronomer Hertzsprung was born on October 8, 1873, in Frederiksberg, near Copenhagen. His father, Severin Hertzsprung, had graduated with a master of science degree in astronomy, but for financial reasons worked as a civil servant in the Department of Finances

American astronomer Harlow Shapley (left) appears in this photograph talking with the Danish astronomer Ejnar Hertzsprung (right), during the 1958 Moscow meeting of the International Astronomical Union. Working independently of the American astronomer Henry Norris Russell, Hertzsprung developed the famous Hertzsprung-Russell diagram—a useful graph that depicts the stars arranged according to their luminosity and spectral classification. *(D. Y. Martynov, courtesy AIP Emilio Segrè Visual Archives, Shapley Collection)*

within the Danish government. Consequently, he encouraged his son to enjoy astronomy as an amateur, believing the field presented very few opportunities for financial security. Hertzsprung responded to his father's well-intended advice and studied chemical engineering at Copenhagen Polytechnic Institute. Upon graduating in 1898, he accepted employment as a chemical engineer in St. Petersburg, Russia, and remained there for two years. He then journeyed to the University of Leipzig, where he studied photochemistry for several months under the German physical chemist Wilhelm Ostwald (1853–1932). The thorough understanding of photochemical processes he acquired from Ostwald allowed Hertzsprung to make his most important contribution to modern astronomy.

He became one of the great observational astronomers of the 20th century, despite the fact that he never received any formal academic training in astronomy. When he returned to

Denmark in 1902, Hertzsprung revived his life-long interest in astronomy by becoming a private astronomer at the Urania and University Observatories in Copenhagen. During this period, he taught himself a great deal about astronomy and used the telescopes available at the two small observatories to make detailed photographic observations of the light from the stars.

Hertzsprung's earliest and perhaps most important contribution to astronomy started with the publication of two papers, both entitled *"Zur Strahlung der Sterne"* (On the radiation of the stars) in a relatively obscure German scientific photography journal. These papers appeared in 1905 and 1907, respectively, and their existence was unknown to the American astronomer Henry Norris Russell, who would soon publish similar observations in 1913. Publication of these two seminal papers marks the beginning of the independent development of the Hertzsprung-Russell (HR) diagram—the famous tool in modern astronomy and astrophysics that graphically portrays the evolutionary processes of visible stars. Its creation is equally credited to both astronomers. Hertzsprung's two papers presented his insightful interpretation of the stellar photography data that revealed the existence of a relationship between the color of a star and its respective brightness. He stated that his photographic data also suggested the existence of both giant and dwarf stars. Hertzsprung then developed these and many other new ideas over the course of his long career as a professional astronomer.

Between 1905 and 1913 Hertzsprung and Russell independently observed and reported that any large sample of stars, when analyzed statistically using a two-dimensional plot of magnitude (or luminosity) versus spectral type (color or temperature), form well-defined groups or bands. Most stars in the sample will lie along an extensive central band called the main sequence, which extends from the upper left corner to lower right corner of the HR diagram. Giant and supergiant stars appear in the upper right portion of the HR diagram, while the white dwarf stars populate the lower left region. Modern astronomers use several convenient forms of the HR diagram to support their visual observations and to describe where a particular style fits in the overall process of stellar evolution. For example, our parent star, the Sun, is a representative yellow dwarf (main sequence) star. Previously, astronomers used the term *dwarf star* to describe any star lying on the main sequence of the HR diagram. Today, the term *main sequence star* is preferred to avoid possible confusion with white dwarf stars—the extremely dense final evolutionary phase of most low-mass stars.

While Hertzsprung was still in Copenhagen, he also started making detailed photographic investigations of star clusters. In 1906 he apparently constructed his first color versus magnitude precursor diagrams based on photographic observations of the Pleiades and subsequently published such data along with companion data for the Hyades in 1911. This activity represents an important milestone in the evolution of the Hertzsprung-Russell diagram. The Pleidaes is a prominent open star cluster in the constellation Taurus about 410 light-years distant. The Hyades is an open cluster of about 200 stars in the constellation Taurus, some 150 light-years away. Astronomers use both clusters as astronomical yardsticks, comparing the brightness of their stars with the brightness of stars in other clusters. Hertzsprung would spend the next two decades making detailed observations of the Pleidaes—a favorite astronomical object used frequently by other astronomers to compare the performance of their telescopes in the Carte du Ciel astrophotography program that began in 1887.

In 1909 the German astronomer KARL SCHWARZSCHILD invited Hertzsprung to visit him in Göttingen, Germany. Schwarzchild quickly recognized that Hertzsprung, despite his lack of formal training in astronomy, possessed exceptional talents in the field and gave Hertzsprung a staff position at the Göttingen Observatory. Later that year Schwarzschild became the director of the Astrophysical Observatory in Potsdam, Germany, and he offered Hertzsprung an appointment as a senior astronomer. Hertzsprung proved to be a patient, exacting observer who was always willing to do much of the tedious work himself. While at the Potsdam Observatory, he investigated the Cepheid stars in the Small Magellanic Cloud. He used the periodicity relationship for Cepheid variables announced in 1912 by HENRIETTA SWAN LEAVITT to help him calculate intergalactic distances. Although the currently estimated distance (195,000 light-years) to the Small Magellanic Cloud is about six times larger than the value of 32,600 light-years that Hertzsprung estimated in 1913, his work still had great importance. It introduced innovative methods for measuring extremely large distances and presented an astronomical distance that was significantly larger than any previously known distance in the universe. This "enlarged view" of the universe encouraged other astronomers, such as HEBER DOUST CURTIS and HARLOW SHAPLEY, to vigorously debate its true extent.

Hertzsprung left the Potsdam Observatory in 1919 and accepted an appointment as associate director of the Leiden Observatory in the Netherlands. He became the director of the observatory in 1935 and served in that position until he retired in 1944. Upon retirement, he returned home to Denmark. He remained active in astronomy, primarily by examining numerous photographs of binary stars and extracting precise position data. The international astronomical community recognized his contributions to astronomy through several prestigious awards. He received the Gold Medal of the Royal Astronomical Society in 1929 and the Bruce Gold Medal from the Astrophysical Society of the Pacific in 1937. After a full life dedicated to progress in observational astronomy, Hertzsprung, the chemical engineer turned astronomer, died at the age of 94 on October 21,

1967, in Roskilde, Denmark. The Hertzsprung-Russell diagram permanently honors his important role in understanding the life cycle of stars.

⊗ **Hess, Victor Francis**
(1883–1964)
Austrian/American
Physicist

As a result of ionizing radiation measurements made during perilous high-altitude balloon flights between 1911 and 1912, Victor Hess discovered the existence of cosmic rays that continually bombard Earth from space. Scientists used his discovery to turn Earth's atmosphere into a giant natural laboratory—a clever research approach that opened the door to many new discoveries in high-energy nuclear physics.

Hess was born in Waldstein Castle, near Peggau, in Steiermark, Austria, on June 24, 1883. His father, Vinzens Hess, was a royal forester in the service of Prince Öttinger-Wallerstein. Hess completed his entire education in Graz, Austria—attending secondary school (the Gymnasium) from 1893 to 1901 and then the University of Graz, as an undergraduate from 1901 to 1905, then as a graduate student in physics, from which he received his Ph.D. degree in 1910. For approximately a decade after earning his doctorate, Hess investigated various aspects of radioactivity while working as a staff member at the Institute of Radium Research of the Viennese Academy of Sciences.

In 1909 and 1910 scientists had used electroscopes (an early nuclear radiation detection instrument) to compare the level of ionizing radiation in high places, such as the top of the Eiffel Tower in Paris, France, or during balloon ascents into the atmosphere. These early studies gave a vague and puzzling, indication that the level of ionizing radiation at higher altitudes was actually greater than the level detected on Earth's surface. As they operated their radiation detectors farther

As a result of his pioneering ionizing radiation detection measurements made during perilous high-altitude balloon flights between 1911 and 1912, the Austrian-American physicist Victor Hess discovered the existence of *Höhenstrahlung* (radiation from above). These cosmic rays, as they were later called, are very energetic atomic particles that continually bombard Earth from space. Cosmic rays proved extremely important in nuclear physics research in the 1920s through the 1950s. They also became a special "window to the universe" that allowed scientists to examine direct evidence from violent astrophysical phenomenon. For his important discovery, Hess shared the 1936 Nobel Prize in physics. *(AIP Emilio Segrè Visual Archives, W. F. Meggers Gallery of Nobel Laureates)*

from Earth's surface and the sources of natural radioactivity within the planet's crust, the scientists expected the observed radiation levels to simply decrease as a function of altitude. They had not anticipated the existence of energetic nuclear particles arriving from outer space.

Hess attacked this mystery by first making considerable improvements in radiation detection instrumentation, which he then took on a number of daytime and nocturnal balloon ascents to heights up to 5.3 kilometers in 1911 and 1912. The results were similar, as they were in 1912 when he made a set of balloon flight measurements during a total solar eclipse. His careful, systematic measurements revealed that there was a decrease in ionization up to about an altitude of one kilometer, but beyond that height the level of ionizing radiation increased considerably, so that at an altitude of five kilometers it had twice the intensity than at sea level. Hess completed analysis of his measurements in 1913 and published his results in the *Proceedings* of the Viennese Academy of Sciences. Carefully examining his measurements, he concluded that there was an extremely penetrating radiation, an "ultra radiation," entering Earth's atmosphere from outer space. Hess had discovered "cosmic rays"—a term coined in 1925 by the American physicist Robert Milikan (1868–1953).

Cosmic rays are very energetic nuclear particles that carry information to Earth from all over the Galaxy. When a primary cosmic ray particle hits the nucleus of an atmospheric atom, the result is a cosmic ray "shower" of secondary particles that gives nuclear scientists a detailed look at the consequences of energetic nuclear reactions.

In 1919 Hess received the Lieben Prize from the Viennese Academy of Sciences for his discovery. The following year he received an appointment as professor of experimental physics at the University of Graz. From 1921 to 1923 he took a brief leave of absence from his position at the university to work in the United States as the director of the research laboratory of the U.S. Radium Company in New Jersey and then as a consultant to the Bureau of Mines of the U.S. Department of the Interior in Washington, D.C.

In 1923 Hess returned to Austria and resumed his position as physics professor at the University of Graz. He moved to the University of Innsbruck in 1931 and became the director of its newly established Institute of Radiology. As part of his activities at Innsbruck, he also founded a research station on Mount Hafelekar to observe and study cosmic rays. In 1932 the Carl Zeiss Institute in Jena awarded him the Abbe Memorial Prize and the Abbe Medal. That same year, Hess became a corresponding member of the Academy of Sciences in Vienna.

But his greatest acknowledgment for his pioneering research came in 1936 when Hess shared that year's Nobel Prize in physics for his discovery of cosmic rays. His fascinating discovery established a major branch of high-energy astrophysics and helped change our understanding of the universe and some of its most violent high-energy phenomena.

In 1938 the Nazis came to power in Austria, and Hess, a Roman Catholic with a Jewish wife, was immediately dismissed from his university position. The couple fled to the United States by way of Switzerland and later that year, Hess accepted a position at Fordham University in the Bronx, New York, as a professor of physics. He became an American citizen in 1944 and retired from Fordham University in 1956.

Victor Hess wrote more than 60 technical papers and published several books, including *The Electrical Conductivity of the Atmosphere and Its Causes* (1928) and *The Ionization Balance of the Atmosphere* (1933). He died on December 17, 1964, in Mount Vernon, New York.

⊠ **Hewish, Antony**
(1924–   )
British
Radio Astronomer

The reception of what were originally thought to be signals from an extraterrestrial civilization led Antony Hewish into a collaborative effort with SIR MARTIN RYLE resulting in the development of radio-wave-based astrophysics. Hewish's efforts involved the discovery of the pulsar, for

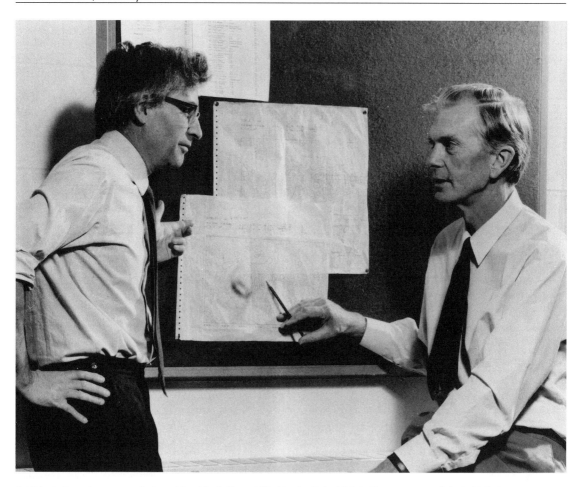

British radio astronomers Antony Hewish (left) and Sir Martin Ryle (right) discuss some of their Nobel Prize–winning extraterrestrial radio wave data. They shared the Nobel Prize in physics in 1974 for their accomplishments in radio astronomy. Hewish received the award primarily for his work identifying the first pulsar—a feat that started during an August 1967 survey of galactic radio waves and the detection of a suspicious "alien" radio signal by his graduate student Susan Joceyln Bell Burnell. *(AIP Emilio Segrè Visual Archives, Physics Today Collection)*

which he shared the 1974 Nobel Prize in physics with Ryle. The fascinating story starts in August 1967, when his graduate student SUSAN JOCELYN BELL BURNELL detected strangely repetitive "alien" radio wave signals from a certain region of space during a survey of galactic radio waves. Subsequent analysis indicated the source as a pulsar, the first every discovered. Through no fault of Hewish, the 1974 Nobel Prize awards

committee inexplicably overlooked Bell's role in this major astronomical discovery.

Hewish was born on May 11, 1924, in Fowey, Cornwall, United Kingdom. He received education at King's College, Taunton, and then matriculated in 1942 at the University of Cambridge. From 1943 to 1946 he performed war service for Great Britain at the Royal Aircraft Establishment in Farnborough and at the Telecommunications Research Establishment in Malvern, Great Britain. During this period, he participated in the development of airborne radar-countermeasure devices and worked with Ryle.

He returned to Cambridge University in 1946 and completed his undergraduate degree in physics at Gonville and Caius College two years later. Upon graduation, he joined Ryle's research team at the Cavendish Laboratory. In 1952 Hewish completed his doctor of philosophy degree in physics at Cambridge University and then lectured at Gonville and Caius College until 1961, when he became the director of studies in physics at Churchill College. For the next three decades, Hewish continued his affiliation with Cambridge University, progressing through positions of increasing academic importance. From 1961–69 he served as university lecturer, from 1969–71 he became a reader, and then from 1971 until his retirement in 1989 he was a professor of radio astronomy. When Ryle became ill in 1977, Hewish assumed leadership of the radio astronomy group at Cambridge and then directed the Mullard Radio Observatory from 1982–88. Upon retirement, Hewish assumed the title of emeritus professor of radio astronomy. As of 2004 he remained actively involved in contemporary astrophysical research from his office in the Cavendish Laboratory.

His experiences with electronics, antennas, and radar during World War II influenced Hewish to conduct research in radio astronomy following the war. His initial area of interest addressed the propagation of radio waves through inhomogeneous media. Specifically, he recognized that the scintillation twinkling of signals from recently discovered cosmic radio sources (that is, radio stars) could be used to investigate conditions within Earth's ionosphere. Earth's ionosphere contains various layers of free electrons and ions and extends from approximately 60 kilometers upward to more than 1,000 kilometers altitude. He developed and set up radio

wave interferometers to exploit this idea. With this new equipment, Hewish was able to perform pioneering measurements that revealed the height and physical extent of plasma clouds within Earth's ionosphere. In 1964 Hewish's radio astronomy group at Cambridge University discovered interplanetary scintillation—the twinkling or variation in brightness of a cosmic radio source due to radio wave scattering by irregularities in the ionized gases within the solar wind. Hewish used this discovery to make the first ground-based measurements of the solar wind, and soon space scientists in other countries adopted his technique.

Between 1965 and 1967 he constructed a large radio telescope at Cambridge with the intention of using the facility to survey about 1,000 radio galaxies using interplanetary scintillation to provide for scintillating radio sources. By a stroke of good luck, this large radio telescope, consisting of a regular array of 2,048 dipoles, operated at 3.7 meters wavelength—precisely the wavelength needed to make astronomical history. On August 6, 1967, one of Hewish's graduate students, Jocelyn Bell, detected an interesting, rapidly fluctuating radio signal that had a periodicity of approximately 1.337 seconds. Hewish initially thought the unusual signal might be a flare star. But subsequent, detailed analysis of the signal indicated it was a regular radio signal from beyond the solar system.

Because of the regularity of the radio wave signal, it was suggested that this could be a message from an intelligent extraterrestrial civilization. So while Hewish puzzled over the nature of this very strange radio signal, both teacher and student decided to call the signal LGM—which stood for "Little Green Man." As they continued to keep track of this signal, Bell detected several more pulsed sources. In February 1968 they resolved the mystery and announced that the strange pulsed radio signal belonged to the first pulsar ever detected.

The pulsar is a rapidly spinning neutron star that had been theoretically postulated in 1934 by WILHELM HEINRICH WALTER BAADE and FRITZ ZWICKY. By 1974, when Hewish received the Nobel Prize in physics for the discovery of the first pulsar (now formerly identified as PSR 1919+21 rather than LGM), astronomers had detected more than 130.

Astrophysicists now believe that a pulsar is a very rapidly rotating neutron star, possibly formed during a supernova explosion. More than 500 radio pulsars are currently known to exist, with periods ranging from 1.56 millisecond to 4 seconds. Jocelyn Bell completed her doctoral degree at Cambridge in 1968 (the same year she married Martin Burnell) and included the discovery of the pulsar as an appendix in her dissertation. Although she did not get any credit from the 1974 Nobel Prize committee for her role in the discovery of the pulsar, Hewish publicly acknowledged his student's "care, diligence, and persistence" in the data collection that led to this discovery, during his Nobel laureate lecture in 1974.

In addition to the 1974 Nobel Prize in physics, Hewish received other awards and honors to commemorate his contributions to radio astronomy and astrophysics. His awards include the Eddington Medal of the Royal Astronomical Society (1969), the Franklin Institute's Michelson Medal (1973), and the Hughes Medal of the British Royal Society in 1976. Hewish became a fellow of the British Royal Society in 1968, a foreign honorary member of the American Academy of Arts and Sciences in 1977, a foreign Fellow of the Indian National Science Academy in 1982, and an associate member of the Belgian Royal Academy in 1989. As an emeritus professor of radio astronomy at Cambridge University, he still enjoys making contributions to astrophysics and giving inspiring lectures to the general public that share the great excitement of scientific discovery.

⊠ Hohmann, Walter
(1880–1945)
German
Engineer, Orbital Mechanics Expert

In his seminal 1925 work, *The Attainability of Celestial Bodies* (*Die Erreichbarkeit der Himmelskörper*), Walter Hohmann described the mathematical principles that govern space vehicle motion, including the most efficient, minimum-energy orbit transfer path between two orbits in the same geometric plane. This widely used orbit transfer technique is now called the Hohmann transfer orbit in his honor.

Hohmann was born to Rudolph and Emma Hohmann on March 18, 1880, in Hardheim, a small German town about 40 kilometers southwest of Würzburg. His father served as a physician and surgeon in the local hospital. The Hohmann family moved to Port Elizabeth in South Africa in 1891 and remained there for about six years before returning to Germany. When his family returned to Germany, he enrolled in high school at Würzburg. Between 1900 and 1904 he studied at the Technical University in Munich and graduated from there as a certified civil engineer with a Diplom-Bauingenieur degree.

Upon graduation and for several years thereafter just prior to the start of World War I, Hohmann worked as a civil engineer for various companies in Austria and Germany. Then, in 1912, he became the combined equivalent of an urban planner and city engineer for the city of Essen, Germany. It was also in 1912 that Hohmann developed his lifelong interest in space flight and the motion of spaceships on interplanetary trajectories. This fascination with astronautics occurred quite by accident when the young engineer read an astronomy book. As the clouds of war descended upon Europe, he dutifully undertook a wartime service position in 1915 and performed noncombat activities for approximately eight months.

He married Luise Juenemann in 1915 and the couple had two sons, Rudolf (born in 1916) and Ernst (born in 1918). In 1916, Hohmann submitted his dissertation on concrete structures to the Technical University of Aachen. However, because of wartime conditions, his work was not formally accepted and approved until 1919, a year after World War I ended. Hohmann could now use the more illustrious academic title of "Dr.-Ing. Walter Hohmann"—that is, "Walter Hohmann, Doctor of Engineering"—in his professional activities.

After World War I Hohmann sought a professorship in Karlsruhe, but did not succeed. So he remained affiliated with Essen as a professional civil engineer for the remainder of his life. He became a distinguished member of Verein deutscher Ingenieure, the society of German engineers. The analytical judgments and mathematical skills he used to discharge his professional duties also served his work in astronautics.

Hohmann's personal interest in space flight was reinforced by the great public interest in science fiction, rocketry, and space travel that permeated the Weimar Republic. Some of this enthusiasm can be attributed to Hermann Julius Oberth, who published an inspirational book, *The Rocket into Interplanetary Space* (Die Rakate zu den Planetenräumen) in 1923. Just after Oberth's book appeared, Hohmann channeled his own enthusiasm for astronautics into another important book. Using his professional mathematical and engineering skills, he related the fundamental principles of orbital mechanics to space travel in the classic book *Die Erreichbarkeit der Himmelskörper* (The Attainability of Celestial Bodies), which was published in 1925. Of particular importance to modern space technology is Hohmann's careful enumeration of the most efficient orbit transfer path between two coplanar circular orbits. Hohmann demonstrated that an interplanetary trajectory of minimum energy is an ellipse that is tangent to the orbits of both planets. This useful principle of orbital mechanics is now called the Hohmann transfer orbit (or sometimes the Hohmann transfer ellipse) in his honor. The technique is not just restricted to traveling between the planets. It is also used to efficiently raise (or lower) the altitude of a spacecraft in a circular orbit around a primary celestial body, like planet Earth. A simple description of Hohmann's widely used minimum energy orbital transfer technique follows below.

Consider a spacecraft in a relatively low altitude, circular orbit around Earth or another celestial body. The Hohmann transfer orbit technique raises the spacecraft to a higher altitude orbit in the same orbital plane with a minimum expenditure of energy and propellant. Hohmann's technique requires two impulsive high-thrust burns, or firings, of the spacecraft's onboard propulsion system. The first burn changes the original circular orbit to an elliptical one whose perigee is tangent with the lower-altitude circular orbit and whose apogee is tangent with the higher-altitude circular orbit. After the spacecraft coasts for half of the elliptical transfer orbit (that is, half of the Hohmann transfer orbit) it reaches a position that is tangent with the destination higher-altitude circular orbit. The spacecraft's propulsion system then performs the second high-thrust firing. When just the right amount of thrust is applied, this second firing circularizes the spacecraft's orbit at the desired new altitude. Hohmann's technique can be used to lower the altitude of a satellite from one circular orbit to another circular orbit of less altitude. The main disadvantage with Hohmann's minimum energy orbit transfer maneuver is that it takes a great deal of time.

Throughout the late 1920s Hohmann examined the orbital mechanics of interplanetary space flight. He actively discussed minimum propellant expenditure, spacecraft design, crew safety, and maneuver analysis with fellow space travel enthusiasts in the German Society for Space Travel (Verein für Raumschiffahrt, or VFR).

Hohmann was a well-respected professional engineer in service to the city of Essen, so his enthusiastic participation in the VFR and its space travel advocacy served as an important endorsement of the society's interplanetary flight concepts. His relationship with the VFR was a mutually beneficial one. Hohmann found a rich fertile environment for his own innovative astronautics concepts within the VFR because society membership included such space travel pioneers as Hermann Oberth, WERNHER MAGNUS VON BRAUN, and Willy Ley (1906–69)—people who quickly recognized the importance of Hohmann's work.

However, his enthusiasm faded when Adolf Hitler (1889–1945) and the Nazi Party seized power in 1933. From that point on Hohmann began to withdraw from any further German rocket or space activity. He did not join many of his former VFR colleagues as they began to develop military rockets for the German army, and he did not become a member of Braun's rocket research team at Peenemünde. Despite Hohmann's desire not to support the military or political objectives of Nazi Germany, the painful consequences of World War II nevertheless adversely affected him and his family. During the war, he lost his son Ernst, a soldier. On March 11, 1945, exploding Allied bombs claimed Hohmann's life just a week before his 65th birthday and less than two months before Germany surrendered. Today, the name of the Hohmann transfer orbit commemorates his pioneering work in orbital mechanics.

⊠　**Hubble, Edwin Powell**
(1889–1953)
American
Astronomer, Physicist, Cosmologist

Edwin Powell Hubble is generally credited with creating the greatest change in the perception of the universe since GALILEO GALILEI. It was

The famous American astronomer Edwin Hubble examines the interesting features of a spiral galaxy. After carefully investigating many galaxies in the 1920s, Hubble announced that the universe was expanding—a premise that now forms the basis for all modern observational cosmology. *(Hale Observatories, courtesy AIP Emilio Segrè Visual Archives)*

Hubble who determined that other galaxies exist outside the Milky Way, and that the universe itself, in fact, was expanding.

Hubble was born on November 20, 1889, in Marshfield, Missouri, the son of Virginia Lee James Hubble of Virginia City, Nevada, and insurance broker John Powell Hubble of Missouri. In 1898 his family moved to Chicago, where Hubble excelled at sports at Wheaton High School, setting the state record for the high jump. At the University of Chicago he lettered in track, boxing, and basketball while working at the school's Yerkes Observatory in Wisconsin. During this time he worked as a laboratory assistant to the renowned Robert Millikan (1868–1953). He earned his B.S. degree in physics in 1910, then won a Rhodes scholarship and moved to England to read Roman and English law at Queens College, Oxford.

When he returned to America in 1913, Hubble passed the bar exam and practiced law in Louisville, Kentucky, for a year. His heart was not in it, however, and in 1914 he returned to the University of Chicago to study astronomy, earning a Ph.D. in 1917. Just before getting his degree, he was invited by GEORGE ELLERY HALE to join the staff at the famous Mount Wilson Observatory, in Pasadena, California, home of the world's largest telescope at 100 inches. Soon, the United States entered World War I, and Hubble's patriotism came to the fore. After cramming for his doctoral degree and taking the oral exams the next morning, he enlisted in the U.S. Army. He was commissioned a captain in the 343rd Infantry, 86th Division, stationed in France, and soon after was promoted to major. After cessation of hostilities, Hubble was mustered out of the service in San Francisco and immediately headed south to Pasadena, where he presented himself in his major's uniform and took the job offered by Hale.

At the time, the accepted theory was that the Milky Way was the boundary of the known universe and that the fuzzy spots of light in the night sky were simply islands of star clusters within that boundary. However, Hubble used the Mount Wilson telescope to measure the distance to the Andromeda Galaxy and determined that it was 100,000 times farther away from the nearest star than was generally thought, and therefore had to be its own galaxy. Hubble went on to measure and classify many galaxies, and proved once and for all that the Milky Way is only one of millions of galaxies in the cosmos.

In what was to become a quantum leap in knowledge of the new field of astrophysics, Hubble then virtually re-created our image of the universe by discovering the "redshift" and announcing to the world in 1929 that the universe was expanding as if it were a gigantic balloon. The redshift was determined to be a change in light toward the red end of the spectrum as galaxies moved away from our own, much like the Doppler effect, which changes the tone of a ringing bell as it is passed in a car or train. It was this groundbreaking first evidence of the big bang theory that caused ALBERT EINSTEIN, who had revised his original calculations of an expanding universe because they were too "far-fetched," to revise his theory again and make the statement that it was "the greatest blunder of my life."

The redshift discovery propelled Hubble to the highest ranks of worldwide astronomers and earned him unexpected celebrity. He became close friends with several prominent movie stars and scientists, and was called by *Time* magazine the "toast of Hollywood." In 1924 Hubble had married Grace Burke, and the couple traveled the world and were honored guests at lavish Hollywod parties during the 1930s and 1940s. When World War II broke out, Hubble wanted to enlist again but became convinced that his scientific knowledge would do the country more good in research on the home front. He ultimately was awarded a Medal of Merit for his ballistics work at the Aberdeen Proving Ground's ultrasonic wind tunnel testing grounds in New Mexico.

After the war, Hubble was instrumental in building the 200-inch Hale telescope at Mount Palomar, California, and was given the high honor of being the first person to use it. He continued his research at Mount Palomar until his death in Pasadena of a cerebral thrombosis on September 28, 1953.

In his distinguished career as one of the 20th century's greatest astronomers, Hubble invented the science of cosmology and completely changed our vision of the universe. His verification of the Big Bang theory has caused scientists and theologians alike to debate the tremendous size of the cosmos and to theorize about what it was that "banged" in the first place.

Such was Hubble's fame and impact on astronomy and cosmology that his name is attached to several parameters and formulae such as Hubble's Law, the Hubble Sequence, Hubble's Zone of Avoidance, the Hubble Galaxy type, the

Hubble Luminosity Law, the Hubble Red-Shift Distance Relation, the Hubble Radius, as well as the Hubble crater on the Moon, and the *Hubble Space Telescope*.

⊠ **Huggins, Sir William**
(1824–1910)
British
Astronomer, Spectroscopist

Sir William Huggins, a skilled amateur astronomer and spectroscopist, helped to revolutionize observational astronomy in the 19th century. Using a private astronomical observatory that he built just outside London, he performed pioneering spectral measurements of the stars and then compared the observed stellar spectra to chemical spectra generated in his laboratory on Earth. In 1868 Huggins profoundly influenced the future direction of cosmology when he attempted to measure a star's radial velocity by using the Doppler shift in its spectrum. Queen Victoria knighted him, and he copublished a major work on stellar spectra in 1899 with his talented young wife, Lady Margaret Murray Huggins.

Huggins was born in London, England, on February 7, 1824. His father was a prosperous merchant, so he provided his son not only a high quality private education but also an important degree of financial independence. Under such fortunate circumstances, Huggins became one of Great Britain's most successful and competent private (amateur) astronomers. Throughout his lifetime, he was able to contribute productively to astronomy without having a formal attachment to either a university or an established observatory. Huggins received his earliest education in the City of London School, but later learned mathematics, the physical sciences, and several languages at home from private teachers. At 18 years of age, he passed up a formal college education to take over the family business when

his father became seriously ill. For more than a decade, Huggins conducted business by day and astronomy at night. He enjoyed both astronomy and microscopy, since both of these disciplines gave him an opportunity to pursue original investigations.

When he was 30, Huggins sold the family business and dedicated the rest of his life to astronomy. To support this decision, he moved to an open region called Upper Tulse Hill, south of London, in 1856 and constructed a large private observatory adjacent to his new house. In 1858 he added a high-quality 8-inch refractor telescope, manufactured by the world-famous American optician ALVAN GRAHAM CLARK.

In January 1862 Huggins began a productive collaboration with W. Allen Miller, a professor of chemistry at King's College in London. Huggins wanted to apply to the Kirchhoff-Bunsen solar spectroscopy experiment to light from the distant stars, but he recognized that this was a challenging effort and that Miller's experience in chemical spectroscopy would prove invaluable. However, Miller himself was initially doubtful that Huggins would be successful in developing stellar spectroscopy. For one thing, the amount of starlight collected by telescope and available for spectral analysis would be only a very tiny fraction (on the order of millionths to billionths) of the amount of sunlight available to Kirchhoff and Bunsen when they solved the mystery of the Fraunhofer lines.

Despite Miller's initial misgivings, one of the great milestones in astronomical spectroscopy took place in 1864 when Huggins announced that the distant stars were composed of the same chemical elements as the Sun. With Miller as a coauthor, Huggins published a piece entitled "On the Lines of Some Fixed Stars," in the *Philosophical Transactions* of the Royal Society. This paper included diagrams of the spectra of stars, such as Sirius, Betelgeuse, and Vega. In one brilliant stroke, spectroscopy became the bridge linking terrestrial chemistry to the chemistry of the "unreachable" stars. Unknown to Huggins, the Italian astronomer PIETRO ANGELO SECCHI in Rome and others were beginning to conduct similarly important efforts in astronomical spectroscopy. This particular effort earned Huggins and Miller the 1867 Gold Medal from the Royal Astronomical Society.

Prior to publishing this work, Huggins also attempted to obtain photographs of the spectra of Sirius and Capella in February 1863, but his efforts were greatly limited by contemporary photographic technology. About a decade later, new photographic materials became available and greatly improved the amateur astronomer's ability to make useful photographic records of stellar spectra.

Huggins orchestrated another great moment in observational astronomy in August 1864 by turning his stellar spectroscope to study the fuzzy celestial objects called nebulas. At the time, astronomers were generally divided concerning the exact nature of nebulas. Some hypothesized that they were great gaseous clouds, while others considered them to be enormous collections of stars too distant to be individually resolved by a telescope.

On the evening of August 29, Huggins observed a single bright line in the spectrum of a nebula in the constellation Draco and immediately concluded that it was a giant cloud of luminous gas. Later, he found a few bright emission lines in certain other nebulas, including the great nebula in Orion. Huggins also correctly noted that certain spiral nebulas, like Messier 31 in Andromeda, showed continuous spectra and must be composed of numerous unresolved stars. As a historic note, many of these distant spiral nebulas did indeed turn out to be enormous collections of stars—that is, other galaxies, or island universes, as originally suggested by Kant and Laplace. However, that particular astronomical knowledge would not clearly emerge until decades later, in the early part of the 20th century.

While solving one nebula mystery, Huggins created another that would remain unresolved for about 60 years. In 1866 he observed the spectra of certain nebulas and discovered previously unidentified bright emission lines that he decided to attribute to a new element that he called "nebulium." It was not until 1927 that the American astrophysicist Ira Sprague Bowen (1898–1973) properly explained this puzzling phenomenon. The bright emission lines did not belong to any new element; rather, Bowen's detailed investigation revealed that the mysterious green lines were actually special (forbidden) transitions of the ionized gases oxygen and nitrogen. In 1866 the Royal Society presented Huggins with its highest award, the Royal Medal, for his research involving astronomical spectroscopy.

During the 1860s Huggins also helped change the direction of modern cosmology when he attempted to measure the radial velocity of Sirius by detecting any Doppler shift in its spectrum. He used his relatively imprecise spectral measurements (acquired without the aid of precision spectral photography) to conclude that there was a redshift in the spectrum of Sirius of sufficient magnitude to suggest that this star was receding from the solar system at about 47 kilometers per second. He described this work in a paper, "On the Spectra of the Fixed Stars," that was coauthored with Miller and belatedly published in 1868 in *Philosophical Transactions* of the Royal Society.

Although this value of radial velocity was not close to the value obtained later, Huggins provided the pioneering concept that encouraged other astronomers, such as EDWIN POWELL HUBBLE, to use the Doppler effect (redshift) in the spectra of distant galaxies. Using this approach, Hubble was eventually able to postulate the cosmological model of an expanding universe. The radial velocity of a star is simply its motion toward (considered negative) or away from (considered positive) the solar system. Contemporary astronomical measurements show the star Sirius has radial velocity of −8 kilometers per second, meaning it is actually approaching the solar system.

On May 18, 1866, Huggins became the first astronomer to observe the spectrum of a nova (Nova Coronae 1866), detecting its bright hydrogen emission lines. He also applied astronomical spectroscopy to comets. In 1868 he observed the spectrum of a comet and reported the presence of luminous carbon gases. He married Margaret Lindsay Murray in 1875 and she immediately became not only a loving wife but also a most talented collaborator. As he aged, Lady Margaret Huggins, being 24 years younger than her husband, did more and more of the actual observing. In 1899 Lord and Lady Huggins published their award-winning *Atlas of Representative Stellar Spectra*. He was knighted by Queen Victoria in 1897 and served as president of the Royal Society from 1900 to 1905. Sir William Huggins died in London on May 12, 1910.

Despite his lack of formal schooling, his epochal astronomical work brought Huggins full recognition from learned societies and universities. He was elected as a member to the Royal Society in 1865 and received several of its prestigious medals, including the Royal Medal (1866), the Rumford Medal (1880), and the Copley Medal (1898). The Royal Astronomical Society gave him its distinguished Gold Medal twice, in 1867 and again in 1885. The French Academy of Sciences presented him its Lalande Gold Medal (1872), the Valz Prize (1883), and the Janssen Gold Medal (1888). Finally, the U.S. National Academy of Sciences awarded him its Henry Draper Medal in 1901. Perhaps the greatest contribution that this wealthy British amateur astronomer made to the development of astronomical spectroscopy was his pioneering effort to interpret stellar observations by using carefully prepared laboratory observations of chemical spectra.

J

⊠ **Jeans, Sir James Hopwood**
(1877–1946)
British
Astronomer, Mathematician, Physicist

At the dawn of 20th-century astronomy, Sir James Hopwood Jeans proposed the controversial and now generally ignored tidal theory to explain the origin of the solar system. In 1928 he made another bold suggestion that matter was being continuously created in the universe. This controversial hypothesis led other scientists to establish a steady-state universe model in cosmology. Big Bang cosmology has now all but eliminated the steady-state model that traces its origins to his initial speculations. Following his knighthood in 1928, he focused much of his creative activity on writing popular books about astronomy.

Jeans was born in Ormskirk, Lancashire, England, on September 11, 1877, but his family moved to London when he was three years old. He was the son of a journalist and this probably influenced his strong inclination toward writing during childhood. Jeans enrolled at Trinity College, Cambridge University, in October 1896. He received a degree in mathematics in 1900 and became a Fellow of Trinity College in 1901. During this period, he was very active, publishing research on a variety of topics in applied mathematics, astronomy, and physics. He was particularly interested in the kinetic theory and thermodynamic behavior of gases, and published his first major textbook, *The Dynamical Theory of Gasses*, in 1904, while recuperating from a bout with tuberculosis.

Jeans began his career as a mathematician, but made important contributions in such diverse fields as molecular physics, astrophysics, and cosmology. In 1904 he became a lecturer in mathematics at Cambridge University, then in 1905 traveled to the United States to serve as a professor of applied mathematics at Princeton University in New Jersey until 1909. While at Princeton he published his second major textbook, *Theoretical Mechanics*, in 1906. He became a Fellow in the British Royal Society in 1907 and also married an American, Charlotte Tiffany Mitchell, a member of the famous Tiffany family from New York. Jeans published another book, *The Mathematical Theory of Electricity and Magnetism*, in 1908. Later the next year he returned to Great Britain and accepted an appointment as the Stokes Lecturer in applied mathematics at Cambridge University. However, he kept this position only until 1912, before devoting himself entirely to technical writing and research in applied mathematics, especially as related to astronomy and cosmology.

In 1917 Jeans won the Adams Prize from Cambridge University for an essay titled

"Problems of Cosmogony and Stellar Dynamics" which was published two years later as a book. In this work Jeans mathematically addressed the stability of a rotating mass and then took direct issue with the nebular hypothesis of Kant and Laplace as the basis for the formation of the planets. He endorsed, instead, a tidal (or catastrophic) theory to explain the formation of the solar system. This interesting theory suggests that a rogue star passed close to the Sun billions of years ago and gravitationally tugged a cigar-shaped wedge of matter out of the Sun that eventually condensed into the planets. For more than a decade the tidal theory remained popular in 20th-century astronomy. This theory implied, however, that planetary systems would be rare because their formation would depend on catastrophic encounters of passing rogue stars. By the mid-1940s, revitalized versions of the nebular hypothesis emerged and displaced tidal theory from its position of dominance in astrophysics. This took place because astronomers once again regarded planet formation as a common, normal part of stellar evolution. The recent discovery of extra solar planets provides compelling observational evidence in support of the refined nebular hypothesis and justifying the abandonment of the tidal hypothesis.

By 1917 his excessively high workload began to stress his heart and adversely influence his health. Therefore, in 1918 Jeans moved his family to more relaxed surroundings in Dorking, Surrey, a place where he could devote more time to recreation and music, while still pursuing technical writing at a less hectic pace. However, his interest in astronomy and cosmology also encouraged him to undertake research at the Mount Wilson Observatory, in Pasadena, California, and Jeans enjoyed an extended appointment there as a research associate from 1923 to 1944. This position exposed him to stimulating new ideas in astronomy and gave him the opportunity to personally interact with many notable American scientists, including Walter Sydney Adams and Arthur Holly Compton.

In 1928 Jeans suggested in his book *Astronomy and Cosmogony* that matter was being continually created in the universe. This interesting conjecture became the foundation of the steady-state theory of cosmology—a theory championed by some scientists in the 1940s and 1950s in opposition to Big Bang cosmology. Today, steady-state cosmology has been displaced by the general acceptance of an expanding universe cosmology based on the Big Bang hypothesis.

Jeans was elevated to knighthood in 1928, and starting in 1929 he focused his diverse intellectual talents on writing books that greatly popularized astronomy and science. His most famous popular books include: *The Universe around Us* (1929), *The Mysterious Universe* (1930), *The New Background of Science* (1933), *Through Space and Time* (1934), and *Physics and Philosophy* (1943).

Jeans received numerous honors and awards. He received the Gold Medal from the Royal Astronomical Society in 1922 and the Franklin Medal in 1931. He served as the president of the Royal Astronomical Society from 1925 to 1927 and as the president of the British Association for the Advancement of Science in 1934. He was also awarded the Order of Merit in 1939.

When his first wife died in 1934, Jeans remarried a year later. His second wife, Suzanne Hock, came from Vienna, Austria, and was an accomplished musician. Their common interest in music served as the source of inspiration for Jeans's popular book *Science and Music* (1938). Jeans died at home in Dorking, England, on September 16, 1946, due to heart failure. His numerous publications and bold hypotheses not only helped shape astrophysics and cosmology in the first half of the 20th century but also made astronomy familiar and popular to many less-technical readers.

Jones, Sir Harold Spencer
(Spencer Jones)
(1890–1960)
British
Astronomer

Sir Harold Spencer Jones was a multitalented individual who used the close passage of the asteroid Eros 433 in 1931 to achieve precise measurements of an important astronomer's "yardstick," the astronomical unit (AU)—that is, the average distance from the Earth to the Sun. From 1933 to 1955 he served Great Britain as the 10th Astronomer Royal. He also wrote a number of books that popularized astronomy, including the pioneering work *Life on Other Worlds* (1940) which opened up all manner of interesting speculations in the exciting field now known as exobiology.

Jones was born in London on March 29, 1890. The son of an accountant, he received his early education at Latymer Upper School and then attended Jesus College at Cambridge University on a scholarship. Following a distinguished undergraduate career, he earned his degree in mathematics in 1911 and in physics in 1912. In 1913 Jones became the chief assistant at the Royal Greenwich Observatory. Except for World War I service with the British Ministry of Munitions, he would hold this position at Greenwich until 1923. He was elected a Fellow of Jesus College in 1914. He married Gladys Mary Owens in 1918 and the couple had two sons. In 1923 he replaced Sir Frank Dyson as the Royal Astronomer at the Cape of Good Hope Observatory in South Africa.

In 1928, as Royal Astronomer at the Cape of Good Hope Observatory, Jones took charge of a cooperative international project under the sponsorship of the International Astronomical Union (IAU) to refine estimates of the astronomical unit using positional data for the asteroid Eros 433. Following a suggestion originally made by JOHANN GOTTFRIED GALLE, Jones led a worldwide effort to refine the estimate of the distance from the Earth to the Sun by triangulating the distance of Eros 433 as it approached close to Earth in 1931. Eros 433 was discovered in 1898 and visited by a NASA spacecraft in 2000. A member of the Mars-crossing Amor group, it is an irregularly shaped minor planet, approximately $33 \times 13 \times 13$ kilometers in size and resembling a fat banana. In 1931 it came within 26 million kilometers of Earth, and astronomers at observatories all over the world photographed the celestial object and recorded accurate data concerning its position. Jones gathered these data and then spent the next decade reviewing more than 3,000 photographs and making extensive calculations in a tremendous effort to refine the estimate of solar parallax—that is, the angle subtended by Earth's radius when viewed from the center of the Sun. Not until the early 1940s, while World War II was raging, did Jones, now back in Greenwich as Astronomer Royal, finally announce his refined value for solar parallax as 8.790 ± 0.001 arc seconds. Because of the global conflict, much of the significance of his great personal effort escaped the immediate attention of the international astronomical community. Today, the IAU defines solar parallax as 8.794148 arc seconds, based on radar measurements.

He left South Africa and returned to Great Britain in 1933 to continue his tenure as Astronomer Royal. In addition to refining solar parallax, Jones made a second major contribution to astronomy by using more accurate quartz clocks to study subtle anomalies in Earth's rotation rate. In 1939 he announced that his measurements indicated that Earth did not rotate regularly, but rather kept time "like an inexpensive clock." Specifically, his measurements indicated that, with respect to other solar system bodies, Earth was rotating slowly by about a second per year—an amount sufficient to account for some of the tiny anomalies recorded in lunar or planetary ephemerides. This work was instrumental in the introduction of ephemeris time in 1958.

After World War II, Jones supervised the plans and activities involved in moving the Royal Observatory from Greenwich to Herstmonceux Castle, in Sussex. Urban light and chemical pollution, detrimental by-products of London's growth, necessitated this change in location. Jones retired in 1955 with the relocation well underway. By 1958 the move of the Royal Observatory to its new location was successfully completed.

While serving as Astronomer Royal, Jones communicated the excitement and wonder of astronomy to both technical and nontechnical readers alike by writing such interesting books

as *Worlds without End* (1935), *Life on Other Worlds* (1940), and *A Picture of the Universe* (1947). He received many honors for his contributions to astronomy. In 1930 he became a Fellow of the Royal Society of London. He was knighted by King George VI in 1943. He also received the Royal Medal of the British Royal Society (1919), the Gold Medal of the Royal Astronomical Society (1943), and the Bruce Gold Medal of the Astronomical Society of the Pacific (1949). As a lasting tribute to his life-long service to Great Britain and to the astronomical sciences, Jones was elevated to the rank of Knight of the British Empire in 1955 by Queen Elizabeth II. He died at age 70 on November 3, 1960, in London.

Kant, Immanuel
(1724–1804)
German
Natural Philosopher, Astronomer

The great German philosopher Immanuel Kant was the first to propose the nebula hypothesis, a concept he introduced in his 1755 book, *General History of Nature and Theory of the Heavens*. In this hypothesis, Kant anticipated modern astrophysics when he suggested that the solar system formed out of a primordial cloud of interstellar matter. Kant also introduced the term *island universes* to describe the distant collections of stars that astronomers now call galaxies. He was a truly brilliant thinker, and his works in metaphysics and philosophy exerted great influence on Western thinking far beyond the 18th century.

Kant was born on April 22, 1724, in Königsberg, East Prussia (now Kaliningrad, Russia). He lived a very orderly and organized life, so precise that his neighbors would set their watches by his behavior. Although he was a brilliant, popular conversationalist, during his entire lifetime Kant never ventured more than about 100 kilometers from his birthplace. His father was a poor, but well-respected, saddler, and his mother created a home environment filled with a strict adherence to Protestantism. At age 16 Kant enrolled at the University of Königsberg,

where he studied philosophy, mathematics, and natural philosophy (physics). He also attended classes in theology. When his father died in 1745, Kant found himself in financial difficulties and served as a private tutor for nine years to earn enough money to finish his studies. In 1755 he finally completed the requirements for his degree at the University of Königsberg and published *General History of Nature and Theory of the Heavens* (Allgemeine naturgeschichte und theorie des himmels) presenting his view of the physical universe. In this book Kant adhered strictly to the principle of mechanical causation (using Newtonian principles) and avoided the probing metaphysical speculations that would later characterize his great works in philosophy.

Although Kant is primarily regarded as a great philosopher and not an astronomer or physicist, his early work addressed Newtonian cosmology and contained several important concepts that, amazingly, anticipated future developments in astronomy and astrophysics. First, *Theory of the Heavens* introduced the nebular hypothesis of the formation of the solar system: namely, that the planets formed from a cloud of primordial interstellar material that slowly collected and began swirling around a protosun under the influence of gravitational attraction and eventually portions of this rotating disc condensed into individual planetary "clumps."

Laplace independently introduced a similar version of Kant's nebular hypothesis in 1796. Second, Kant suggested that the Milky Way Galaxy was a lens-shaped collection of many stars that orbited around a common center, similar to the way the rings of Saturn orbited around the gaseous giant planet. Third, he speculated that certain distant spiral nebulas were actually other "island universes"—a term he coined to describe the modern concept of other galaxies. Fourth, Kant suggested that tidal friction was slowing down Earth's rotation just a bit—a bold scientific speculation confirmed by precise measurements in the 20th century. Finally, he also suggested the tides on Earth are caused by the Moon and that this action is responsible for the fact the Moon is locked in a synchronous orbit around Earth, always keeping the same "face" (called the nearside) presented to observers on Earth. Not too bad, for a person who had just completed his university degree and was about to embark on a long and productive academic career best known for its important philosophical contributions to Western civilization.

Upon graduating from the University of Königsberg, Kant became a lecturer (*Privatdozent*) at the university and served in that capacity for the next 15 years. During that time, his fame as a brilliant thinker and wonderful teacher grew. In 1770 university officials appointed him to the chair of logic and metaphysics. This promotion marked the beginning of a period in which Kant wrote a powerful series of books that shaped Western thinking for the next two centuries. The most important of these works was his *Critique of Pure Reason (Kritik der reinen Vernunft)*—a treatise on metaphysics that is regarded as a classic contribution to philosophy. A full and detailed discussion of Kant's other great works in metaphysics, ethics, judgment, and reason is well beyond the scope of this book.

Although Kant never married and did not venture far from home, he was not a recluse.

On the contrary, he was a brilliant conversationalist and excellent teacher, who enjoyed a wide circle of friends who constantly sought his views on contemporary intellectual and political issues. He retired from the university in 1796 and died on February 12, 1804, in Königsberg.

⌧ **Kapteyn, Jacobus Cornelius**
(1851–1922)
Dutch
Astronomer

Toward the end of the 19th century, Jacobus Kapteyn made significant contributions to the emerging new field of astrophotography—that is, taking analysis-quality photographs of celestial objects. He also pioneered the use of statistical methods to evaluate stellar distributions and the radial and proper motions of stars. Between 1886 and 1900, Kapteyn carefully scrutinized numerous photographic plates collected by SIR DAVID GILL at the Cape of Good Hope Royal Observatory, in South Africa. He then published these data as a three-volume catalog that contained the positions and photographic magnitudes of more than 450,000 Southern Hemisphere stars.

Kapteyn was born on January 19, 1851, at Barneveld in the Netherlands. From 1869 to 1875, he studied physics at the University of Utrecht. After graduating with his doctorate, Kapteyn accepted a position as a member of the observing staff at the Leiden Observatory. During his studies, he had not given any special attention to astronomy, but this employment opportunity convinced him that studying stellar populations and the structure of the universe was how he wanted to spend his professional life.

In 1878 Kapteyn received an appointment as the chair of astronomy and theoretical mechanics at the University of Groningen. Since the university did not have its own observatory,

Kapteyn decided to establish a special astronomical center dedicated to the detailed analysis of photographic data collected by other astronomers. The year following this brilliant career maneuver, he married Catharina Karlshoven, with whom he had three children. In the mid-1880s Kapteyn interacted with Sir David Gill, Royal Astronomer at the Cape of Good Hope Observatory. The two astronomers developed a cordial relationship and Kapteyn volunteered to make an extensive catalog of the stars in the Southern Hemisphere, using the photographs taken by Gill at the Cape Observatory.

In 1886 Kapteyn's collaboration with Gill began. Underestimating the true immensity of the task for which he had just volunteered, Kapteyn worked for more than a decade in two small rooms at the University of Groningen. During his labor-intensive personal crusade in photographic plate analysis, he enjoyed support from just one permanent assistant, several part-time assistants, and occasional convict labor from the local prison.

The end product of Kapteyn's massive personal effort was the *Annals of the Cape Observatory*—an immense three-volume work published between 1896 and 1900 that contained the position and photographic magnitudes of 454,875 Southern Hemisphere stars. His pioneering effort led to the creation of the Astronomical Laboratory at the University of Groningen, and the establishment of the highly productive Dutch school of 20th-century astronomers.

In 1904 Kapteyn noticed that the proper motion of the stars in the Sun's neighborhood did not occur randomly, but rather moved in two different, opposite streams. However, he did not recognize the full significance of his discovery. Years later, one of his students, JAN HENDRIK OORT, demonstrated that the star-streaming phenomenon first detected by Kapteyn was observational evidence that the Galaxy was rotating.

Along with statistical astronomy, Kapteyn was interested in the general structure of the Galaxy. In 1906 he attempted to organize a cooperative international effort in which astronomers would determine the population of stars of differing magnitude in 200 selected areas of the sky and also measure their radial velocities and proper motions. The radial velocity of a star is a measure of its velocity along the line of sight with Earth. Astronomers use the Doppler shift of a star's spectrum to indicate whether it is approaching Earth (blueshifted spectrum) or receding from Earth (redshifted spectrum). The proper motion of a star is the gradual change in the position of a star due to its motion relative to the Sun. The proper motion (symbol μ) is the apparent angular motion per year of a star as observed on the celestial sphere—that is, the change in its position in a direction that is perpendicular to the line of sight.

Kapteyn gathered enormous quantities of data, but World War I interrupted this multinational cooperative effort. Nevertheless, he examined the data that was collected but overlooked the starlight-absorbing impact of interstellar dust. So, in the tradition of SIR WILLIAM HERSCHEL, he postulated that the Milky Way must be lens-shaped, with the Sun located close to its center. The American astronomer HARLOW SHAPLEY disagreed with the Dutch astronomer's erroneous model and, in 1918, conclusively demonstrated that the Galaxy was far larger than Kapteyn had estimated and that the Sun actually resided far away from the galactic center.

Kapteyn's model of the Galaxy suffered when applied to the galactic plane because he lacked sufficient knowledge about the consequences of interstellar absorption, not because he was careless in his evaluation of observational data. On the contrary, he was a very skilled and careful analyst who helped found the important field of statistical astronomy. He died in Amsterdam, the Netherlands, on June 18, 1922. During his lifetime he received many international awards and medals, including the Gold Medal of the Royal Astronomical Society of London in 1902, the Watson Medal from the American Academy of Sciences in 1913, and the Bruce Gold Medal from the Astronomical Society of the Pacific, also awarded in 1913. Perhaps his greatest overall contribution to astronomical sciences was the creation of the Dutch school of astronomy, which provided many great astronomers who continued his legacy of excellence in data analysis throughout the 20th century.

⊠ Kepler, Johannes (Johann Kepler)
(1571–1630)
German
Astronomer, Mathematician

Johannes Kepler was a brilliant contemporary of GALILEO GALILEI who helped start the Scientific Revolution. While Galileo's pioneering use of the telescope provided the observational foundations for Copernican cosmology, Kepler's innovative mathematical contributions to astronomy provided the theoretical foundations. Through painstaking and arduous analyses, Kepler developed his three laws of planetary motion—important physical principles that describe the elliptical of orbits of planets around the Sun. His work not only provided the empirical basis for the early acceptance of NICOLAS COPERNICUS's heliocentric hypothesis but also the mathematical starting point from which SIR ISAAC NEWTON developed his law of gravitation. Kepler gave astronomy its modern, mathematical foundation.

Kepler was born on December 27, 1571, in the small town of Weil der Stadt, Württemberg, Germany (then part of the Holy Roman Empire). His father was a mercenary soldier and his mother was the daughter of an innkeeper. A sickly child, Kepler suffered from smallpox at the

Johannes Kepler combined painstaking and arduous analyses with simple mathematics to describe the elliptical movement of the known planets around the Sun. With his famous three laws of planetary motion, Kepler established the mathematical foundations for heliocentric (Copernican) cosmology. His work not only provided the empirical basis for the early acceptance of the heliocentric hypothesis, it also provided the mathematical starting point from which Sir Isaac Newton could develop his law of gravitation. *(AIP Emilio Segrè Visual Archives)*

age of three. The disease impaired the use of his hands and affected his eyesight for the remainder of his life. When Kepler was five, his father left to fight in one of the many local conflicts ravaging Europe at the time, and he never returned home. So Kepler spent the remainder of his childhood living with his mother at his grandfather's inn.

As a child, his mother took him outside the walls of the city so he could witness the great comet of 1577. This special event appears to have encouraged his lifelong interest in astronomy.

After completing local schooling, he pursued a religious education at the University of Tübingen in the hopes of becoming a Lutheran minister. He graduated in 1588 with his baccalaureate degree and then went on to complete a master's degree in 1591.

While Kepler was a student at the University of Tübingen, his professor of astronomy, Michael Maestlin (1550–1632) introduced him to the Copernican hypothesis. Kepler immediately embraced the new heliocentric model of the solar system. As his interest in the motion of the planets grew, Kepler's skill in mathematics also emerged. By 1594 partially because he alienated church officials after expressing his personal disagreement with certain aspects of Lutheran doctrine, he totally abandoned any plans for the Lutheran ministry and became a mathematics instructor in Gratz (Graz), Austria.

As well as an accomplished astronomer and mathematician, Kepler also maintained a strong interest in mysticism throughout his life. He extracted many of his mystical notions from the early Greeks, including the "music of the celestial spheres" that was originally suggested by Pythagoras in the sixth century B.C.E. As was common for many 17th-century astronomers, Kepler also dabbled in astrology. He often cast horoscopes for important benefactors, such as Emperor Rudolf II (1552–1612) and Duke (Imperial General) Albrecht von Wallenstein. Astrology provided the frequently impoverished Kepler with a supplemental source of income. His well-prepared horoscopes also earned Kepler the political protection of many satisfied high-ranking officials in the Holy Roman Empire. As a result, he was often the only prominent Protestant allowed to live in a German city under control of a Catholic ruler.

In 1596 Kepler published *Mysterium Cosmographicum* (The cosmographic mystery)—an intriguing work in which he unsuccessfully tried to analytically relate PLATO's five basic geometric solids (as found in early Greek philosophy and mathematics) to the distances of the six known planets from the Sun. This work, often considered the first outspoken defense of the Copernican hypothesis, also attracted the attention of TYCHO BRAHE, the most famous (pretelescope) astronomer of the period.

Kepler married his first wife in 1597. She was a wealthy widow named Barbara Mueller. However, when all the Protestants were forced to leave the Catholic-controlled city of Gratz in 1598, Kepler found it extremely difficult to liquidate her property holdings. Any nuptial financial comfort quickly dissipated, as the Kepler family hastily departed Gratz and moved to Prague in response to Brahe's invitation.

Kepler joined the elderly Danish astronomer in 1600 as his assistant. When Brahe died the following year, Kepler succeeded him as the imperial mathematician to the Holy Roman Emperor Rudolf II. As a result, Kepler acquired all the precise astronomical data Brahe had collected. These data, especially Brahe's precise observations of the motion of Mars, played an important role in the development of Kepler's famous three laws of planetary motion.

In 1604 Kepler wrote the book *De Stella Nova* (The new star), in which he described a bright new star (now known as a supernova) in the constellation Ophiuchus. He first observed the event on October 9, 1604. Modern astronomers sometimes refer to this supernova event, the remnants of which form radio source 3C 358, as Kepler's Star. According to early 17th-century astronomical observations that took place in Europe and Korea, this particular supernova remained visible for approximately one year.

From 1604 until 1609 Kepler's main interest was a detailed study of the orbital motion of Mars. Before his death, Brahe had challenged his young assistant with the task of explaining the puzzling motion of the Red Planet. Even accepting the Copernican hypothesis, which Kepler did but Brahe did not, a circular orbit around the Sun did not properly fit Brahe's carefully observed orbital position data. With youthful optimism, Kepler told the older astronomer he would have an answer for him in a week, but it took Kepler eight long and hard years to finally obtain the solution. The movement of Mars could not be explained and predicted unless Kepler assumed the orbit was an ellipse with the Sun located at one focus. This revolutionary assumption produced a major advance in solar system astronomy and provided the first empirical evidence of the validity of the Copernican model.

Kepler recognized that the other observable planets also followed elliptical orbits around the Sun. He published this important discovery in 1609 in the book *Astronomia Nova* (New astronomy). The book, dedicated to Emperor Rudolf II, confirmed the Copernican model and permanently shattered 2,000 years of geocentric Greek astronomy. Kepler became the first scientist to present a well-written demonstration of the scientific method. This work clearly acknowledged the errors and imperfections in the observational data and then compensated for these inaccuracies by creating a new scientific law (model) that used mathematics to accurately predict the natural phenomenon of interest (here the orbital position of Mars). In this important document, Kepler announced that the orbits of the planets are ellipses with the Sun as a common focus. Today, astronomers call this statement Kepler's First Law of Planetary Motion. Kepler also introduced his Second Law of Planetary Motion in this book.

German church officials (Lutheran or Catholic) never officially attacked Kepler for advocating the Copernican hypothesis. This was probably due to his powerful political benefactors. However, the same benign neglect did not befall his fiery Italian contemporary, Galileo, who shared the same passion for the Copernican hypothesis and corresponded with Kepler regularly until about 1610.

The year 1611 proved extremely difficult for Kepler in that his first wife, Barbara, and their seven-year-old son died. Then his royal patron, Emperor Rudolf II, abdicated the throne in favor of his brother Matthias. Unlike his brother Rudolf, Matthias did not believe in tolerance for the Protestants living in his realm. So Kepler, along with many other Lutherans, left Prague to avoid the start of an impending civil war between Catholics and Protestants. After burying his wife and son together, he moved his surviving children to Linz, where he accepted a position as district mathematician and teacher.

In 1613, desperately needing someone to care for his children, he married Susanna Reuttinger, herself an orphan. Although his second marriage was a generally happy one, Kepler continued to suffer from misfortune. He had chronic financial troubles, experienced the deaths of two infant daughters, and had to return to Württemberg to defend his mother, Katharina Kepler, who was being put on trial as a witch. This trial dragged on for three years and Kepler used all his legal wit and political capital to get her released. Her torture and death at the stake were the very likely outcomes that Kepler struggled so hard to prevent. Despite these personally draining problems, he remained a diligent mathematician in Linz until 1628 and used his available time to write several books. He moved his family in 1628 to Sagan (in Silesia) in order to work for the Imperial general Albrecht von Wallenstein (1583–1632) as his court mathematician. With the Thirty Years' War (1618–48) raging in Germany, in 1630 he had to flee Sagan to avoid religious persecution as Wallenstein fell from power.

When he published *De Harmonica Mundi* (Concerning the harmonies of the world) in 1619, Kepler continued his great work involving the orbital dynamics of the planets. Although this book extensively reflected Kepler's fascination with mysticism, it also provided a very significant insight that connected the mean distances of the planets from the Sun with their orbital periods. This discovery became known as Kepler's Third Law of Planetary Motion.

Between 1618 and 1621, despite constant relocations Kepler summarized all of his planetary studies in an important seven-volume effort entitled *Epitome Astronomica Copernicanae* (Epitome of Copernican astronomy). This work presents all of Kepler's heliocentric astronomy in a systematic way. As a point of scientific history, Kepler actually based his second law (the law of equal areas) on the mistaken (but reasonable for the time) physical assumption that the Sun exerted a strong magnetic influence on all the planets. Later in the 17th century, Newton provided the right physical explanation when he developed the universal law of gravitation to identify the force that causes the planetary motion so correctly described by Kepler's second law.

Kepler's three laws of planetary motion are: (1) The law of ellipses: the planets move in elliptical orbits with the Sun as a common focus; (2) The law of areas: as a planet orbits the Sun, the radial line joining the planet to the Sun sweeps out equal areas within the ellipse in equal times; and (3) The harmonic law: the square of the orbital period (P) of a planet is proportional to the cube of its mean distance (a) from the Sun. The third law states that there is a fixed ratio between the time it takes a planet to complete an orbit around the Sun and the size of the orbit. Astronomers often express this ratio as P^2/a^3, where "a" is the semimajor axis of the ellipse and "P" is the period of revolution around the Sun. Kepler's three laws established modern observational astronomy on a solid mathematical foundation.

In 1627 Kepler's *Tabulae Rudolphinae* (Rudolphine tables), named after his benefactor Emperor Rudolf and dedicated to Brahe, provided astronomers detailed planetary position data. The tables remained in use until the 18th century. Kepler used the logarithm to help perform the tedious calculations, which was the first important scientific application of the logarithm, a new mathematical function invented by the Scottish mathematician, John Napier (1550–1617).

Kepler also made important contributions in the field of optics. While in Prague, he wrote *Optica* (Optics) in 1604. Prior to 1610, Galileo and Kepler communicated with each other, although they never met. According to one historic account of their relationship, in 1610 Kepler refused to believe that Jupiter had four moons that behaved like a miniature solar system, unless he personally observed them. When a Galilean telescope arrived at his doorstep. Kepler promptly used the device and immediately called the four major moons discovered by Galileo "satellites"—a term Kepler derived from the Latin word *satelles* meaning "the people who escort or loiter around a powerful person." In 1611 Kepler improved the design of Galileo's original telescope by introducing two convex lenses in place of the one convex lens and one concave lens arrangement used by the great Italian astronomer. Kepler's final published work in Prague was his *Dioptrice* (Dioptics), which he completed in 1611. This book is the first scientific work in geometrical optics.

Before his death in 1630, Kepler wrote a novel called *Somnium* (The dream), about an Icelandic astronomer who travels to the Moon, involving demons and witches who help get the hero to the Moon's surface in a dream state. Kepler's description of the lunar surface was quite accurate. Therefore, many historians treat this story, published after his death in 1634, as the first genuine piece of science fiction.

Kepler married twice, fathered 13 children, constantly battled financial difficulties, and endured the turmoil of numerous relocations caused by religious persecution. He died on November 15, 1630, in Regensburg, Bavaria, of a fever contracted while journeying to see the emperor. Kepler was trying to collect the payment owed him for his service as imperial mathematician and for his effort in producing the *Rudolphine Tables*. Newton regarded both Galileo and Kepler as those "giants" upon whose shoulders he (Newton) stood to see farther into the mysteries and workings of the mechanical universe.

⊠ al-Khwarizmi, Abu (Abu'Jafar Muhammad ibn-Musa al-Khwarizmi)
(ca. 780–850)
Arabian
Mathematician, Astrologer, Astronomer, Geographer

The foundation for practical mathematics as it is known and used today is derived from the work of Abu al-Khwarizmi, who is most renowned as "the father of algebra." But as a scholar of his time, broadly estimated between the years of 770 to 850, al-Khwarizmi was a student, practitioner, and teacher of many scientific disciplines—mathematics, astrology, astronomy, and geography—because during this time the study of the stars usually involved an understanding of all of these sciences.

Science first made its appearance in a significant way in Baghdad under the rule of caliphs. During this time, many aspects of life were conducted according to astronomy. Physicians practiced medicine based on the stars, seeking stellar guidance for the best time to perform medical treatments. Agriculture was tied to astronomy, as is often the case today when almanacs are consulted for the best time to plant and harvest. Calendars needed to be devised, and the passing of time using a moon-calendar could only be told by looking at the night sky. Religious rituals, which required praying at precise times and facing an exact direction (geography), depended on the expert interpretation of the heavens. And the lives and future destinies of rulers were decided amongst the stars.

In 813 the caliph Al-Ma'mun, who ruled from 813 to 833, ordered the translation of the work of a man from India who reportedly could predict eclipses, and al-Khwarizmi became the first person to publish astronomical tables based on the Indian's work. The Indian books, however, were strictly numerical, with limited written explanations. But because India had a numerical system that included the concept of zero, al-Khwarizmi was able to translate the tables and move into a realm of mathematics that he devised for the average person to use in everyday life.

Al-Ma'mun's reign was one that both supported and promoted scientific study, and his excellent skills during a peace negotiation with the Byzantines made it possible for him to acquire PTOLEMY's original work. Arab scientists again went to work, this time translating Ptolemy, and the result was the *Almagest*, created in about 827. Ptolemy's influence, and indeed some of his exact measurements, can be found in the latter-day translations of al-Khwarizmi's writings.

During this time, translating involved more than changing the information from one language to another. Recalculation of measurements was equally important, to assure an accurate reading. In essence, al-Khwarizmi was both compiling information and creating new astronomical tables for his time and region.

Applying geography to the understanding of astronomy, then, was a natural extension of developing the tables. This led to the creation of the first map of the known world, completed in 830, under the leadership of al-Khwarizmi with the help of 70 other geographers. Determining the volume and circumference of the newly mapped Earth—a scientific blending of geography and astronomy—was another task undertaken by al-Khwarizmi for his caliph.

A gifted scientist, al-Khwarizmi worked within many disciplines as they related to astronomy and produced significant writings, the fundamentals of which have withstood centuries. Aside from his most noteworthy work in algebra, *Al-Jabrwa-al-Muqabilah*, al-Khwarizmi's scientific writings include his work on geography, *Kitab Surat-al-Ard*, which is based on Ptolemy's *Geography* and provides 2,402 distinct longitude and latitude calculations that are generally agreed to be more accurate than Ptolemy's calculations; his writing on the Jewish calendar, *Istikhraj Tarikh al-Yuhad*; his work on sundials, *Kitab al-Tarikh* and *Kitab al-Rukhmat*; two works on the astrolabe, an instrument that was used in early times to measure the angle between the horizon and a celestial body; a horoscope of prominent members of society that was written as a political history; and his title on astronomy, *Sindhind zij*, which covers astrological tables, calendars, and calculations relating to the Sun, the Moon, and planets, as well as eclipses. Of this work, nearly all of the originals are lost.

Abu al-Khwarizmi began his life's work translating science. It is ironic, then, that all of our knowledge of his scientific discovery comes from translations of his work by other men.

Gustav Kirchhoff was the talented German physicist who collaborated with Robert Bunsen in demonstrating the principles of spectroscopy. A major breakthrough took place when Kirchhoff applied spectroscopy to study the chemical composition of the Sun—especially the production of the Fraunhofer lines in the solar spectrum. His pioneering work contributed to the development of astronomical spectroscopy—one of the major tools in modern astronomy. *(AIP Emilio Segrè Visual Archives, W.F. Meggers Gallery of Nobel Laureates)*

⊠ Kirchhoff, Gustav Robert
(1824–1887)
German
Physicist, Spectroscopist

This gifted German physicist collaborated with ROBERT WILHELM BUNSEN in developing the fundamental principles of spectroscopy. While investigating the phenomenon of blackbody radiation, Kirchhoff applied spectroscopy to study the chemical composition of the Sun—especially the production of the Fraunhofer lines in the solar spectrum. His pioneering work contributed to the development of astronomical spectroscopy—one of the major tools in modern astronomy.

Gustav Robert Kirchhoff was born on March 12, 1824, in Königsberg, Prussia (now Kaliningrad, Russia). He was a student of Carl Friedrich Gauss (1777–1855) at the University of Königsberg and graduated in 1847. While still a student, he extended the work of Georg Simon Ohm (1787–1854) by introducing his own set of physical laws (now called Kirchhoff's laws) to describe the network relationship between current, voltage, and resistance in electrical circuits. Following graduation, Kirchhoff taught as an unsalaried lecturer (*Privatdozent*) at the University of Berlin and remained there for approximately three years before joining the University of Breslau as a physics professor. Four years later, he accepted a more prestigious appointment as a professor of physics at the University of Heidelberg and remained with that institution until 1875. At Heidelberg he collaborated with the German chemist Bunsen in a series of innovative experiments that significantly changed observational astronomy through the introduction of astronomical spectroscopy.

Prior to his innovative work in spectroscopy, Kirchhoff produced a theoretical calculation in 1857 at the University of Heidelberg demonstrating the physical principle that an alternating electric current flowing through a zero-resistance conductor would flow through the circuit at the speed of light. His work became a key step for the Scottish physicist James Clerk Maxwell (1831–79) during his formulation of electromagnetic wave theory.

Kirchhoff's most significant contributions to astrophysics and astronomy were in the field of spectroscopy. Fraunhofer's work had established the technical foundations of the science of spectroscopy, but undiscovered was the fact that each chemical element had its own characteristic spectrum. That critical leap of knowledge was necessary before physicists and astronomers could solve the puzzling mystery of the Fraunhofer lines and make spectroscopy an indispensable tool in both observational astronomy and numerous other applications. In 1859 Kirchhoff achieved that giant step in knowledge, while working in collaboration with Bunsen at the University of Heidelberg. In his breakthrough experiment, Kirchhoff sent sunlight through a sodium flame and, with the primitive spectroscope he and Bunsen had constructed, observed two dark lines on a bright background just where the Fraunhofer D-lines in the solar spectrum occurred. He and Bunsen immediately concluded

that the gases in the sodium flame were absorbing the D-line radiation from the Sun, producing an absorption spectrum.

After additional experiments, Kirchhoff also realized that all the other Fraunhofer lines were actually absorption lines—that is, gases in the Sun's outer atmosphere were absorbing some of the visible radiation coming from the solar interior thereby creating these dark lines, or "holes" in the solar spectrum. By comparing solar spectral lines with the spectra of known elements, Kirchhoff and Bunsen detected a number of elements present in the Sun, with hydrogen being the most abundant. This classic set of experiments, performed in Bunsen's laboratory in Heidelberg with a primitive spectroscope assembled from salvaged telescope parts, gave rise to the entire field of spectroscopy, including astronomical spectroscopy.

Bunsen and Kirchhoff then applied their spectroscope to resolving elemental mysteries on Earth. In 1861 they discovered the fourth and fifth alkali metals, which they named cesium (from the Latin *caesium*, meaning sky blue) and rubidium (from the Latin *rubidus*, meaning darkest red). Today scientists use spectroscopy to identify individual chemical elements from the light each emits or absorbs when heated to incandescence.

Modern spectral analyses trace their heritage directly back to the pioneering work of Bunsen and Kirchhoff. Astronomers and astrophysicists use spectra from celestial objects in a number of important applications, including composition evaluation, stellar classification, and radial velocity determination.

In 1875 Kirchhoff left Heidelberg because the cumulative effect of a crippling injury he sustained in an earlier accident now prevented him from performing experimental research. Confined to a wheelchair or crutches, he accepted an appointment to the chair of mathematical physics at the University of Berlin. In the new less physically demanding position, he pursued numerous topics in theoretical physics for the remainder of his life. During this period he made significant contributions to the field of radiation heat transfer. He discovered that the emissive power of a body to the emissive power of a blackbody at the same temperature is equal to the absorptivity of the body. His work in blackbody radiation was fundamental in the development of quantum theory by MAX KARL PLANCK.

Recognizing his great contributions to physics and astronomy, the British Royal Society elected Kirchhoff as a fellow in 1875. Failing health forced him to retire prematurely in 1886 from his academic position at the University of Berlin. He died in Berlin on October 17, 1887, and some of his scientific work appeared posthumously. While at the University of Berlin, he prepared his most comprehensive publication, *Lectures on Mathematical Physics* (Vorlesungen über mathematische Physik)—a four-volume effort that appeared between 1876 and 1894. His collaborative experiments with Bunsen and his insightful interpretation of their results started a new era in astronomy.

⊠ Korolev, Sergei Pavlovich
(1907–1966)
Ukrainian/Russian
Rocket Engineer

Sergei Korolev is often called the Russian WERNHER VON BRAUN, because he was the driving technical force behind the initial intercontinental ballistic missile (ICBM) and the space exploration programs of the Soviet Union. In 1954 he started work on the first Soviet ICBM, called the R-7. It was a powerful rocket designed to carry a massive nuclear warhead across continental distances. As part of cold war politics, Soviet premier Nikita Khrushchev (1894–1971) gave Korolev permission to use that military rocket to place the world's first artificial satel-

lite, *Sputnik 1*, into orbit around Earth. On October 4, 1957, Korolev became the rocket engineer who started the Space Age.

Korolev was born on January 12, 1907, in Zhitomir, the Ukraine—at the time part of czarist Russia. Korolev's birth date sometimes appears as December 30, 1906, a date corresponding to an obsolete czarist calendar system. As a young boy, Korolev obtained his first ideas about space travel in the inspirational books of KONSTANTIN EDUARDOVICH TSIOLKOVSKY. After discovering the rocket, Korolev decided on a career in engineering. He entered Kiev Polytechnic Institute in 1924 and two years later transferred to Moscow's Bauman High Technical School, where he studied aeronautical engineering under such famous Russian aircraft designers as Andrey Tupolev. Korolev graduated as an aeronautical engineer in 1929.

He began to champion rocket propulsion in 1931, when he helped to organize the Moscow Group for the Investigation of Reactive Motion, (Gruppa Isutcheniya Reaktvnovo Dvisheniya) (GIRD). Like its German counterpart, the Verein für Raumschiffahrt (VFR), (the German Society for Space Travel), this Soviet technical society began testing liquid-propellant rockets of increasing size. In 1933 GIRD merged with a similar group from Leningrad (St. Petersburg) to form the Reaction Propulsion Scientific Research Institute (RNII). Korolev was very active in this new organization and encouraged its members to develop and launch a series of rocket-propelled missiles and gliders during the mid-1930s. The crowning achievement of Korolev's early aeronautical engineering efforts was his creation of the RP-318, Russia's first rocket-propelled aircraft.

In 1934 the Soviet Ministry of Defense published Korolev's book *Rocket Flight into the Stratosphere*. Between 1936 and 1938, he supervised a series of rocket engine tests and winged-rocket flights within RNII. However, Soviet dictator Joseph Stalin (1879–1953) was eliminating many intellectuals through a series of brutal purges. Despite his technical brilliance, Korolev, along with most of the staff at RNII, found themselves imprisoned in 1938. During World War II Korolev remained in a scientific labor camp. His particular prison design bureau, called Sharashka TsKB-29, worked on jet-assisted take-off (JATO) systems for aircraft.

Freed from the labor camp after the war, Korolev resumed his work on rockets. He accepted an initial appointment as the chief constructor for the development of a long-range ballistic missile. At this point in his life, Korolev essentially disappeared from public view, and all his rocket and space activities remained a tightly guarded state secret.

Following World War II Stalin became more preoccupied with developing an atomic bomb than with exploiting captured German V-2 rocket technology. But he apparently sanctioned some work on long-range rockets and also approved the construction of a ballistic missile test range at Kapustin Yar, near the city of Volgograd. It was this approval that allowed Korolev to form a group of rocket experts to examine captured German rocket hardware and to resume the Soviet rocket research program. In late October 1947 Korolev's group successfully test-fired a captured German V-2 rocket from the new launch site at Kapustin Yar. By 1949 Korolev had developed a new rocket, the Pobeda (Victory-class) ballistic missile. He used Russian-modified German V-2 rockets and Pobeda rockets to send instruments and animals into the upper atmosphere. In May 1949 one of his modified V-2 rockets lifted a 120-kilogram payload to an altitude of 110 kilometers.

As cold war tensions mounted between the Soviet Union and the United States in 1954, Korolev began work on the first Soviet ICBM. Soviet leaders were focusing their nation's defense resources on developing very powerful rockets to carry the country's much heavier, less design-efficient nuclear weapons. Responding to this emphasis, Korolev designed the R-7. This powerful rocket was capable of carrying a 5,000-kilogram payload more than 5,000 kilometers.

With the death of Stalin in 1953, a new leader, Nikita Khrushchev, decided to use Soviet technology accomplishments to emphasize the superiority of Soviet communism over Western capitalism. Under Khrushchev, Korolev received permission to send some of his powerful military missiles into the heavens on missions of space exploration, as long as such space missions also had high-profile political benefits.

In summer 1955 construction began on a secret launch complex in a remote area of Kazakhstan north of a town called Tyuratam. This central Asian site is now called the Baikonur Cosmodrome and lies within the political boundaries of the Republic of Kazakhstan. In August and September 1957 Korolev successfully launched the R-7, on long-range demonstration flights from this location. Encouraged by the success of these test flights, Khrushchev allowed Korolev to use an R-7 military missile as a space launch vehicle in order to beat the United States into outer space with the first artificial satellite.

On October 4, 1957, a mighty R-7 rocket roared from its secret launch pad at Tyuratam, placing *Sputnik 1* into orbit around Earth. Korolev, the "anonymous" engineering genius, had propelled the Soviet Union into the world spotlight and started the Space Age. To Khrushchev's delight, a supposedly technically inferior nation won a major psychological victory against the United States. With the success of *Sputnik 1*, space technology became a key instrument of cold war politics and superpower competition.

The provocative and boisterous Khrushchev immediately demanded additional high-visibility space successes from Korolev. The rocket engineer responded on November 3, 1957, by placing a much larger satellite into orbit. *Sputnik 2* carried the first living space traveler into orbit around Earth. The passenger was a dog named Laika. Korolev continued to press the Soviet Union's more powerful booster advantage by developing the Vostok (one-person) spacecraft to support human space flight. On April 12, 1961, another of Korolev's powerful military rockets placed the *Vostok 1* spacecraft, carrying cosmonaut Yuri Gagarin (1934–68), into orbit around Earth. Gagarin's brief flight (only one orbital revolution) took place just before the United States sent its first astronaut, Alan B. Shepard Jr. (1923–99), on a suborbital mission from Cape Canaveral, Florida, on May 5, 1961.

Using Korolev's powerful rockets like so many trump cards, Khrushchev continued to reap international prestige for the Soviet Union, which had just become the first nation to place a person into orbit around Earth. This particular Soviet accomplishment forced U.S. president John F. Kennedy (1917–63) into a daring response. In May 1961 Kennedy announced his bold plan to land U.S. astronauts on the Moon in less than a decade. From the perspective of history, Korolev's space achievements became the political catalyst by which Braun received a mandate to build more powerful rockets for the United States. Starting in July 1969, Braun's new rockets would carry U.S. explorers to the surface of the Moon—a technical accomplishment that soundly defeated the Soviet Union in the great space race of the 1960s.

Following the success of Sputnik, Korolev used his rockets to propel large Soviet spacecraft to the Moon, Mars, and Venus. One of his spacecraft, *Lunik 3*, took the first photographic images of the Moon's far side in October 1959. These initial images of the Moon's long-hidden hemisphere excited both astronomers and the general public. Appropriately, one of the largest features on the Moon's far side now bears Korolev's name. He also planned a series of

interesting follow-on Soyuz spacecraft for future space projects involving multiple human crews. However, he was not allowed to pursue these developments in a logical, safe fashion.

Premier Khrushchev kept demanding other space mission "firsts" from Korolev's team. The Soviet leader wanted to fly a three-person crew in space before the United States could complete the first flight of its new two-person *Gemini* capsule. The first crewed *Gemini* flight occurred on March 23, 1965. Korolev recognized how dangerous it would be to convert his one-person *Vostok* spacecraft design into an internally stripped-down "three-seater." But he also remembered the painful time he spent in a political prison camp for perceived disloyalty to Stalin's regime.

So, despite strong opposition from his design engineers, Korolev removed the *Vostok* spacecraft's single ejection seat and replaced it with three couches. Without an ejection seat, the cosmonaut crew could not leave the space capsule during the final stages of reentry descent, as had been done during previous successful *Vostok* missions. To accommodate the demands of this mission, Korolev came up with several clever ideas, including a new retro-rocket system and a larger reentry parachute. To quell the vehement safety objections from his design team, he also made them an offer they could not refuse: If they could design this modified three-person spacecraft in time to beat the Americans, one of the engineers would be allowed to participate in the flight as a cosmonaut.

On October 12, 1964, a powerful military booster sent the "improvised" *Voskhod 1* space capsule into a 170 kilometer by 409 kilometer orbit around Earth. *Voshkod* is the Russian word for "sunrise." The flight lasted just one day. The retrofitted *Vostok* spacecraft carried three cosmonauts without spacesuits under extremely cramped conditions. Cosmonaut Vladimir Komarov commanded the flight; a medical expert, Boris Yegorov, and a design engineer, Konstantin Feoktistov, accompanied him. Korolev won an extremely high-risk technical gamble. His engineering skill and luck not only satisfied Khrushchev's insatiable appetite for politically oriented space accomplishments, but he also kept his promise of a ride in space to one of his engineers.

Khrushchev responded to Kennedy's moon race challenge by ordering Korolev's Experimental Design Bureau No. 1 (OKB-1) to accelerate its plans for a very large booster. As originally envisioned by Korolev, the Russian N-1 rocket was to be a very large and powerful booster for placing extremely heavy payloads into Earth's orbit, including space stations, nuclear-rocket upper stages, and various military payloads. But after Korolev's untimely death in 1966, the highly secret N-1 rocket became the responsibility of another rocket design group. It suffered four catastrophic failures between 1969 and 1972, and then the project was quietly canceled.

From 1962 to 1964, Khrushchev's continued political use of space technology seriously diverted Korolev's creative energies from much more important projects, such as new boosters, the *Soyuz* spacecraft, a Moon-landing mission, and the space station. His design team was just beginning to recover from Khrushchev's constant interruptions when disaster struck. On January 14, 1966, Korolev died during a botched routine surgery at a hospital in Moscow. He was only 58 years old. Some of Korolev's contributions to space technology include the powerful, legendary R-7 rocket (1956), the first artificial satellite (1957), pioneering lunar spacecraft missions (1959), the first human space flight (1961), a spacecraft to Mars (1962), and the first space walk (1965). Even after his death, the Soviet government chose to hide Korolev's identity by publicly referring to him only as the "Chief Designer of Carrier Rockets and Spacecraft." Despite this official anonymity, Chief Designer and Academician Korolev is now properly recognized as the rocket engineer who started the Space Age.

⊠ **Kuiper, Gerard Peter (Gerrit Pieter)**
(1905–1973)
Dutch/American
Astronomer

In 1951 Gerard Kuiper, the Dutch-American planetary astronomer, boldly postulated the presence of thousands of icy planetesimals in an extended region at the edge of the solar system beyond the orbit of Pluto. Today, this region of frigid, icy objects has been detected and is called the Kuiper belt in his honor. In 1944 Kuiper discovered that Saturn's largest moon, Titan, had an atmosphere. He continued to revive interest in planetary astronomy by discovering Miranda, the fifth-largest moon of Uranus, in 1948, and Nereid, the outermost moon of Neptune, in

The Dutch-American planetary astronomer Gerard Kuiper was the exceptionally skilled observer who in 1944 discovered that Saturn's largest moon, Titan, had an atmosphere. In the late 1940s he helped revive interest in planetary astronomy, when he found Miranda, the fifth largest moon of Uranus in 1948, and Nereid, the outermost moon of Neptune, in 1949. He also postulated the existence of a large reservoir of small icy bodies beyond the orbit of Pluto—now called the Kuiper belt in his honor. *(AIP Emilio Segrè Visual Archives, Physics Today Collection)*

1949. Transitioning to Space Age astronomy, he served with distinction as a scientific adviser to the U.S. National Aeronautics and Space Administration (NASA) on the early lunar and planetary missions in the 1960s, especially NASA's Ranger Project and Surveyor Project.

Gerard Kuiper was born on December 7, 1905, in Harenkarspel, the Netherlands. When he was young, two factors influenced Kuiper's interest in astronomy. First, his father gave him a gift of a small telescope that he used to great advantage because of his exceptional visual acuity. Second, he was drawn to astronomy by the cosmological and philosophical writings of the great French mathematician René Descartes (1596–1650).

While enrolling at the University of Leiden in September 1924, Kuiper made the acquaintance of a fellow student, BARTHOLOMEUS JAN BOK, who would remain a lifelong friend. He completed a bachelor of science degree in 1927 and immediately pursued his graduate studies. One of Kuiper's professors at the University of Leiden was EJNAR HERTZSPRUNG, under whom he did his doctoral thesis on the subject of binary stars. After receiving his doctoral degree in 1933, Kuiper traveled to the United States under a fellowship and conducted postdoctoral research on binary stars at the Lick Observatory, near San Jose, California.

In August 1935 he left California and spent a year at the Harvard College Observatory. While there, he met and later married (in June 1936) Sarah Parker Fuller, whose family had donated the property on which the Harvard Oak Ridge Observatory stood. The couple had two children: a son and a daughter. In 1936 Kuiper joined the Yerkes Observatory of the University of Chicago. He became a naturalized American citizen the following year. He joined other members of the Yerkes staff who worked with the University of Texas in planning and developing the McDonald Observatory, near Fort Davis, Texas. The 2.1-meter reflector of this important astronomical facility was dedicated and began operation in 1939. At the University of Chicago, Kuiper progressed up through the academic ranks, becoming a full professor of astronomy in 1943.

In an important 1941 technical paper, Kuiper introduced the concept of "contact binaries"—a term describing close, mass-exchange binary stars characterized by accretion disks. During World War II, he took a leave of absence from the University of Chicago to support various defense-related projects. His duties included special technical service missions as a member of the U.S. War Department's ALSOS Mission, which assessed the state of science in Nazi Germany during the closing days of the war. In one of his most interesting adventures, he helped rescue MAX KARL PLANCK from the advancing Soviet armies. Kuiper, an astronomer turned commando, took charge of a military vehicle and driver, dashed across war-torn Germany from the U.S. lines to Planck's location in eastern Germany, and then spirited the aging German physicist and his wife away to a safe location in the western part of Germany near Göttingen.

While performing his wartime service, Kuiper still managed to make a major contribution to modern planetary astronomy, the astronomical area within which he would become the recognized world leader. Taking a short break from his wartime research activities in winter 1943–44, Kuiper conducted an opportunistic spectroscopic study of the giant planets Jupiter and Saturn and their major moons at the McDonald Observatory. To his great surprise, early in 1944 he observed methane on the largest Saturnian moon, Titan—making the large moon the only satellite in the solar system with an atmosphere. This fortuitous discovery steered Kuiper into his very productive work in solar system astronomy, especially the study of planetary atmospheres.

Through spectroscopy he discovered in 1948 that the Martian atmosphere contained carbon dioxide. In 1948 he discovered the fifth moon of Uranus, which he named Miranda, and then Neptune's second moon, Nereid, in 1949. He went on to postulate the existence of a region of icy planetesimals and minor planets beyond the orbit of Pluto. The first members of this region, now called the Kuiper belt in his honor, would be detected in the early 1990s.

Kuiper left the University of Chicago and joined the University of Arizona in 1960. At Arizona, he founded and directed the Lunar and Planetary Laboratory, making major contributions to planetary astronomy at the dawn of the Space Age. He produced major lunar atlases for both the U.S. Air Force and NASA. His research concerning the Moon in the 1950s and 1960s provided strong support for the impact theory of crater formation. Prior to the Space Age, most astronomers held that the craters on the Moon had been formed by volcanic activity.

Kuiper promoted a multidisciplinary approach to the study of the solar system, and this emphasis stimulated the creation of a new astronomical discipline called planetary science. He assisted in the selection of the superior ground-based observatory site at Mauna Kea, on the island of Hawaii. Because of his reputation as an outstanding planetary astronomer, he served as a scientific adviser on many of NASA's 1960 and 1970 space missions to the Moon and the inner planets. Kuiper was the chief scientist for NASA's Ranger Project (1961–65), which sent robot spacecraft equipped with television cameras crashing into the lunar surface. Later, he helped identify suitable landing sites for NASA's *Surveyor* robot-lander spacecraft as well as the Apollo human-landing missions.

He served as principal author or general editor on several major works in modern astronomy, including *The Solar System*, a four-volume series published between 1953 and 1963; *Stars and Stellar Systems*, a nine-volume series appearing in 1960; the famous *Photographic Atlas of the Moon* (1959); the *Orthographic Atlas of the Moon* (1961); the *Rectified Lunar Atlas* (1963); and the *Consolidated Lunar Atlas* (1967).

Kuiper was also a pioneer in applying infrared technology to astronomy and played a very influential role in the development of airborne infrared astronomy in the 1960s and early 1970s. Starting in 1967 Kuiper used NASA's Convair 990 jet aircraft, which was equipped with a telescope system capable of performing infrared spectroscopy. As the aircraft flew above 40,000 feet, Kuiper and his research assistants performed trend-setting infrared spectroscopy measurements of the Sun and other stars, as well as planets, discovering many interesting phenomena that could not be observed by ground-based astronomers because of the blocking influence of the denser portions of Earth's atmosphere.

Kuiper died on December 24, 1973, while on a trip to Mexico City with his wife and several family friends. As one tribute to his numerous contributions to astronomy, in 1975 NASA named its newest airborne infrared observatory, a specially outfitted C-141 jet transport aircraft, the Gerard P. Kuiper Observatory (KAO). This unique airborne astronomical facility operated out of Moffett Field, California, for more than two decades until it was retired from service in October 1995.

Kuiper received the Janssen Medal in 1947 from the Astronomical Society of France and the Kepler Medal in 1971 from the American Association for the Advancement of Science and the Franklin Institute. He was also elected as a member of the American Academy of Sciences in 1950. No astronomer did more in the 20th century to revitalize planetary astronomy. Celestial objects carry his name or continue his legacy of discovery across the entire solar system, from a specially named crater on Mercury, to Miranda around Uranus, to Nereid around Neptune, to an entire cluster of icy minor planets beyond Pluto.

L

Lagrange, Comte Joseph-Louis
(Lodovico Lagrange, Luigi Lagrange, Giuseppe Luigi Lagrangia)
(1736–1813)
Italian/French
Mathematician, Celestial Mechanics Expert

The 18th-century mathematician Joseph Lagrange made significant contributions to celestial mechanics. In about 1772 he identified certain special regions in outer space—now called the Lagrangian libration points—that mark the five equilibrium points for a small celestial object moving under the gravitational influence of two much larger bodies. Other astronomers used his discovery of the Lagrangian libration points to find new objects in the solar system, such as the Trojan Group of asteroids. His influential book, *Analytical Mechanics*, represented an elegant compendium of the mathematical principles that described the motion of heavenly bodies.

Lagrange was born on January 25, 1736, in Turin, Italy. His father, Giuseppe Francesco Lodovico Lagrangia, was a prosperous public official in the service of the king of Savoy, but impoverished his family through unwise financial investments and speculations. Lagrange, a gentle individual with a brilliant mathematical mind, recalled his childhood poverty by later quipping: "If I had been rich, I probably would not have devoted myself to mathematics." Although given an Italian family name at baptism, throughout his life Lagrange preferred to emphasize his French ancestry, derived from his father's side of the family. He would, therefore, sign his name "Luigi Lagrange" or "Lodovico Lagrange"—combining an Italian first name with a French family name.

Lagrange studied at the College of Turin, enjoying instruction in classical Latin and ignoring Greek geometry. But after reading Sir Edmond Halley's treatise on the application of algebra in optics, Lagrange changed his earlier career plans to become a lawyer and embraced the study of higher mathematics. In 1755 he was appointed professor of mathematics at the Royal Artillery School in Turin. At the time, Lagrange was just 19 years old and already exchanging impressive mathematical notes on his calculus of variations with the famous Swiss mathematician Leonhard Euler. In 1757 Lagrange became a founding member of a local scientific group that eventually evolved into the Royal Academy of Science of Turin. He published many of his elegant papers on the calculus of variations in the society's journal, *Mélanges de Turin*.

By the early 1760s Lagrange had earned a reputation as one of Europe's most gifted mathematicians. Yet he was a humble person who did

The gifted 18th-century Italian-French mathematician Joseph Lagrange made important contributions to celestial mechanics, extending the work begun by Sir Isaac Newton. In about 1772 Lagrange identified several special regions in outer space—now called the Lagrangian libration points—that correspond to the five equilibrium points for a small celestial object moving under the gravitational influence of two much larger bodies. Later observational astronomers, such as Max Wolf, used knowledge of the Lagrangian libration points to find new objects in the solar system, such as the Trojan Group of asteroids. *(AIP Emilio Segrè Visual Archives, E. Scott Barr Collection)*

not seek fame or position. He just wanted to pursue the study of mathematics with a modest amount of financial security. In 1764 Lagrange received a prize from the Paris Academy of Sciences for his brilliant mathematical paper on the libration of the Moon. The Moon's libration is the phenomenon by which 59 percent of the

lunar surface is visible to an observer on Earth over a period of 30 years. This results from a complicated collection of minor perturbations in the Moon's orbit as it travels around Earth. Soon after that award, he won another prize from the Paris Academy in 1776 for his theory on the motion of Jupiter's moons.

When Euler left Berlin in 1766 to return to St. Petersburg, he recommended to King Frederick II of Prussia that Lagrange serve as his replacement. Therefore, in November 1766 Frederick II invited Lagrange to succeed Euler as mathematical director of the Berlin Academy of Science of the Prussian Academy of Sciences. A little less than a year after his arrival in Berlin, Lagrange married Vittoria Conti, a cousin who had lived for an extended time with his family in Turin. The couple had no children.

For two decades Lagrange worked in Berlin and during this period produced a steady number of top quality, award-winning mathematical papers. He corresponded frequently with PIERRE-SIMON MARQUIS DE LAPLACE, a contemporary French mathematician living in Paris. Lagrange won prizes from the Paris Academy of Sciences for his mathematics in 1772, 1774, and 1780. He shared the 1772 Paris Academy prize with Euler for his superb work on the challenging three-body problem in celestial mechanics—an effort that involved the existence and location of the Lagrangian libration points. These five points (usually designated as L_1, L_2, L_3, L_4, and L_5) are the locations in outer space where a small object can experience a stable orbit in spite of the force of gravity exerted by two much more massive celestial bodies when they all orbit about a common center of mass. He won the 1774 prize for another brilliant paper on the motion on the Moon. His 1780 award-winning effort discussed the perturbation of cometary orbits by the planets.

However, during his 20 years in Berlin, his health began to fail. His wife also suffered from poor health, and she died in 1783 after an extended illness. Her death plunged Lagrange into a state of depression. After the death of Frederick II, Lagrange departed from Berlin to accept an invitation from the French king Louis XVI. In May 1787 Lagrange became a member of the Paris Academy of Sciences, a position he retained for the rest of his career despite the turmoil of the French Revolution. Upon Lagrange's arrival in Paris, King Louis XVI offered Lagrange apartments in the Louvre, from which comfortable surroundings he published in 1788 his elegant synthesis of mechanics, entitled *Mécanique analytique* (Analytical mechanics), which he had written in Berlin.

In May 1790 Lagrange became a member of and eventually chaired the committee of the Paris Academy of Sciences that standardized measurements and created the international system (Système Internationale, or SI) of units currently used throughout the world. In 1792 Lagrange married his second wife, a much younger woman, named Renée Le Monnier, the daughter of one of his astronomer colleagues at the Paris Academy of Sciences.

Following the Reign of Terror (1793), Lagrange became the first professor of analysis at the famous École Polytechnique, founded in March 1794. Though brilliant, the aging mathematician was not an accomplished lecturer, and much of what he discussed sailed over the heads of his audience of inattentive students. Despite his shortcomings, the notes from his calculus lectures were collected and published as the *Theory of Analytic Functions* (1797) and *Lessons on the Calculus of Functions* (1804)—the first textbooks on the mathematics of real analytical functions.

In 1791 the Royal Society in London made him a Fellow. Napoleon, the emperor of France, bestowed many honors upon the aging mathematician, making him a senator, a member of the Legion of Honor, and a count of the empire. Despite the lavish political attention, Lagrange remained a quiet academician who preferred to keep himself absorbed in his thoughts about mathematics. He died in Paris on April 10, 1813. His numerous works on celestial mechanics prepared the way for 19th-century mathematical astronomers to make discoveries using the subtle perturbations and irregularities in the motion of solar system bodies.

Laplace, Pierre-Simon, Marquis de
(1749–1827)
French
Mathematician, Astronomer, Celestial Mechanics Expert

Called the "French Newton," Pierre-Simon Laplace's work in celestial mechanics established the foundations of 19th-century mathematical astronomy. Laplace extended SIR ISAAC NEWTON's gravitational theory and provided a more complete mechanical interpretation of the solar system, including the subtle perturbations of planetary motions. His mathematical formulations supported the discovery of Uranus (18th century), Neptune (19th century), and Pluto (20th century). In 1796, apparently independent of IMMANUEL KANT, Laplace introduced his own version of the nebula hypothesis—suggesting that the Sun and the planets had condensed from a primeval interstellar cloud of gas.

Laplace was born on March 23, 1749, at Beaumont-en-Auge, in Normandy, France. The son of a moderately prosperous farmer, he attended a Benedictine priory school in Beaumont, a preparatory school for future members of the clergy or military. At age 16 Laplace enrolled in the University of Caen, where he studied theology in response to his father's wishes. However, after two years at the university, Laplace discovered his great mathematical abilities and abandoned any plans for an ecclesiastical career.

At 18 he left the university without receiving a degree and went to live in Paris. There

Marquis Laplace is often called the "French Newton." A brilliant mathematician, his work in celestial mechanics established the foundations of 19th-century mathematical astronomy. Laplace extended Newton's gravitational theory to provide a more complete mechanical interpretation of the solar system. He specifically addressed those subtle perturbations in planetary motions that appeared to threaten the long-term stability of the solar system. His mathematical formulations proved that the solar system was stable and supported the discovery of Uranus (18th century), Neptune (19th century), and Pluto (20th century). (*AIP Emilio Segrè Visual Archives*)

Laplace met Jean d'Alembert (1717–83) and greatly impressed the French mathematician and philosopher. D'Alembert secured an appointment for Laplace at the prestigious École Militaire. By age 19 Laplace had become the professor of mathematics at the school.

In 1773 he began to make his important contributions to mathematical astronomy. Working independently of, but cooperatively (through correspondence) with COMTE JOSEPH-LOUIS LAGRANGE, who then resided in Berlin, Laplace

applied and refined Newton's law of gravitation to bodies throughout the solar system. Careful observations of the six known planets and the Moon indicated subtle irregularities in their orbital motions. This raised a very puzzling question: How could the solar system remain stable? Newton, for all his great scientific contributions, considered this particular problem far too complicated to be treated with mathematics. He simply concluded that "divine intervention" took place on occasion to keep the entire system in equilibrium.

Laplace, however, was determined to solve this problem with his own excellent skills in mathematics. Eighteenth-century astronomers could not explain why Saturn's orbit seemed to be continually expanding, while Jupiter's orbit seemed to be getting smaller. Laplace was able to mathematically demonstrate that while there certainly were measured irregularities in the orbital motions of these giant planets, such anomalies were not cumulative, but rather periodic and self-correcting. His important discovery represented a mathematically and physically sound conclusion that the solar system was actually an inherently stable system.

As a result of this brilliant work, Laplace became recognized as one of France's great mathematicians. In 1773 Laplace became an associate member of the French Academy of Sciences; he became a full member in 1785. However, as his fame grew, he more frequently acted the part of a political opportunist rather than a great scientist, tending to forget his humble roots and rarely acknowledging his many benefactors and collaborators. Good fortune came to Laplace in 1784, when he became an examiner at the Royal Artillery Corps. He evaluated young cadets who would soon serve in the French army. While fulfilling this duty in 1785, he had the opportunity to grade the performance of and pass a 16-year-old cadet named Napoleon Bonaparte. The examiner position allowed Laplace to make direct contact with many of the powerful men who would run France during its next four politically turbulent decades, including the monarchy of Louis XVI, the revolution, the rise and fall of Napoleon, and the restoration of the monarchy under Louis XVIII. While other great scientists, such as Antoine de Lavoisier (1743–94), would quite literally lose their heads around him, Lagrange maintained an ability to correctly change his views to suit the changing political events.

In 1787 Lagrange left Berlin and joined Laplace in Paris. Despite a rivalry between them, the great mathematical geniuses benefited from their constant exchange of ideas. They both served on the commission of the French Academy of Sciences that developed the metric system, although Laplace was eventually discharged from the commission during the Reign of Terror. Laplace married Marie-Charlotte de Courty de Romanges in May 1788. His wife was 20 years younger than Laplace and bore him two children. Just before the Reign of Terror began, Laplace departed Paris with his wife and two children, and they did not return until after July 1794.

Laplace published *Exposition du système du monde* (The system of the world) in 1796. This book was a five-volume popular (that is, nonmathematical) treatment of astronomy and celestial mechanics. From a historic perspective, it also served as the precursor for his more detailed mathematical work on the subject. Laplace used this book to introduce his version of the nebular hypothesis, in which he suggested that the Sun and planets formed from the cooling and gravitational contraction of a large, flattened and slowly rotating gaseous nebula. Apparently, Laplace was unaware that Kant had proposed a similar hypothesis about 40 years earlier. Laplace wrote this book in such exquisite prose that in 1816 he won admission to the French Academy—an honor only rarely bestowed upon an astronomer or mathematician. In 1817 he served as president of this distinguished and exclusive literary organization.

Between 1799 and 1825, Laplace summed up gravitational theory and completed Newton's pioneering work in a monumental five-volume work, entitled *Mécanique Céleste* (Celestial mechanics). Laplace provided a complete mechanical interpretation of the solar system and his application of the law of gravitation. Central to this work is Laplace's great discovery of the invariability of the mean motions of the planets. With this finding, he demonstrated the inherent stability of the solar system and gained recognition throughout his country and the rest of Europe as the "French Newton."

Laplace maintained an extraordinarily high scientific output despite the tremendous political changes taking place around him. He always demonstrated political adroitness and changed sides whenever it was to his advantage. Under Napoleon he became a member and then chancellor of the French senate, received the Legion of Honor in 1805, and became a count of the empire in 1806. Yet when Napoleon fell from power, Laplace quickly supported restoration of the monarchy and was rewarded by King Louis XVIII, who named him a marquis in 1817.

Laplace made scientific contributions beyond his great work in celestial mechanics. He provided the mathematical foundation for probability theory in two important books. In 1812 he published his *Théorie analytique des probabilités* (Analytic theory of probability) in which he presented many of the mathematical tools he invented to apply probability theory to games of chance and events in nature, including astronomy and collisions of comets and planets. He also wrote a popular discussion of probability theory, *Essai philosophique sur les probabilités* (A philosophical essay on probability) that appeared in 1814.

Laplace also explored other areas of physical science and mathematics between 1805 and 1820. These areas included heat transfer, capillary action, optics, the behavior of elastic fluids, and the velocity of sound. However, as other physical theories began to emerge with their intelligent young champions, Laplace's dominant position in French science came to an end. He died in Paris on March 5, 1827. Throughout France and Europe, Laplace was highly regarded as one of the greatest scientists of all time. He held membership in many foreign societies, including the British Royal Society, which elected him a Fellow in 1789.

⊠ Leavitt, Henrietta Swan
(1868–1921)
American
Astronomer

It is a startling coincidence that Henrietta Swan Leavitt, another one of the "computers" hired at the beginning of the 20th century by professor EDWARD CHARLES PICKERING to make complex data analyses at the Harvard Observatory, contributed to astronomy while suffering almost total deafness, the same condition as her famous coworker, ANNIE JUMP CANNON.

Leavitt was born on July 4, 1868, in Cambridge, Massachusetts, the daughter of a Congregational minister. She attended Oberlin College and then Radcliffe College, which was then called the Society for Collegiate Instruction of Women. She graduated with a B.A. degree from Radcliffe in 1892, having become interested in astronomy in her senior year, and, to pursue her new interest, enrolled in further courses in astronomy.

Leavitt suffered a serious illness, however, which left her nearly deaf, forcing her to spend a few years at home. Eventually she volunteered

In about 1912, while working at the Harvard College Observatory, the talented American astronomer Henrietta Leavitt discovered the period-luminosity relationship for Cepheid variable stars. Her important finding permitted other astronomers, such as Harlow Shapley, to make more accurate estimates of distances in the Milky Way Galaxy. *(Photo courtesy Margaret Harwood, AIP Emilio Segrè Visual Archives, Shapley Collection)*

to work at the Harvard College Observatory as a research assistant, seeking to gather experience as an astronomer. Seven years later, in 1902, Professor Pickering hired her at 30 cents an hour to be one of his computers. In this capacity she drifted toward the photometric side of the astronomy department, and became specialized in photographing stars as a method of determining their magnitude.

In 1904, after spending almost all of her time searching photographic plates for Cepheid variables, variable stars whose brightness changes due to alternate expansions and contractions in volume, Leavitt discovered 152 variables in the Large Magellanic Cloud (LMC) and 59 in the Small Magellanic Cloud (SMC). The following year she reported 843 new variables in the SMC.

In 1912 Leavitt devised the "period-luminosity" relationship, the direct correlation between the time it took a star to go from brightest to dimmest, which allowed astronomers to determine the exact brightness of a star. Today astronomers use the period-luminosity relationship to determine the distance of galaxies from the Earth.

With this knowledge, she studied 299 plates from 13 different telescopes, using logarithmic equations to classify stars over 17 degrees of magnitude and developing a standard of photographic measurements, eventually known as the Harvard Standard. In 1913 the International Committee on Photographic Magnitudes accepted the standard that Leavitt described. In the course of her studies, Leavitt also discovered four novae and more than 2,400 variables, approximately half of the total known to exist at the time.

Leavitt was a member of Phi Beta Kappa, the American Association of University Women, the American Astronomical and Astrophysical Society, the American Society for the Advancement of Science, and an honorary member of the American Association of Variable Star Observers.

Without Leavitt's pioneering work, modern astronomers would not be able to calculate the distance from the Earth to faraway galaxies. She was working on a new photographic magnitude scale when she died of cancer in 1921. Four years later, the Swedish Academy of Sciences recognized her contribution to science by nominating her for the Nobel Prize. A crater on the moon was named Leavitt to honor the work of deaf astronomers.

⊠ Lemaître, Abbé Georges-Édouard
(1894–1966)
Belgian
Astrophysicist, Cosmologist

Georges-Édouard Lemaître was the innovative cosmologist who suggested in 1927 that a violent explosion might have started an expanding universe. He based this hypothesis on his interpretation of ALBERT EINSTEIN's general relativity theory and upon EDWIN POWELL HUBBLE's contemporary observation of galactic redshifts—an observational indication that the universe was indeed expanding. Other physicists, such as RALPH ASHER ALPHER under the direction of GEORGE GAMOW, built upon Lemaître's pioneering work and developed it into the widely accepted big bang theory of modern cosmology. Central to Lemaître's model is the idea of an initial superdense primeval atom, his "cosmic egg," that started the universe in a colossal ancient explosion.

Lemaître was born on July 17, 1894, in Charleroi, Belgium. Prior to World War I, he studied civil engineering at the University of Louvain. At the outbreak of war in 1914, he volunteered for service in the Belgian army and earned the Belgian Croix de Guerre for his combat activities as an officer in the artillery corps. Following the war, he returned to the University of Louvain where he pursued additional studies in physics and mathematics. In the early 1920s he also responded to another vocational calling and became an ordained priest in the Roman Catholic Church in 1923. After ordination, Abbé Lemaître pursued a career devoted to science, especially the area of cosmology, in which he would make a major 20th-century contribution.

Taking advantage of an advanced studies scholarship from the Belgian government, Lemaître traveled to England to study (1923–24) with Sir Arthur Stanley Eddington at the solar physics laboratory of the University of Cambridge. He continued his travels and came to the United States where he studied at the Massachusetts Institute of Technology (MIT) from 1925–27, earning his Ph.D. in physics. While at MIT, Lemaître became familiar with the expanding-universe concepts just being developed by the American astronomers Harlow Shapley and Edwin Hubble.

In 1927 Lemaître returned to Belgium and joined the faculty of the University of Louvain as a professor of astrophysics, a position he held for the rest of his life. That year he published his first major paper related to cosmology. Unaware of similar work by the Russian mathematician Alexander Friedmann (1888–1925), Lemaître blended Einstein's general relativity and Hubble's early work on expanding galaxies to reach the conclusion that an expanding-universe model is the only appropriate explanation for observed redshifts of distant galaxies. Lemaître suggested the distant galaxies serve as "test particles" that clearly demonstrate the universe is in a state of expansion. Unfortunately, Lemaître's important insight attracted little attention within the astronomical community, mainly because he published this paper in a rather obscure scientific journal in Belgium.

But in 1931 Eddington discovered Lemaître's paper and, with his permission, had the paper translated into English and then published in the *Monthly Notices of the Royal Astronomical Society*. Eddington also provided a lengthy commentary to accompany the translation of Lemaître's paper. By this time Hubble had formally announced his famous law (Hubble's law) that related the

distance of galaxies to their observed redshift. So Lemaître's paper, previously ignored, now created quite a stir in the astrophysical community. He was invited to come to London to lecture about his expanding-universe concept. During this visit, he not only presented an extensive account of his original theory of the expanding universe, he also introduced his idea about a "primitive atom" (sometimes called Lemaître's "cosmic egg"). He was busy thinking not only about how to show that the universe was expanding but also about what caused the expansion and how this great process began.

Lemaître cleverly reasoned that if the galaxies are now everywhere expanding, then in the past they must have been much closer together. In his mind, he essentially ran time backward to see what conditions were as the galaxies came closer and closer together in the early universe. He hypothesized that eventually a point would be reached when all matter resided in a superdense primal atom. He further suggested that sum instability within this primal atom would result in an enormous explosion that would start the universe expanding. Big Bang cosmology results directly from his concepts, although it took several decades to fully develop and establish the Big Bang theory as the currently preferred cosmological model.

Lemaître's other significant papers include "Discussion on the Evolution of the Universe" (1933) and "Hypothesis of the Primeval Atom" (1946). His contributions to astrophysics and modern cosmology were widely recognized. In 1934 he received the Prix Francqui (Francqui Award) directly from the hands of Einstein, who had personally nominated Lemaître for this prestigious award. In 1936 he became a member of the Pontifical Academy of Sciences and then presided over this special papal scientific assembly as its president from 1960 until his death. In 1941, the Royal Belgian Academy of Sciences and Fine Art voted him membership.

The Belgian government bestowed its highest award for scientific achievement upon him in 1950, and the British Royal Astronomical Society awarded him the society's first Eddington Medal in 1953. Lemaître died at Louvain, Belgium, on June 20, 1966. However, he lived long enough to witness the detection by ARNO ALLEN PENZIAS and ROBERT WOODROW Wilson of the microwave remnants of the big bang—an event he had boldly hypothesized almost four decades earlier.

⊠ **Leverrier, Urbain-Jean-Joseph**
(1811–1877)
French
Astronomer, Mathematician, Orbital Mechanics Expert

Urbain Leverrier was a skilled celestial mechanics practitioner who mathematically predicted in 1846 (independent of JOHN COUCH ADAMS) the possible location of an eighth, as yet undetected, planet in the outer regions of the solar system. Leverrier sent his calculations to the Berlin Observatory and Neptune was quickly discovered by telescopic observation.

Leverrier was born on March 11, 1811, in Saint-Lô, France. His father, a local government official, made a great financial sacrifice so that his son could receive a good education at the prestigious École Polytechnique. After graduation, Leverrier began his professional life as a chemist. He investigated the nature of certain chemical compounds under the supervision of the French chemist Joseph Gay-Lussac (1778–1850), who was a professor of chemistry at the École Polytechnique. In 1836 Leverrier accepted an appointment as a lecturer in astronomy at the same institution. Consequently, with neither previous personal interest in nor extensive formal training for astronomy, Leverrier suddenly found himself embarking on an astronomically oriented academic career. The process

In 1846 the French celestial mechanics practitioner Urbain J. J. Leverrier mathematically predicted the location of the planet Neptune so well that the German astronomer Johann Galle was able to quickly discover Neptune by telescopic observation on the evening of September 23 of that year. Leverrier's computational effort, accomplished independently of John Couch Adam's, is considered to be one of the great triumphs of mathematical astronomy in the 19th century. *(Courtesy of NASA)*

all came about rather quickly, when the opportunity for academic promotion at the École Polytechnique presented itself and Leverrier seized the moment.

As he settled into this new academic position, Leverrier began investigating lingering issues in celestial mechanics. He decided to focus his attention on continuing the work of PIERRE-SIMON MARQUIS DE LAPLACE in mathematical astronomy and demonstrate with even more precision the inherent stability of the solar system. Following the suggestion given him in 1845 by DOMINIQUE-FRANÇOIS-JEAN ARAGO, he began investigating the subtle perturbations in the orbital motion of Uranus, the outer planet discovered by SIR WILLIAM HERSCHEL some 50 years prior. In 1846 Leverrier, following the suggestions of other astronomers (among them FRIEDRICH WILHELM BESSEL), hypothesized that an undiscovered planet lay beyond the orbit of Neptune. He then examined the perturbations in the orbit of Uranus and used the law of gravitation to calculate the approximate size and location of this outer planet. Unknown to Leverrier, the British astronomer Adams had made similar calculations. However, during this period, Sir George Biddell Airy (1801–92), Britain's Astronomer Royal, basically ignored Couch's calculations and correspondence about the possible location of a new outer planet.

In mid-September 1846 Leverrier sent a detailed letter to the German astronomer JOHANN GOTTFRIED GALLE at the Berlin Observatory telling Galle where to look in the night sky for a planet beyond Uranus. On the evening of September 23, 1846, Galle found Neptune after about an hour of searching. The French-German team of mathematical and observational astronomers had beaten the British astronomical establishment to one of the 19th century's greatest discoveries. In France Leverrier experienced the celebrity that often accompanies a widely recognized scientific accomplishment. He received appointment to a specially created chair of astronomy at the University of Paris, and the French government made him an officer in the Legion of Honor. Leverrier received international credit for mathematical discovery of Neptune, which was immediately followed by controversy. The Royal Society of London awarded him its Copley Medal in 1846 and elected him a Fellow in 1847. As a point of scientific justice, Adams received the Copley Medal in 1848 for his discovery of Neptune.

Bristling with success, fame, and good fortune, Leverrier pressed his luck in 1846 by applying mathematical astronomy to explain minor perturbations in the solar system. This time, he looked inward at the puzzling behavior in the orbital motion of Mercury, the innermost known planet. Unfortunately, Leverrier failed quite miserably with his hypothesis that the observed perturbations in Mercury were due to the presence of an inner undiscovered planet he called Vulcan, or possibly even a belt of asteroids orbiting closer to the Sun. Many 19th-century observational astronomers, including SAMUEL HEINRICH SCHWABE, searched in vain for Vulcan. ALBERT EINSTEIN's general relativity theory in the early part of the 20th century explained the perturbations of Mercury without resorting to Leverrier's hypothetical planet.

In 1854 Leverrier succeeded Arago as the director of the Paris Observatory, founded in 1667 by King Louis XIV who appointed GIOVANNI DOMENICO CASSINI as its first director. This observatory served as the national observatory of France and was the first such national observatory established in the era of telescopic astronomy. Leverrier focused his efforts from 1847 until his death on producing more accurate planetary data tables and on creating a set of standard references for use by astronomers.

Unfortunately, Leverrier was an extremely unpopular director, who ran the Paris Observatory like a tyrant. By 1870 his coworkers had had enough and he was replaced as director. However, he was reinstated to the directorship in 1873 when his successor, Charles-Eugene Delaunay (1816–72) died suddenly. This time, however, he served under the very watchful authority of a directing council.

Leverrier died in Paris on September 23, 1877. His role in celestial mechanics and the predictive discovery of Neptune represents one of the greatest triumphs of mathematical astronomy.

⊠ **Lippershey, Hans (Hans Lippersheim, Jan Lippersheim)**
(ca. 1570–ca. 1619)
Dutch
Optician

Hans Lippershey is the Dutch optician generally credited with the invention of the telescope, because he was the first among his contemporaries to attempt to patent a simple, two-lens viewing instrument in 1608. However, once news of this new type of optical instrument circulated through western Europe, creative persons such as GALILEO GALILEI and JOHANNES KEPLER quickly recognized the telescope's important role in observational astronomy and spearheaded a litany of design improvements that continues up to modern times.

The Dutch lens maker (optician) Lippershey was born in approximately the year 1570 in the town of Wesel, Germany. He moved to Middelburg, the capital city of the province of Zeeland, in what is now the modern Kingdom of the Netherlands. Practicing his craft as a lens and spectacle-maker, he married there in 1594, and became a citizen. At the time, Middelburg was a prosperous city experiencing an influx of Dutch Protestant refugees who were fleeing from the Spanish invasion of the southern provinces of the Netherlands (now modern Belgium). In the late 16th century this region, like much of western Europe, was in a state of political and religious turmoil. The Dutch people living in northern provinces of the region known as the Seventeen Provinces of the Netherlands had embraced Protestantism and were striving for more political freedom from the Catholic king of Spain. Through the Union of Utrecht (1579), William the Silent, prince of Orange, led a confederation of seven northern provinces called estates (Holland, Zeeland, Utrecht, Gelderland, Overrijssel, Groningen, and Friesland). The estates retained individual sovereignty, but were represented jointly in the States-General, a governing political body that had control of foreign affair and defense. By 1581 the estates repudiated allegiance to Spain and formed the Republic of the Seven United Netherlands—the nucleus of the Dutch republic that grew into a 17th-century naval and economic power.

Several historic anecdotes describe how the telescope came into being. Hans Lippershey generally appears at the center of most of these stories by virtue of his being the first person to actually apply for a patent for a *kijker*, or "looker." One of the most popular stories suggests that an apprentice in Lippershey's lens-making shop was tinkering with various arrangements of lenses and noticed that a special combination two short-focus convex lenses placed at the opposite ends of a long tube would make distant objects appear nearer. Lippershey immediately realized the potential military significance of this device and applied for a patent from the States-General through the local government of Zeeland.

The oldest surviving record concerning the invention of the telescope is a letter, dated September 25, 1608, from the government of the Estate of Zeeland to the States-General of the Seven United Netherlands, requesting that the States-General assist the bearer of the letter (Lippershey) "who claims to have a certain device by means of which all things at a very great distance can be seen as if they were nearby." Members of the States-General discussed Lippershey's patent application on October 2, 1608, and then denied the application, stating that the basic concept for such a device could not be kept secret.

Members of the assembly were quite correct on this point. Upon learning about the Dutch telescope from his son, Galileo immediately fashioned his own devices in 1609, looked into the night sky, and started the field of telescopic astronomy. However, the States-General of the Seven Netherlands did reward Lippershey for his efforts. The governing body awarded him a generous fee (900 florins) to modify his telescope into a binocular device, which they considered more practical for military applications.

Other Dutch lens makers also claimed credit for inventing the telescope. As the States-General was reviewing Lippershey's patent request, another patent request arrived from Jacob Metius (1580–1628), an instrument maker from Alkmaar, a city in the northern part of the Netherlands. Metius sought a patent for his "perspicilla" in October 1608, barely two weeks after Lippershey initiated his request. Several years later, Zacharias Janssen (1580–ca. 1638), another Dutch optician from Middelburg, claimed credit for inventing the telescope. At this point, on the basis of surviving historic records, it is impossible to establish exactly who really "invented" the telescope or precisely when the creative moment took place. Lippershey holds the favored position because his request for patent is the oldest surviving record of the period. Sometime in 1619, he died in Middelburg.

⊗ **Lockyer, Sir Joseph Norman**
(1836–1920)
British
Physicist, Astronomer, Spectroscopist

Helium is the second most abundant element in the universe, yet its existence was totally unknown prior to the late 19th century. In 1868 the British physicist Sir Joseph Norman Lockyer collaborated with the French astronomer Pierre-Jules-César Janssen (1824–1907) and discovered the element helium through spectroscopic studies of solar prominences. However, it was not until 1895 that the Scottish chemist Sir William Ramsay (1852–1916) finally detected the gas in Earth's atmosphere. Lockyer also founded the prestigious scientific periodical *Nature* and pioneered the field of archaeological astronomy.

Lockyer was born on May 17, 1836, in Rugby, England. Following a traditional postsecondary education outside of Great Britain on the European Continent, Lockyer began a career as a civil servant in the War Office of the British government. While serving in this capacity from approximately 1857–69, he also became an avid amateur astronomer, enthusiastically pursuing this scientific hobby in every spare moment. Following observation of a solar eclipse in 1858, he constructed a private observatory at his home. In 1861 while working at the War Office in London, he married and settled in Wimbledon. Lockyer made several important astronomical discoveries in the 1860s and turned to astronomy and science on a full-time basis at the end of that decade.

While still an amateur astronomer in the 1860s, Lockyer decided to apply the emerging field of spectroscopy to study the Sun. This decision enabled Lockyer to join two other early astronomical spectroscopists, Janssen and SIR WILLIAM HUGGINS, in extending the pioneering work of ROBERT WILHELM BUNSEN and GUSTAV ROBERT KIRCHHOFF. Lockyer's efforts directly contributed to development of modern astrophysics. In 1866 he started spectroscopic observation of sunspots. He observed Doppler shifts in their spectral lines and suggested that strong convective currents of gas existed in the outer regions of the Sun—a region he named the chromosphere.

By 1868 he devised a clever method of acquiring the spectra of the solar prominences without waiting for an eclipse of the Sun to take place. A prominence is a cooler, cloudlike feature found in the Sun's corona. Prominences often appear around the Sun's limb during total solar eclipses. Lockyer discovered that he could observe the spectra of prominences without eclipse conditions if he passed light from the very edge of the Sun through a prism.

Almost simultaneously, but independent of Lockyer, the French astronomer Janssen also developed this method of obtaining the spectra of solar prominences. Janssen and Lockyer reported their identical discoveries at the same meeting of the French Academy of Sciences. Officials of the academy wisely awarded, 10 years later, both men a special "joint medallion" to honor their simultaneous and independent contribution to solar physics.

In 1868 Janssen used his newly developed spectroscopic technique to investigate prominences during a solar eclipse expedition to Guntur, India. Janssen reported an unknown bright orange line in the spectral data he collected. Lockyer reviewed Janssen's report and then compared the position of the reported unknown orange line to the spectral lines of all the known chemical elements. When he could not find any correlation, Lockyer concluded that this line was the spectral signature of a yet undiscovered element. He named the element helium, after Helios, the sun god in Greek mythology. Because spectroscopy was still in its infancy, most scientists waited for almost three decades before accepting Lockyer's identification of a new element in the Sun. Wide scientific acceptance of Lockyer's discovery of helium took place only after Ramsay detected the presence of this elusive gas within Earth's atmosphere in 1895.

In 1869 Lockyer won election as a fellow to the Royal Society. He decided to abandon his civil service career so he could pursue astronomy and science full time. That year he founded the internationally acclaimed scientific journal *Nature*; he would serve as its editor for almost 50 years. Between 1870 and 1905, Lockyer remained an active solar astronomer and personally participated in eight solar eclipse expeditions.

In 1873 Lockyer turned his attention to stellar spectroscopy and introduced his theory of atomic and molecular dissociation in an attempt to explain puzzling green lines in the spectra of certain nebulas. He disagreed with fellow British spectroscopist Huggins who suggested that the green lines came from an unknown element he called "nebulium." Instead, Lockyer hypothesized that they might actually be from terrestrial

compounds that had dissociated into unusual combinations of simpler substances, making their spectral signatures unrecognizable. An American astrophysicist, Ira Bowen (1898–1973) solved the mystery of "nebulium" in the 1920s, when he demonstrated that the green emission lines from certain nebulas were actually caused by forbidden electron transitions taking place in oxygen and nitrogen under various excited states of ionization. Lockyer's theory of dissociation was definitely on the right track.

Lockyer was an inspiring lecturer. In 1881 he developed an interesting new course in astrophysics at the Royal College of Science (today part of the Imperial College). By 1885 he received an academic promotion to professor of astrophysics, making him the world's first professor in this discipline. The Royal College also constructed the Solar Physics Observatory at Kensington to support his research and teaching activities. Lockyer served as the observatory's director until approximately 1913, when the facility moved to a new location at Cambridge University.

During travel to Greece and Egypt in the early 1890s, Lockyer noticed how many ancient temples had their foundations aligned along an east-west axis—a consistent alignment that suggested to him some astronomical significance with respect to the rising and setting Sun. To pursue this interesting hypothesis, Lockyer visited Karnack, one of the great temples of ancient Egypt. He discussed the hypothesis in his 1894 book, *The Dawn of Astronomy*. This book is often regarded as the beginning of archaeological astronomy—the scientific investigation of the astronomical significance of ancient structures and sites.

As part of this effort, Lockyer studied Stonehenge, an ancient site located in south England. However, he could not accurately determine the site's construction date. As a result, he could not confidently project the solar calendar back to a sufficiently precise moment in history that would reveal how the curious circular ring of large vertical stones topped by capstones might be connected to some astronomical practice of the ancient Britons. Lockyer's visionary work clearly anticipated the results of modern studies of Stonehenge—results that suggest the site could have served as an ancient astronomical calendar around 2000 B.C.E.

Lockyer was a prolific writer. His best-known published works include *Studies in Spectrum Analysis* (1872), *The Chemistry of the Sun* (1887), and *The Sun's Place in Nature* (1897). In 1874 the Royal Society of London recognized his important achievements in solar physics and astronomical spectroscopy and awarded him its Rumford Medal.

In 1897 Lockyer joined the Order of Knight Commander of the Bath (KCB), a royal honor bestowed upon him by Britain's Queen Victoria for his discovery of helium and lifetime contributions to science. After the Solar Physics Observatory at the Royal College was moved from Kensington to Cambridge University, Sir Lockyer remained active in astronomy by constructing his own private observatory in 1912 in Sidmouth, Devonshire. This facility, called the Norman Lockyer Observatory, currently operates under the supervision of an organization of amateur astronomers.

Lockyer, the British civil servant who entered astronomy in the 1850s as a hobby and became one of the great pioneers in astrophysics and stellar spectroscopy, died on August 16, 1920, at Salcombe Regis, Devonshire.

⊠ Lowell, Percival
(1855–1916)
American
Astronomer

Late in the 19th century Percival Lowell established a private astronomical observatory, the Lowell Observatory, near Flagstaff, Arizona, primarily to support his personal interest in Mars and his aggressive search for signs of an intelligent civilization there. Driving Lowell was his misinterpretation of GIOVANNI VIRGINIO SCHIAPARELLI's use of the word *canali* in an 1877 technical report in which the Italian astronomer discussed his telescopic observations of the Martian surface. Lowell took this report as early observational evidence of large, water-bearing canals built by intelligent beings. Lowell wrote books, such as *Mars and Its Canals* (1906) and *Mars As the Abode of Life* (1908) to communicate his Martian civilization theory to the public. While his nonscientific (but popular) interpretation of observed surface features on Mars proved quite inaccurate, his astronomical instincts were correct for another part of the solar system. Based on perturbations in the orbit of Neptune, Lowell predicted in 1905 the existence of a planet-sized, trans-Neptunian object. In 1930 CLYDE WILLIAM TOMBAUGH, working at the Lowell Observatory, discovered Lowell's Planet X and called the tiny planet Pluto.

Lowell was born on March 13, 1855, in Boston, Massachusetts, into an independently wealthy family. His brother Abbott Lowell became president of Harvard University and his sister Amy became an accomplished poet. Following graduation with honors from Harvard University in 1876, Lowell devoted his time to business and to traveling throughout the Far East. Based on his experiences between 1883 and 1895, Lowell published several books about the Far East, including: *Chosön* (1886), *The Soul of the Far East* (1888), *Noto* (1891) and *Occult Japan* (1895).

He was not especially attracted to astronomy until later in life, when he discovered an English translation of Schiaparelli's 1877 Mars observation report that presented the Italian word *canali*. As originally intended by Schiaparelli, *canali* simply meant "channels." In the early 1890s Lowell unfortunately became erroneously inspired by the thought of artificially constructed canals on Mars, the presence of which he then extended to imply the existence of an advanced alien civilization. From this point on, Lowell decided to become an astronomer and dedicate himself to a detailed study of Mars.

Unlike other observational astronomers, however, Lowell was independently wealthy and already had a general idea of what he was searching for—evidence of an advanced civilization of the Red Planet. To support this quest, he spared no expense and sought the assistance of professional astronomers. William Henry Pickering (1858–1938) from the Harvard College Observatory, and his assistant ANDREW ELLICOTT DOUGLASS helped Lowell find an excellent "seeing" site upon which to build a private observatory for the study of Mars, which Lowell constructed near Flagstaff, Arizona. Called the Lowell Observatory, it was opened in 1894 and housed a top quality 24-inch refractor telescope that allowed Lowell to perform some excellent planetary astronomy. However, his "observations" tended to anticipate the things he reported, like oases and seasonal changes in vegetation. Other astronomers, including Douglass, would label these blurred features simply indistinguishable natural markings. As Lowell more aggressively embellished his Mars observations, Douglass began to question Lowell's interpretation of these data. Perturbed by Douglass's scientific challenge, Lowell simply fired him in 1901 and then hired Vesto M. Slipher to fill the vacancy.

In 1902 the Massachusetts Institute of Technology (MIT) gave Lowell an appointment as a nonresident astronomer. He was definitely an accomplished observer, but often could not resist the temptation to greatly stretch his interpretation of generally fuzzy and optically distorted surface features on Mars into observational evidence of artifacts from an advanced civilization. With such books as *Mars and Its Canals* Lowell became popular with the general public, which drew excitement from his speculative (but scientifically unproven) theory of an

intelligent civilization on Mars struggling to distribute water from the planet's polar regions with a series of elaborate giant canals. While most planetary astronomers shied away from such unfounded speculation, science fiction writers flocked to Lowell's hypothesis—a premise that survived in various forms until the dawn of the Space Age. On July 14, 1965, NASA's *Mariner 4* spacecraft flew past the Red Planet and returned images of its surface that shattered all previous speculations and romantic myths about a series of large canals built by a race of ancient Martians.

Since the *Mariner 4* encounter with Mars, an armada of other robot spacecraft have also studied Mars in great detail. No cities, canals, nor intelligent creatures have been found on the Red Planet. What has been discovered, however, is an interesting "halfway" world. Part of the Martian surface is ancient, like the surfaces of the Moon and Mercury, while part is more evolved and Earthlike. In the 21st century, robot spacecraft and eventually human explorers will continue Lowell's quest for Martians—but this time they will hunt for tiny microorganisms *possibly* living in sheltered biological niches or else *possible* fossilized evidence of ancient Martian life-forms from the time when the Red Planet was a milder, kinder and wetter world.

While Lowell's quest for signs of intelligent life on Mars may have lacked scientific rigor by a considerable margin, his astronomical instincts about "Planet X"—his name for a suspected icy world lurking beyond the orbit of Neptune—proved technically correct. In 1905 Lowell began to make detailed studies of the subtle perturbations in Neptune's orbit and predicted the existence of a planet-sized trans-Neptunian object. He then began an almost decade-long telescopic search, but he failed to find this elusive object. In 1914, near the end of his life, he published the negative results of his search for Planet X and bequeathed the task to some future astronomer. Lowell died in Flagstaff, Arizona, on November 12, 1916.

In 1930, Tombaugh, a farm boy turned amateur astronomer, fulfilled Lowell's quest. Hired by Slipher to work as an observer at the Lowell Observatory and search for Planet X, Tombaugh made use of the blinking comparator technique to find the planet Pluto on February 18, 1930. The Lowell Observatory still functions today as a major private astronomical observatory. An often forgotten milestone in astronomy took place at the Lowell Observatory in 1912, when Slipher made early measurements of the Doppler shift of distant nebulas. He found many to be receding from Earth at a high rate of speed. Slipher's work provided EDWIN POWELL HUBBLE the basic direction he needed to discover the expansion of galaxies.

M

⊠ **Messier, Charles**
(1730–1817)
French
Astronomer

Charles Messier, the first French astronomer to spot comet Halley during its anticipated return in 1758, was an avid "comet hunter." He personally discovered at least 13 new comets and assisted in the codiscovery of at least six others. In 1758 he began compiling a list of nebulas and star clusters that eventually became the famous noncomet *Messier Catalogue*. He assembled this list of "unmoving" fuzzy nebulas and star clusters primarily as a tool for astronomers engaged in comet searches.

Messier was born in Badonviller, Lorraine, France, on June 26, 1730. His father died when Messier was a young boy and the family was thrust into poverty. The appearance of a multi-tailed comet and the solar eclipse of 1748 stimulated his childhood interest in astronomy. Messier went to Paris in 1751 and became a clerk for Joseph Nicolas Delisle (1688–1768), who was then the astronomer of the French navy. Messier's ability to carefully record data attracted Delisle's attention, and soon the young clerk was cataloging observations made at Hôtel de Cluny, including the transit of Mercury in May 1753. Delisle, like many other astronomers, anticipated the return of comet Halley sometime in 1758 and made his own calculations in an attempt to be the first to detect its latest visit.

Delisle assigned the task of searching for Halley's comet to his young clerk. Messier searched diligently throughout 1758 but mostly in the wrong location, because Delisle's calculations were in error. However, in August of that year, Messier discovered and tracked a different comet for several months. He detected a distant fuzzy object that he thought was the anticipated comet Halley. However, to Messier's frustration, the object did not move, so he realized it had to be a nebula. On September 12, 1758, he carefully noted its position in his observation log. This particular "fuzzy patch" was actually the Crab Nebula—a fuzzy celestial object previously discovered by another astronomer.

Cometography, that is, "comet hunting," was one of the great triumphs of 18th-century astronomy. Messier enjoyed the thrill and excitement that accompanied a successful quest for a new "hairy star"—the early Greek name (κομετεζ) for a comet. The young clerk-astronomer also realized that fuzzy nebulas and star clusters caused time-consuming false alarms. He had the solution—prepare a list of noncomet objects, such as nebulas and star clusters. In the 18th century, astronomers used the word *nebula* to describe any fuzzy, blurry, luminous celestial object that could not be sufficiently resolved by available telescopes. The fuzzy object he mistook for a comet in late August 1758 eventually became object number one (M1) in the *Messier Catalogue*—namely, the great Crab Nebula in the constellation Taurus.

Despite false alarms caused by nebulas and his discovery of another comet, Messier kept vigorously searching for comet Halley throughout 1758. Then, on the evening of December 25, Christmas Day, an amateur German astronomer named Georg Palitzch became the first observer to catch a glimpse of the greatly anticipated comet, as it returned to the inner portions of the solar system. About a month later, on the evening of January 21, 1759, Messier succeeded in viewing this comet and became the first French astronomer to have successfully done so during the comet's 1758–59 journey through perihelion. His work earned him recognition at the highest levels. The French king Louis XV fondly referred to Messier as "the comet ferret."

After the passage of comet Halley, Messier continued to search for new comets and to assemble his noncomet list of nebulas and star clusters. By 1764 this list contained 40 objects, 39 of which he personally verified through his own observations. Within the resolution limits of the telescopes of his day, Messier's objects included nebulas, star clusters, and distant galaxies.

By 1771 he had completed the first version of his famous noncomet list and published it as the *Catalogue of Nebulae and of Star Clusters* in 1774. This initial catalog contained 45 Messier objects. He completed the final version of his noncomet list of celestial objects in 1781. It contained 103 objects (seven additional were added later) and was published in 1784. Messier designated each of the entries in the list by a separate number, prefixed by the letter M. For example, Messier object M1 represented the Crab Nebula, and object M31 represented the Andromeda Galaxy. His designations persist to the present day, although the Messier designa-

tions have generally been superseded by the designations presented in the *New General Catalogue of Nebulae and Clusters of Stars* (NGC) published in 1888 by the Danish astronomer Johan Dreyer (1852–1926).

In 1764 Messier was elected as a foreign member of the Royal Society of London. In 1770 he became a member of the Paris Academy of Sciences, and the astronomer of the French navy, officially assuming the position formerly held by Delisle, who had retired six years earlier. However, Messier lost this position during the French Revolution and his observatory at the Hôtel de Cluny fell into general disrepair.

In 1806 Napoleon Bonaparte (1769–1821) awarded Messier the cross of the Legion of Honor—possibly because Messier had openly and rather unscientifically suggested in a technical paper that the appearance of the great comet of 1769 correlated with Napoleon's birth. This politically inspired act of astrology is the last known attempt by an otherwise knowledgeable astronomer to tie the appearance of a comet to a significant historic event on Earth. Messier died in Paris on April 11, 1817. Most of the comets he so eagerly sought have been forgotten, but his famous noncomet list of celestial objects serves as a permanent tribute to one of the 18th century's most successful "comet hunters."

⊠ **Michelson, Albert Abraham**
(1852–1931)
German/American
Physicist

How fast does light travel? Physicists have pondered that challenging question since GALILEO GALILEI's time. In the late 1880s Albert Michelson provided the first very accurate answer—a velocity slightly less than 300,000 kilometers per second. He received the 1907 Nobel Prize in physics for his innovative optical measurements and precision measurements.

The Nobel laureate Albert Michelson had a lifelong passion for precision optical instruments (like the type he is inspecting in the photograph) and extremely accurate measurements of the velocity of light. He was the first American scientist to receive a Nobel Prize in physics and experienced this great honor in 1907 for his innovative optical experiments and precision measurements. *(AIP Emilio Segrè Visual Archives)*

Michelson was born in Strelno, Prussia (now Strzelno, Poland) on December 19, 1852. When he was two years old, his family immigrated to the United States, settling first in Virginia City, Nevada, and then in San Francisco, California, where he received his early education. In 1869 Michelson received an appointment from President Grant to the U.S. Naval Academy in Annapolis, Maryland. More skilled as a scientist than as a seaman, he graduated from Annapolis in 1873 and received his commission as an ensign in the U.S. Navy. Following two years of sea duty, Michelson became an instructor in physics and chemistry at the academy, then in 1879 an assignment to the navy's Nautical Almanac Office in Washington, D.C. There he met and worked with SIMON NEWCOMB who, among many other things, was attempting to measure the velocity of light.

After about a year at the Nautical Almanac Office, Michelson took a leave of absence to study advanced optics in Europe. In Germany he visited the University of Berlin and the University of Heidelberg. He also studied at the Collège de France and the École Polytechnique in Paris. In 1881 he resigned from the U.S. Navy and then returned to the United States to accept an appointment as a professor of physics at the Case School in Applied Science in Cleveland, Ohio. Michelson had conducted some preliminary experiments in measuring the speed of light in 1881 at the University of Berlin. Once settled in at Case, he refined his experimental techniques and obtained a value of 299,853 kilometers per second. This value remained a standard within physics and astronomy for more than two decades and changed only when Michelson himself improved the value in the 1920s.

In the early 1880s, with financial support from the Scottish-American inventor Alexander Graham Bell (1847–1922), Michelson constructed a precision optical interferometer—an instrument that splits a beam of light into two paths and then reunites the two separate beams. Should either beam experience travel across different distances or at different velocities (due to passage through different media), the reunited beam would appear out of phase and produce a distinctive arrangement of dark and light bands called an interference pattern. Using an interferometer in 1887, Michelson collaborated with the American physicist William Morley (1838–1923) to perform one of the most important "failed" experiments ever undertaken in science.

Now generally referred to as the Michelson-Morley experiment, this experiment used an interferometer to test whether light traveling in the same direction as Earth through space moves more slowly than light traveling at right angles to Earth's motion. Their failure to observe velocity differences in the perpendicular beams of light dispelled the prevailing concept that light traveled through the universe using some sort of invisible "cosmic ether" as the medium. Michelson had reasoned that when rejoined in the interferometer the two beams of light should be out of phase due to Earth's motion through the postulated ether. But their very careful measurements failed to detect any interfering influence of the hypothetical, all-pervading ether. Of course, the real reason their classic experiment failed is now very obvious—there isn't any "cosmic ether." Michelson's work was exact, precise, and provided a very correct, albeit null, result. The absence of ether gave ALBERT EINSTEIN important empirical evidence upon which he constructed his special relativity theory in 1905. The Michelson-Morley experiment provided the first direct evidence that light travels at a constant speed in space—the very premise upon which all of special relativity is built.

From 1889 to 1892 Michelson served as a professor of physics at Clark University in Worcester, Massachusetts. He then left Clark University to accept a position in physics in the newly created University of Chicago. He remained affiliated with this university until his retirement in 1929. In 1907 Michelson became the first American scientist to receive the Nobel Prize in physics. He received this distinguished international award because of his excellence in the development and application of precision optical instruments.

During World War I, he rejoined the U.S. Navy. Following this wartime service, Michelson returned to the University of Chicago and turned his attention to astronomy. Using a sophisticated improvement of his optical interferometer, he measured the diameter of the star Betelgeuse in 1920. His achievement represented the first accurate determination of the size of a star, excluding the Sun. In 1923 he resumed his lifelong quest to improve the measurement of the velocity of light. This time he used a 35-kilometer-long

pathway between two mountain peaks in California that he had carefully surveyed to an accuracy of less than 2.5 centimeters. With a specially designed revolving eight-sided mirror, he measured the velocity of light as 299,798 kilometers per second. After retiring from the University of Chicago in 1929, Michelson joined the staff of the Mount Wilson Observatory in California. In the early 1930s he bounced light beams back and forth in an evacuated tube to produce an extended 16-kilometer pathway to measure optical velocity in a vacuum. The final result, announced in 1933, after his death, was a velocity of 299,794 kilometers per second. To appreciate the precision of Michelson's work, the currently accepted value of the velocity of light in a vacuum is 299,792.5 kilometers per second.

Michelson died in Pasadena, California, on May 9, 1931. His most notable scientific works include *Velocity of Light* (1902), *Light Waves and Their Uses* (1899–1903) and *Studies in Optics* (1927). In addition to being the first American scientist to win a Nobel Prize, Michelson received numerous other awards and international recognition for his contributions to physics. He became a member (1888) and served as president of (1923–27) the American National Academy of Sciences. The Royal Society of London made him a foreign Fellow in 1902 and presented him its Copley Medal in 1907. He also received the Draper Medal from the American National Academy of Sciences in 1912, the Franklin Medal from the Franklin Institute in Philadelphia in 1923, and the Gold Medal of the Royal Astronomical Society in 1923. His greatest scientific accomplishment was the accurate measurement of the velocity of light—a physical "yardstick" of the universe and an important constant throughout all of modern physics.

N

Newcomb, Simon
(1835–1909)
Canadian/American
Astronomer, Mathematician

Simon Newcomb was one of the 19th century's leading mathematical astronomers. While working for the Nautical Almanac Office of the U.S. Naval Observatory, he prepared extremely accurate tables that predicted the position of solar system bodies. Before the arrival of navigation satellites, sailors found their way at sea using an accurate knowledge of the positions of natural celestial objects, such as the Sun, the Moon, and other planets. The more accurate the available nautical tables, the more precisely sea captains could chart their voyages. In 1860 Newcomb presented a trend-busting astronomical paper in which he correctly speculated that the main-belt asteroids did not originate from the disintegration of a single ancient planet, as was then commonly assumed.

Newcomb was born on March 12, 1835, in Wallace, Nova Scotia, Canada. As the son of an itinerant teacher, he received little formal schooling, and what education he did experience in childhood took place privately. Newcomb moved to the United States in 1853 and, much like his father, worked as a teacher in Maryland between 1854 and 1856. While proximate to the libraries in Washington, D.C., Newcomb began to study mathematics extensively on his own. By 1857 Newcomb's mathematical aptitude caught the attention of the great American physicist Joseph Henry (1797–1878), who helped him find employment as a computer (that is, a person who does precise astronomical calculations) at the U.S. Nautical Office, then located at Harvard University in Cambridge, Massachusetts. In 1858 Newcomb graduated from Harvard University with a bachelor of science degree; he continued there for three years afterward as a graduate student.

Newcomb received an appointment as a professor of mathematics for the U.S. Navy in 1861 and was assigned to duty at the U.S. Naval Observatory in Washington, D.C. A year earlier he had made his initial mark on mathematical astronomy when he presented a paper demonstrating that the orbits of the asteroids (minor planets) did not diverge from a single point. His careful calculations dismantled the then-popular hypothesis in solar system astronomy that the main belt of minor planets between Mars and Jupiter was the remnants of a single ancient planet that tore itself apart. Unfortunately, Newcomb's interesting paper was quickly overshadowed by the start of the Civil War in April 1861 and the pressing needs of the U.S. Navy for more accurate nautical charts.

This photograph shows one of the 19th century's leading mathematical astronomers, Simon Newcomb, at his desk. As the superintendent of the American Nautical Almanac Office until his retirement in 1897, Newcomb prepared extremely accurate tables that predicted the position of solar system bodies. His efforts led to a global standardization of astronomical reference data and supported great improvements in navigation at sea. *(AIP Emilio Segrè Visual Archives, W. F. Meggers Gallery of Nobel Laureates)*

At the U.S. Naval Observatory, Newcomb focused all his mathematical talents and energies on preparing the most accurate nautical charts ever made (before the era of electronic computers). One of his first major responsibilities was to negotiate the contract for the Naval Observatory's new 26-inch telescope, which had recently been authorized by Congress. Newcomb also planned the tower and dome, and supervised construction. Later, he assisted in the development of the Lick Observatory in California.

Although primarily a mathematical astronomer rather than a skilled observer, Newcomb participated in a number of astronomical expeditions while serving the U.S. Naval Observatory. He traveled to Gibraltar in 1870–71 to observe an eclipse of the Sun and to the Cape of Good Hope, at the southern tip of Africa, in 1882 to observe a transit of Venus. In 1877 he received promotion to senior professor of mathematics in the U.S. Navy (a position with the equivalent naval rank of captain) and was in charge of the office of an important publication, the *American Ephemeris and Nautical Almanac*. Newcomb had a number of military and civilian assistants helping him produce the *Nautical Almanac*, including ASAPH HALL, the American astronomer who discovered the two tiny moons of Mars in 1877 while a staff member at the U.S. Naval Observatory.

By the 1890s, as American naval power reached around the globe, Newcomb supported its needs for improved navigation by producing the highest-quality nautical almanac yet developed. To accomplish this, he supervised a large staff of astronomical computers (that is, human beings) and enjoyed control of the largest observatory budget in the world. To support the production of a high-quality *Nautical Almanac*, Newcomb pursued, produced, and promoted a new and more accurate system of astronomical constants that became the world standard by 1896. He retired from his position as superintendent of the U.S. Nautical Almanac Office in 1897 and received promotion to the rank of rear admiral in 1905.

Newcomb was also a professor of mathematics and astronomy at Johns Hopkins University between 1884 and 1893. He served as editor of the *American Journal of Mathematics* for many years and as president of the American Mathematical Society from 1897–98. He was a founding member and the first president (1899–1905) of the American Astronomical Society.

He wrote numerous technical papers, including "An Investigation of the Orbit of Uranus, with General Table of Its Motion" (1874) and "Measurement of the Velocity of Light" (1884)—a paper that brought him in contact with a brilliant young scientist named ALBERT ABRAHAM MICHELSON. A gifted writer, Newcomb published books on various subjects for audiences of varying levels, including *Popular Astronomy* (1877), *Principles of Political Economy* (1886), and *Calculus* (1887).

As a gifted mathematical astronomer, Newcomb received many honors and won election to many distinguished societies. In 1874 he received the Gold Medal of the Royal Astronomical Society in Great Britain. He also became a Fellow of the Royal Society of London in 1877 and received that society's Copley Medal in 1890. The Astronomical Society of the Pacific made him the first Bruce Gold Medallist in 1898.

He died in Washington, D.C., on July 11, 1909, and was buried with full military honors at Arlington National Cemetery. To commemorate his great contributions to the United States in the field of mathematical astronomy and nautical navigation, the navy named a surveying ship the *USS Simon Newcomb*.

Newton, Sir Isaac
(1642–1727)
British
Mathematician, Physicist

Only one scientist has been simultaneously thought of in such diverse ways as "the greatest genius that ever existed," and "as a man, he was a failure; as a monster he was superb." This genius monster was the famous mathematician Sir

When the brilliant English physicist and mathematician Isaac Newton published his great work *The Principia* in 1687, he single-handedly transformed the practice of physical science and completed the scientific revolution started by Copernicus, Galileo, and Kepler. He is considered one the greatest scientific minds that ever lived. *(Original engraving by unknown artist, courtesy AIP Emilio Segrè Visual Archives, Physics Today Collection)*

Isaac Newton, who made perhaps the biggest contribution of all time to the field of astronomy with the scientific notions he devised as a 22-year-old on his family's farm.

December 25, 1642, Christmas Day, was a day to be remembered in astronomy. In Italy, the famous mathematician and scientist GALILEO GALILEI passed away, blind from too many observations of the Sun, and under house arrest by the Roman Catholic Church—courtesy of the Inquisition for his beliefs in a heliocentric universe, which he was forced to recant 10 years earlier to avoid being tortured to death. As one great life ended, another began in Woolsthorpe, England, where on this day Isaac Newton was born—the first child of the recently widowed Hannah Ayscough. Newton was named for his father, an illiterate but wealthy farmer who had died three months earlier.

From age two, the young Newton was passed around among his relatives after his mother remarried, living mostly with his grandparents. Newton hated his stepfather, Barnabus Smith, once even threatening to burn his house down with his mother and stepfather in it. After Smith died when Newton was 10, the boy was reunited with his mother and her three children by Smith, and then promptly sent away to school. His first attempt at education was brief—he was pulled out of school after just a few months because he lacked any visible ability to learn. But at age 13, after failing miserably at taking over the task of running his mother's estate, Newton was sent back to school, and this time he enjoyed learning so much that he wrote it must be "a sin."

At the very late age of 18, in 1661, Newton entered Trinity College at Cambridge. With no financial backing from his mother, Newton was obliged to work as a "sizar"—a servant to the wealthier students in exchange for financial aid to attend the university. In 1663 he got his first look at Euclid's *Elements,* and he suddenly became entranced with mathematics. He graduated in 1665, the university closed because of the plague, and Newton went back to his family's farm where he spent the next year and a half doing nothing spectacular—nothing but devising the laws that changed the world's understanding of the universe.

As a 22-year-old on the farm, Newton had time to let his scientific mind conjure up the creative ideas for which he is known today, including calculus, the laws of motion, and the Universal Law of Gravity. Newton's laws of motion consist of three laws that deal with mass and force and are the basis for Newtonian mechanics, both on Earth and in space.

The First Law of Motion is based on inertia, a concept Galileo introduced around a half-century before Newton was born. It states that an object at rest stays at rest, and an object in motion stays in motion in a straight line and at a constant speed unless an outside influence or force acts upon that object. More scientifically, it is written, "In the absence of outside forces, the momentum of a system remains constant." Momentum can be simply defined as the product of mass and velocity: *momentum = mass × velocity.*

The Second Law of Motion deals with force, and so is about change in momentum. It states, "If a force acts upon a body, the body accelerates in the direction of the force, its momentum changing at a rate numerically equal to that force." As with the first law, if there is no force, there is no change in momentum. Acceleration is the rate at which velocity changes, and so force is often the product of mass and acceleration: *force = mass × acceleration.* But if the mass of a body changes—for example, if the weight of a rocket diminishes as fuel burns out of it—then the formula is not quite so simple.

If the acceleration to a body occurs in the same direction as the velocity of the body, then the object speeds up. An example of this is the force of gravity accelerating the momentum of a falling object, which occurs at a rate of 32 feet per second^2. If a force is accelerating in the opposite direction of the momentum of a body, then it slows the object down, for example, the force of friction. Newton also explained the computations for two or more forces acting on one object; and if two forces cancel each other, then the body does not accelerate but instead stays in a state of equilibrium.

Newton's Third Law of Motion is the famous "reaction" law: "For every action, there is an equal and opposite reaction." This was an original concept in which Newton states that all forces happen in pairs that are mutually equal and opposite to one another.

Newton also needed to explain the idea of circular momentum, because if a force on a body causes the object to accelerate in a straight line, how could anything travel in a circular path? For this to happen, there has to be a constant force of acceleration toward the center of the circle. This is called centripetal acceleration, or centripetal force. Both Newton and a Dutch physicist, Christiaan Huygens (1629–95), arrived at this idea independently—Newton in 1666, which of course he did not publish, and Huygens in 1673, which he published. The equation is that force is equal to the product of mass and volume squared, divided by the radius of the circle: $F = mv^2/r$.

The Universal Law of Gravitation is the famous "apple falling off the tree" law that children are taught when they first learn about gravity in school. The laws of motion, the formula for centripetal force, and JOHANNES KEPLER's laws helped Newton arrive at the Law of Gravitation. The formula shows that the force between two bodies is the constant of gravitation times the mass of each of the two bodies separated by the distance squared: $F = G(m_1m_2/d^2)$. Newton stated it as "Between any two objects anywhere in space there exists a force of attraction that is in proportion to the product of the masses of the objects and in inverse proportion to the square of the distance between them."

Newton's laws are generally acceptable for most, but not all, natural physics. The two exceptions are relativity, explained by ALBERT EINSTEIN, and quantum theory, which Newtonian mechanics cannot describe.

In order to give all of the necessary computations for his laws, Newton created calculus during his stay on the farm. Then, in 1667, he returned to Cambridge to work on his master's degree, and two years later became Lucasian Professor in mathematics, a position that would also be held 300 years later by SIR STEPHEN WILLIAM HAWKING. Newton made little mention of his calculus or any other work when he returned to Cambridge.

Instead, Newton was interested in optics, and he soon discovered that white light is made of a spectrum of colors—red, orange, yellow, green, blue, indigo, and violet—and can be seen when a ray of light is bent, or refracted, for example through a glass prism. This discovery concerned Newton about the optics of refracting telescopes, so he invented a reflecting telescope. Then, in 1672, he gave one of his new telescopes to the Royal Society. In honor of his gift, he was elected as a member to the society and was allowed to publish his first paper on optics and the properties of light.

Trouble soon followed. Mathematician Robert Hooke had discussed his ideas on light with Newton, and when Hooke saw these concepts in print he accused Newton of stealing his work. Huygens didn't agree with Newton's science. And the Jesuits saw his ideas as an attack on Christianity and sent him a barrage of violent letters. After years of badgering, Newton could no longer stand the pressure, and in 1678 he suffered his first nervous breakdown.

After such disastrous results with publishing, it is surprising that any of Newton's later work ever made it into print. But the famous astronomer SIR EDMUND HALLEY, who sought Newton's help to work on the problem of orbits, discovered that Newton had developed the laws of motion and gravity and had the calculus to back it up. He pushed Newton to publish his work, and Newton agreed only after Halley offered to pay for the printing with his own money. In 1687 the first edition of 2,500 copies of *Philosophiae Naturalis Principia Mathematica* (Mathematical principles of natural philosophy) was published, and *Principia,* as it became known, put Newton in the ranks of the greatest scientific minds ever to grace the planet.

But Newton was busy with the politics of the university and the state, publicly fighting King James II's decree to fill the university's teaching positions with academically unqualified Catholics. When William of Orange ousted James II, Newton was rewarded for his stance with a seat in Parliament. From 1689 to 1693 Newton filled his time with both the university work and his new work in government, but again the pressures of life took hold of him, and he suffered another nervous breakdown. It took several years for him to get back on his feet, but when he did, in 1696, he became extremely wealthy as the head of the Royal Mint.

The Royal Society elected Newton as president in 1703, and in 1704, after Hooke's death, he published his entire work on optics, part of which had caused the scandal with Hooke earlier in his career. For Newton's now famous work in science, Queen Anne knighted him in 1705, giving him the honor of being the first man ever knighted for scientific achievement.

In 1711 controversy struck again, this time in a disastrous scandal involving the famous mathematician Gottfried Wilhelm Leibniz (1646–1716), who was commonly regarded as the inventor of calculus. An article published by the Royal Society, of which Newton was then president, accused Leibniz of plagiarism and declared Newton as the inventor of calculus. Newton was known for being a horrid man, and in a position of power he could do whatever he wanted to guarantee that things went his way. He announced that he was appointing an "impartial committee" to determine who really invented calculus, then wrote an anonymous decision on behalf of the Royal Society in favor of himself, and promptly closed the case, officially pronouncing himself the true inventor. Leibniz was never allowed to give any evidence in his defense, and the issue became a hotly debated international incident.

Newton died in 1727, and two years later, 42 years after its original publication, *Principia* was translated from its original Latin into English. Newton was buried in Westminster Abbey, with a Latin inscription on his tomb that reads, "Mortals! Rejoice at so great an ornament to the human race." The Leibniz incident followed both Newton and Leibniz all the way to their graves, and remained an international debate years after their deaths (Leibniz and Newton are now considered to have independently invented calculus).

The French mathematician COMTE JOSEPH-LOUIS LAGRANGE called Newton "the greatest genius that ever existed," and SUBRAHMANYAN CHANDRASEKHAR considered his brilliance second to none, not even Einstein. But in Newton's case, scientific brilliance had its "equal and opposite" dark side in a mean-spirited and evil temperament. Newton ordered several counterfeiters executed by the state, and he had nasty public fights with other important figures, not the least of whom was the Astronomer Royal John Flamsteed. Newton was superbly monstrous, an observation written by the literary philosopher Aldous Huxley (1894–1963) and a view shared by many, both during and after Newton's lifetime. Yet his professional contributions to astronomy were nothing short of genius. To his credit, Newton has said, "If I have seen farther than others, it is because I was standing on the shoulders of giants."

Oberth, Hermann Julius
(1894–1989)
Romanian/German
Mathematician, Physicist, (Theoretical) Rocket Engineer

The Transylvanian-born German physicist Hermann Oberth was a theoretician who helped establish the field of astronautics early in the 20th century. He originally worked independently of KONSTANTIN EDUARDOVICH TSIOLKOVSKY and ROBERT HUTCHINGS GODDARD in his advocacy of rockets for space travel, but later discovered and acknowledged their prior efforts. Unlike Tsiolkovsky and Goddard, however, Oberth made human beings an integral part of the space travel vision. His inspirational 1923 publication *The Rocket into Interplanetary Space* provided a comprehensive discussion of all the major aspects of space travel, and his 1929 award-winning book, *Roads to Space Travel*, popularized the concept of space travel for technical and nontechnical readers alike. His technical publications and inspiring lectures exerted a tremendous career influence on many young Germans, including the legendary WERNHER MAGNUS VON BRAUN.

Oberth was born on June 25, 1894, in the town of Hermannstadt in a German enclave within the Transylvanian region of Romania (part of the Austro-Hungarian Empire). His parents were members of the historically German-speaking community of the region. His father, Julius, was a prominent physician who tried to influence his son in a career in medicine. But Hermann was a free-spirited thinker who preferred to ponder the future while challenging the ideas held by the current scientific establishment.

At age 11 Oberth discovered the works of Jules Verne, especially *De la terre à la lune* (From the Earth to the Moon). In his famous 1865 novel, Verne was first to provide young readers like Oberth a somewhat credible account of a human voyage to the Moon. Verne's travelers are blasted on a journey around the Moon in a special hollowed-out capsule fired from a large cannon. Verne correctly located the cannon at a low-latitude site on the west coast of Florida called "Tampa Town." By coincidence, this fictitious site is about 120 kilometers to west of Launch Complex 39 at the NASA Kennedy Space Center from which Apollo astronauts actually left for journeys to the Moon between 1968 and 1972.

Oberth read this novel many times and then, while remaining excited about space travel, questioned the story's technical plausibility. He soon discovered that the acceleration down the barrel of this huge cannon would have

Hermann Oberth, one of the founding fathers of astronautics, appears in the forefront of this 1956 photograph, accompanied by officials of the U.S. Army Ballistic Missile Agency at Huntsville, Alabama. Included behind Oberth (from left to right) are Ernst Stuhlinger (seated), Major General H. N. Toftoy (uniformed, standing), Wernher von Braun (seated on table), and Eberhard Rees (standing, far right). General Toftoy was the American military officer responsible for "Project Paperclip"—the operation that took rocket scientists and engineers (including Oberth, von Braun, and Stuhlinger) out of Germany after World War II and relocated them in the United States so they could help design U.S. rockets during the cold war. *(Courtesy of NASA)*

crushed the intrepid explorers and that the capsule itself would have burned up in Earth's atmosphere. But Verne's story made space travel appear technically possible, and this important idea thoroughly intrigued Oberth. So, after identifying the technical limitations in Verne's fictional approach, he started searching for a more practical way to travel into space. That search quickly led him to the rocket.

In 1909, at age 15, Oberth completed his first plan for a rocket to carry several people. His design for a liquid-propellant rocket that burned hydrogen and oxygen came three years later. He graduated from the *Gymnasium* (high school) in June 1912, receiving an award in mathematics. The next year he entered the University of Munich with the intention of studying medicine, but World War I interrupted his studies. As a soldier, he was wounded and transferred to a military medical unit. To the great disappointment of his father, Oberth's three years of duty in this medical unit reinforced his decision not to study medicine.

During World War I Oberth tried to interest the Imperial German War Ministry in developing a long-range military rocket. In 1917 he submitted his specific proposal for a large liquid-fueled rocket. He received a very abrupt response for his efforts. Certain Prussian armaments "experts" within the ministry quickly rejected his plan and reminded him that their experience clearly showed that military rockets could not fly farther than seven kilometers. Of course, these officials were limited by their own experience with contemporary military rockets that used inefficient black powder for propellant. They totally missed Oberth's breakthrough idea involving a better-performing liquid-fueled rocket.

Undaunted, Oberth returned to the university and continued to investigate the theoretical problems of rocketry. In 1918 he married Mathilde Hummel and a year later went back to the University of Munich to study physics. He studied at the University of Göttingen then at the University of Heidelberg before becoming certified as a secondary-school mathematics and physics teacher in 1923. At this point in his life, he was unaware of the contemporary rocket theory work by Tsiolkovsky in Russia and Goddard in the United States. In 1922 he presented a doctoral dissertation on rocketry to the faculty at the University of Heidelberg, but the university committee rejected his dissertation.

Still inspired by space travel, he revised this work and published it in 1923 as *Die Rakete zu den Planetenräumen* (The rocket into interplanetary space). This modest book provided a thorough discussion of the major aspects of space travel, and its contents inspired many young German scientists and engineers to explore rocketry, including WERNHER MAGNUS VON BRAUN. Oberth worked as a teacher and writer in the 1920s. He discovered and acknowledged the rocketry work of Goddard and Tsiolkovsky in the mid-1920s and became the organizing figure around which the practical application of rocketry developed in Germany. He served as a leading member of Verein für Raumschiffahrt (VFR), the German Society for Space Travel. Members of this technical society conducted critical experiments in rocketry in the late 1920s and early 1930s, until the German army absorbed their efforts and established a large military rocket program.

Fritz Lang (1890–1976), the popular Austrian filmmaker, hired Oberth as a technical adviser during the production of his 1929 film *Die Frau im Mond* (The woman in the moon). Collaborating with Willy Ley (1906–69), Oberth provided for the film an exceptionally prophetic two-stage cinematic rocket that startled and delighted audiences with its impressive blastoff. Lang, ever the showman, also wanted Oberth to build and launch a real rocket as part of a publicity stunt to promote the new film. Unfortunately, with little engineering experience, Oberth accepted Lang's challenge, but was soon overwhelmed by the arduous task of turning theory into practical hardware. Shocked and injured by a nearly fatal liquid-propellant explosion, Oberth abandoned Lang's publicity rocket and retreated to the comfort and safety of writing and teaching mathematics in his native Romania.

Oberth was a much better theorist and technical writer than nuts-and-bolts rocket engineer. In 1929 Oberth expanded his ideas concerning the rocket for space travel and human space flight in a book entitled *Wege zur Raumschiffahrt* (Roads to space travel). The Astronomical Society of France gave Oberth an award for this book, which helped popularize the concept of space travel for both technical and nontechnical audiences. As newly elected president of the VFR, Oberth used some of his book's prize money to fund rocket engine research within the society. Young engineers like Braun had a chance to experiment with liquid-propellant engines, including one of Oberth's concepts, the Kegeldüse (conic) engine design. In this visionary book, Oberth also anticipated the development of electric rockets and ion propulsion systems.

Throughout the 1930s, Oberth continued to work on liquid-propellant rocket concepts and on the idea of human space flight. In 1938 Oberth joined the faulty at the Technical University of Vienna. There, he participated briefly in a rocket-related project for the German air force. In 1940 he became a German citizen. The following year, he joined Braun's rocket development team at Peenemünde.

Oberth worked only briefly with Braun. In 1943 he transferred to another location to work on solid-propellant antiaircraft rockets. At the end of World War II, Allied forces captured him and placed him in an internment camp. Upon release, he left a devastated Germany and sought rocket-related employment as a writer and lecturer in Switzerland and Italy. In 1955 he joined Braun's team again, this time at the U.S. Army's Redstone Arsenal. He worked there for several years before returning to Germany in 1958 and retiring.

Of the three founding fathers of astronautics, only Oberth lived to see some of his pioneering visions come true. These visions included the dawn of the Space Age (1957, with *Sputnik 1*), human space flight (1961), the first human landing on the Moon (1969), the first space station (1971), and the first flight of a reusable launch vehicle—NASA's *Space Shuttle* (1982). He died in Nuremberg, Germany, on December 29, 1989. Oberth studied the theoretical problems of rocketry and outlined the technology needed for people to live and work in space. The last paragraph of his 1954 book, *Man into Space*, addresses the important question: "Why space travel?" His eloquently philosophical response is, "This is the goal: To make available for life every place where life is possible. To make inhabitable all worlds as yet uninhabitable, and all life purposeful."

⊠ Olbers, Heinrich Wilhelm Matthäus
(1758–1840)
German
Astronomer (Amateur)

Sometimes a person is primarily remembered for asking an interesting question that baffles the best scientific minds of the time. The German astronomer Heinrich Wilhelm Matthäus Olbers was just such a person. To his credit as an observational astronomer, he discovered the main-belt asteroids Pallas in 1802 and Vesta in 1807. Today he is best remembered for formulating the interesting philosophical question now known as Olber's paradox. In about 1826 he challenged his fellow astronomers by posing the following seemingly innocent question: "Why isn't the night sky with its infinite number of stars not as bright as the surface of the Sun?"

Olbers was born on October 11, 1758, at Ardbergen (near Bremen), Germany. The son of a Lutheran minister, he entered the University of Göttingen in 1777 to study medicine, but also took courses in mathematics and astronomy. By 1781 he became qualified to practice medicine and established a practice in Bremen. By day, he served his community as a competent physician; in the evening he pursued his hobby of astronomy. He converted the upper floor of his home into an astronomical observatory and used several telescopes to conduct regular observations.

"Comet hunting," or cometography, was a favorite pursuit of 18th-century astronomers, and Olbers, a dedicated amateur astronomer, eagerly engaged in this quest. In 1796 he discovered a comet and then used a new mathematical technique he had developed to compute its orbit. His technique, often referred to as Olber's method, proved computationally efficient and was soon adopted by many other astronomers. At the start of the 19th century, Olbers continued to successfully practice medicine in Bremen, but he also became widely recognized as a skilled astronomer.

At the end of the 18th century, astronomers began turning their attention to the interesting gap between Mars and Jupiter. At the time, popularization of the Titius-Bode law—actually just an empirical statement of relative planetary distances from the Sun—caused astronomers to speculate that a planet should have been in this gap. Responding to this wave of interest, GIUSEPPE PIAZZI discovered the first and largest of the minor planets, which he named Ceres, on the first of January in 1801. Unfortunately, Piazzi soon lost track of Ceres. However, based on Piazzi's few observations, the brilliant German mathematician Carl Friedrich Gauss (1777–1855) was able to calculate its orbit. Consequently, a year later, on January 1, 1802, Olbers relocated Ceres. While observing Ceres, Olbers also discovered another minor planet, Pallas. He continued looking in the gap and found the third main-belt asteroid, Vesta, in 1807. Anticipating the discovery of solar radiation pressure, he

suggested in 1811 that a comet's tail points away from the Sun during perihelion passage because the material ejected by the comet's nucleus is influenced (pushed) by the Sun. He remained an active observational astronomer and discovered four more comets, including one in 1815 that now bears his name.

Olbers is best remembered, however, for popularizing the question "Why is the night sky dark?" He presented a technical paper in 1826 that attempted to solve this interesting puzzle. Previous astronomers in the 17th and 18th century were certainly aware of this question, which is sometimes traced back to the writings of the 17th-century astronomer JOHANNES KEPLER, or to a 1744 publication by the Swiss astronomer Jean-Philippe Loys de Chéseaux. However, Olbers's paper represents the most enduring formulation of the question, which is now known as Olber's paradox.

Within the perspective of 19th-century cosmology, the basic problem can be summarized as follows: If the universe is infinite, unchanging, homogeneous, and therefore filled with stars (in the early 20th century "galaxies" was included), then the entire night sky should be filled with light and as bright as daytime. Of course, Olbers saw that night sky was obviously not bright as the day—so what accounted for this inconsistency or paradox between observation and theory? Olbers explanation suggested that the dust and gas in interstellar space blocked much of the light traveling to Earth from distant stars. Unfortunately, he was wrong in this particular suggestion, because within his cosmological model, such absorbing gas would have heated up to incandescent temperatures and glowed.

To solve this apparent contraction, Olbers needed to discard the prevalent infinite, static, homogeneous model of the universe and discover Big Bang cosmology, with its expanding, finite, and inhomogeneous universe. Modern astronomers now provide a reasonable explanation of Olber's paradox with the assistance of an expanding, finite universe model. Some suggest that as the universe expands the very distant stars and galaxies become obscure due to extensive redshift—a phenomenon that weakens their apparent light and makes the night sky dark. Other astrophysicists declare that the observed redshift to lower energies is not a sufficient reason to darken the sky. They suggest that the finite extent of the physical, observable universe and its changing time-dependent nature provide the reason the night sky is truly dark—despite the faint amounts of starlight reaching Earth.

Today, modern cosmology explains that the universe is dynamic and not statically filled up with unchanging stars and galaxies. Finally, as inherently implied within Big Bang cosmology, since the universe is finite in extent and the speed of light is assumed constant, there has not even been a sufficient amount of time for the light from those galaxies beyond a certain range (called the observable universe) to reach Earth.

Olbers died in Bremen, Germany, on March 2, 1840. The chief astronomical legacy of this successful German physician and accomplished amateur astronomer is Olber's paradox—a coherent response to which helps scientists highlight some of the most important aspects and implications of modern cosmology.

⊠ Oort, Jan Hendrik
(1900–1992)
Dutch
Astronomer

An accomplished astronomer, Jan Oort made pioneering studies of the dimensions and structure of the Milky Way Galaxy in the 1920s. However, he is now most frequently remembered for the interesting hypothesis he extended in 1950 when he postulated that a large swarm of comets circles the Sun at a great distance—somewhere between 50,000 and 100,000 astronomical units. Today astronomers refer to this

The Dutch astronomer Jan Oort, shown here at his desk in 1986, made pioneering studies of the dimensions and structure of the Milky Way Galaxy in the 1920s. He also boldly speculated in 1950 that a large swarm of comets (now called the Oort cloud) encircles the Sun at a great distance—somewhere between 50,000 and 100,000 astronomical units. *(Photo by Ron Doel, courtesy AIP Emilio Segrè Visual Archives)*

very distant, postulated reservoir of icy bodies as the Oort cloud in his honor.

Oort was born in the township of Franeker, in Friesland Province, the Netherlands, on April 28, 1900. At age 17 he entered the University of Groningen where he studied stellar dynamics under JACOBUS CORNELIUS KAPTEYN. He completed his doctoral coursework in 1921 and remained at the university as a research assistant for Kapteyn for about one year. He then performed astrometry-related research at Yale University between 1922 and 1924. Upon returning to Holland in 1924, he accepted an appointment as a staff member at Leiden Observatory and remained affiliated with the University of Leiden in some capacity for the remainder of his life.

Following completion of all the requirements for his doctoral degree in 1926, Oort began making important contributions to our understanding of the structure of the cosmos. During seven decades of productive intellectual activity, Oort made significant contributions to the rapidly emerging body of astronomical knowledge. At the start of his career, the universe was considered to be contained within just the Milky Way Galaxy—a vast collection of billions of stars bound within some rather poorly defined borders. When he published his last paper in 1992, Oort was busy describing some of the interesting characteristics of an expanding universe now considered comprised of billions of galaxies whose space and time dimensions extending in all directions to the limits of observation.

In 1927 he revisited Kapteyn's star streaming data, while also building upon the recent work of the Swedish astronomer Bertil Lindblad (1895–1965). This approach produced a classic paper in which Oort demonstrated that the differential systematic motions of stars in the solar neighborhood could be explained in terms of a rotating galaxy hypothesis. The paper, "Observational Evidence Confirming Lindblad's Hypothesis of a Rotation of the Galactic System," served as Oort's platform for introducing his concept of differential galactic rotation.

Differential rotation occurs when different parts of a gravitationally bound gaseous system rotate at different speeds. This implies that the various components of a rotating galaxy share in the overall rotation around the common center to varying degrees. Oort continued to pursue evidence for differential galactic rotation and was able to establish the mathematical theory of galactic structure.

Starting in 1935, Oort served the University of Leiden as both an associate professor of as-tronomy and a joint director of the observatory. In 1945, at the end of World War II, he received a promotion to full professor and the position of the director of its observatory. He continued his university duties until retirement in 1970. However, after retirement from the university, he still continued to work regularly at the Leiden Observatory and produced technical papers until just before his death on November 5, 1992.

Despite the research-limited conditions that characterized German-occupied Holland during World War II, Oort saw the important galactic research potential of the newly emerging field of radio astronomy. He therefore strongly encouraged his graduate student Hendrik van de Hulst (1918–2000) to investigate whether hydrogen clouds might emit a useful radio frequency signal. In 1944 van de Hulst was able to theoretically predict that hydrogen should have a characteristic radio wave emission at a wavelength of 21 centimeters.

After Holland recovered from World War II and the University of Leiden returned to its normal academic functions, Oort was able to form a Dutch team, including van de Hulst, which discovered in 1951 the predicted 21-centimeter wavelength radio frequency emission from neutral hydrogen in outer space. By measuring the distribution of this telltale radiation, Oort and his colleagues mapped the location of hydrogen gas clouds throughout the Galaxy. Then, they used this application of radio astronomy to find the large-scale spiral structure of the Galaxy, the location of the galactic center, and the characteristic motions of large interstellar clouds of hydrogen.

Throughout his career as an astronomer, comets remained one of Oort's favorite topics. Oort supervised completion of research by a postwar doctoral student who was investigating the origin of comets and the statistical distribution of the major axes of their elliptical orbits. Going beyond this student's research, Oort decided to investigate the consequences that passing rogue stars might have on cometary orbits.

As early as 1932, the Estonian astronomer Ernst Öpik (1893–1985) had suggested that the long-period comets that occasionally dashed through the inner solar system might reside in some gravitationally bound cloud at a great distance from the Sun. In 1950 Oort revived and extended the concept of a heliocentric reservoir of icy bodies in orbit around the Sun. This reservoir is now called the Oort cloud.

In refining his comet reservoir model, Oort first suggested that in the early solar system these icy bodies formed at a distance of less than 30 astronomical units from the Sun—that is, within the orbit of Neptune. But then, they diffused outward as a result of gravitational interaction with the giant gaseous planets Jupiter and Saturn. According to Oort's hypothesis, these deflected comets then collected in a swarm or cloud that extended in distance from the Sun of between 30,000 and 100,000 astronomical units. At this extreme range of distances (from about 20 to 30 percent of the way to the nearest star, Proxima Centuri) the perturbations within other stars and gas clouds helped shape the heliocentric orbits so that only an occasional comet now pops out of the cloud on a visit back through the inner solar system. Oort's theory has become a generally accepted model for the origin of very long-period comets.

He received numerous awards for his theory of galactic structure, including the Bruce Medal of the Astronomical Society of the Pacific in 1942 and the Gold Medal of the Royal Astronomical Society in 1946. However, it is the Oort cloud—his comet cloud hypothesis—that gives his work a permanent legacy in astronomy.

P

⊠ Penzias, Arno Allen
(1933–)
German/American
Physicist, Radio Astronomer

Sometimes the world's greatest scientific discoveries take place when least expected. While collaborating in the mid–1960s with ROBERT WOODROW WILSON, a fellow radio astronomer at Bell Laboratories, Arno Penzias went about the task of examining natural sources of extraterrestrial radio wave noise that might interfere with transmissions from communication satellites. To his great surprise, he and Wilson quite by accident stumbled upon the "Holy Grail" of modern cosmology—the cosmic microwave background. This all-pervading microwave background radiation resides at the very edge of the observable universe and is considered the cooled remnant (about three kelvins) of the ancient big bang explosion. They carefully confirmed the data that rocked the world of physics and gave cosmologists the first empirical evidence pointing to a very hot, explosive phase at the beginning of the universe. Penzias and Wilson shared the 1978 Nobel Prize in physics for this most important discovery.

Penzias was born in Munich, Germany, on April 26, 1933. Of Jewish heritage, his family was one of the last to successfully flee Nazi Germany before the outbreak of World War II and escaped almost certain death in one of Hitler's concentration camps. In 1939 his parents placed Arno and his younger brother on a special refugee train to Great Britain and then traveled there under separate exit visas to join their sons. Once reunited, the family sailed for the United States in December and arrived in New York City in January 1940. Penzias followed the education route taken by many thousands of upwardly mobile immigrants. He used hard work and the New York City public school system as his pathway to a better life. In 1954 he graduated from the City College of New York (CCNY) with a bachelor of science degree in physics. After graduation he married, then joined the U.S. Army, where he served in the Signal Corps for two years. He then enrolled at Columbia University in 1956 and graduated from that institution with a master of arts degree in 1958 and a Ph.D. degree in 1962.

His technical experience in the army's Signal Corps allowed him to obtain a research assistantship in the Columbia Radiation Laboratory. There he met and studied under Charles Townes (b. 1915), the American physicist and Nobel laureate who developed the first operational maser (microwave amplification by stimulated emission of radiation)—the forerunner to the laser. For a doctoral research project,

Shown here in front of their historic radio telescope at the Bell Laboratories in New Jersey are the Noble laureates Robert Woodrow Wilson (left, background) and Arno Penzias (right, foreground). In the mid-1960s, while performing other experiments with this facility, the two radio astronomers serendipitously detected the cosmic microwave background—the cooled remnant of the "big bang" at the edge of the observable universe. Their discovery provided cosmologists the first direct evidence of a very hot, explosive phase at the beginning of the universe. For this great achievement, they shared the 1978 Nobel Prize in physics. *(Lucent Technologies' Bell Laboratories, courtesy AIP Emilio Segrè Visual Archives, Physics Today Collection)*

Professor Townes assigned Penzias the challenging task of building a maser amplifier suitable for use in a radio astronomy experiment of his own choosing.

Upon completion of his thesis work in 1961, Penzias sought temporary employment at Bell Laboratories in Holmdel, New Jersey. To his surprise, the director of the Radio Research Laboratory offered him a permanent position. Penzias quickly accepted the radio astronomer position and subsequently remained a member of Bell Laboratories in various technical and management capacities until his retirement in May 1998.

In 1963 Penzias met another radio astronomer, Wilson, who had recently been hired to work at Bell Laboratories after graduating from the California Institute of Technology (Caltech). The two scientists embarked on the task of using a special horn-reflector antenna 20 feet in diameter that had just been constructed at Bell Laboratories to serve as an ultra-low-noise receiver for signals bounced from NASA's *Echo One* satellite—the world's first passive communications satellite experiment, launched in August 1960. At the time, this ultrasensitive six-meter horn-reflector was no longer needed for satellite work, so both Penzias and Wilson began to modify the instrument for radio astronomy. As modified with traveling-wave maser amplifiers, their modest-sized horn reflector became the most sensitive radio telescope in the world at the time. Both radio astronomers were eager to extend portions of their doctoral research with this newly converted radio telescope. Their initial objective was to measure the intensity of several interesting extraterrestrial radio sources at a radio frequency wavelength of 7.53 centimeters—a task with potential value in both the development of satellite-based telecommunications and radio astronomy.

On May 20, 1964, Wilson and Penzias made a historic measurement that later proved to be the very first to clearly indicated the presence of the cosmic microwave background—the remnant "cold light" from the dawn of creation. At the time, neither was sure why this strange background "noise" at around three degrees Kelvin equivalent temperature kept showing up in all the measurements. As excellent scientists do, they checked every potential source for this unexplained signal. They ruled out equipment error (including antenna noise due to faulty joints), the lingering effects of a previous nuclear test in outer space, the presence of human-made radio frequency signals (including those emanating from New York City), sources within the Milky Way Galaxy, and even the possible echoing effects of pigeon droppings from a pair of birds that made a portion of the antenna their home. Nothing explained this persistent, omnidirectional, uniform microwave signal. By early 1965 Penzias and Wilson had completed their initial data collection with their antenna, but were still no closer to unraveling the true physical nature of this persistent signal. Their careful analysis indicated the signal was definitely "real"—but what was it?

Then the mystery unraveled quickly when, during a meeting with another physicist in spring 1965, Penzias casually mentioned his recent radio astronomy work and the interesting noise-like signal he and Wilson had collected. The physicist suggested that Penzias contact Robert Dicke (1916–97) and members of his group at Princeton University who were exploring the physics of the universe and something called the big bang hypothesis. Penzias contacted Dicke who soon visited Bell Laboratories. After he reviewed their precise work, Dicke immediately realized that Penzias and Wilson had, totally by accident, beaten all the other astrophysicists and cosmologists, including Dicke, in detecting evidence of the hypothesized microwave remnant of the big bang. Recognizing the scientific importance of this work, they agreed to publish letters in the *Astrophysical Journal*. As a result, Volume 142 of the *Astrophysical Journal* contains two historic letters that appear side by side. In one letter, Dicke and his team at Princeton University discuss the cosmological implications of the cosmic microwave background; in the other letter, Penzias and Wilson present their cosmic microwave background measurements without elaborating on the cosmological significance of the data.

Once the Penzias and Wilson data began to circulate within the astronomical community, its true significance quickly emerged. Scientists revisited the earlier big bang hypothesis work of RALPH ASHER ALPHER and engaged in a series of corroborating measurements. For providing the first definitive evidence of the cosmic microwave background, Penzias and Wilson shared the 1978 Nobel Prize in physics with the Russian physicist Pyotr Kapitsa, who received his award for unrelated achievements in low-temperature physics.

While responding to the research needs and priorities of Bell Laboratories in the 1960s and 1970s, Penzias and Wilson also continued to participate in pioneering radio astronomy research. In 1973, while investigating molecules in interstellar space, they discovered the presence of the deuterated molecular species DCN and were then able to use this molecular species to trace the distribution of deuterium in the Galaxy. In 1972 Penzias became head of the Radio Physics Research Department at Bell Laboratories, and early in 1979 his management responsibilities increased when he received a promotion to head the laboratory's Communications Sciences Research Division. At the end of 1981, he received another promotion and became vice president of research as the Bell System experienced a major transformation. While discharging his responsibilities in this new executive position, Penzias began to concentrate on the creation and effective use of new technologies. In 1989 he published a book, *Ideas and Information*, which discussed the impact of computers and other new technologies on society. In May 1998 Penzias retired from his position as chief scientist at Bell Laboratories, now a division of Lucent Technologies.

Best known as the radio astronomer who codiscovered the cosmic microwave background, Penzias received many awards and honors in addition to his 1978 Nobel Prize. Some of his other awards include the Henry Draper Medal of the American National Academy of Sciences (1977) and the Herschel Medal of the Royal Astronomical Society (1977). He also received membership in many distinguished organizations, including the National Academy of Sciences, the American Academy of Arts and Sciences, and the American Physical Society.

⊠ Piazzi, Giuseppe
(1746–1826)
Italian
Astronomer

In the late 18th century this intellectually gifted Italian monk was a professor of mathematics at the academy in Palermo and also served as the director of the new astronomical observatory constructed in Sicily in 1787. Piazzi was a skilled observational astronomer and discovered the first asteroid on New Year's Day in 1801. Following prevailing astronomical traditions, he named the newly found celestial object Ceres, after the patron goddess of agriculture in Roman mythology.

Giuseppe Piazzi was born at Ponte, in Valtellina, Italy, on July 16, 1746. Early in his life he decided to enter the Theatine Order, a religious order for men founded in southern Italy in the 16th century. Upon completing his initial religious training, Piazzi accepted the order's strict vow of poverty. As a newly ordained monk, he pursued advanced studies at colleges in Milan, Turin, Genoa, and Rome. Piazzi emerged as a professor of theology and a professor of mathematics. In the late 1770s he taught briefly at the university on the island of Malta and then returned to a college post in Rome to serve with distinction as a professor of dogmatic theology. His academic colleague in Rome was another monk named Luigi Chiaramonti, who was later elected Pope Pius VII (1800–23). In 1780 his religious order sent Piazzi to the academy in Palermo to assume the chair of higher mathematics.

While working at this academy in Sicily, Piazzi obtained a grant from the viceroy of Sicily, Prince Caramanico, to construct an observatory. As part of the development of this new observatory, he traveled to Paris in 1787 to study astronomy with Joseph Lalande (1732–1807) and then went on to Great Britain the following year to work with England's fifth Astronomer Royal,

Nevil Maslelyne (1732–1811). As a result of his travels, Piazzi gathered new astronomical equipment for the observatory in Palermo. Upon returning to Sicily, he set up the new equipment on top of a tower in the royal palace between 1789 and 1790.

All was soon ready and Piazzi began to make astronomical observations in May 1791. He published his first astronomical reports in 1792. His primary goal was to improve the stellar position data in existing star catalogs. Piazzi paid particular attention to compensating for errors induced by such subtle phenomenon as the aberration of starlight, discovered in 1728 by the British astronomer JAMES BRADLEY. The monk-astronomer was a very skilled observer who was soon able to publish a refined positional list of approximately 6,800 stars in 1803. He followed this initial effort with the publication of a second star catalog containing about 7,600 stars in 1814. Both of these publications were well received within the astronomical community and honored by prizes from the French academy.

However, Piazzi is today best remembered as the person who discovered the first asteroid (or minor planet). He made this important discovery on January 1, 1801, from the observatory that he founded and directed in Palermo. To please his royal benefactors while still maintaining astronomical traditions, he called the new celestial object Ceres, after the ancient Roman goddess of agriculture who was once widely worshiped on the island of Sicily. Ceres (asteroid 1) is the largest asteroid, with a diameter of about 935 kilometers. However, because of its very low albedo (about 10 percent), Ceres cannot be observed by the naked eye. After just three good telescopic observations in early 1801, Piazzi lost the asteroid in the Sun's glare. Yet Piazzi's carefully recorded positions enabled the brilliant German mathematician Carl Friedrich Gauss (1777–1855), to accurately predict the asteroid's orbit. Gauss's calculations allowed another early

asteroid hunter, HEINRICH WILHELM MATTHÄUS OLBERS, to relocate the minor planet on January 1, 1802. Olbers then went on to find the asteroids Pallas (1802) and Vesta (1807).

When Piazzi died in Naples, Italy, on July 22, 1826, only four asteroids were known to exist in the gap between Mars and Jupiter. But is his pioneering discovery of Ceres in 1801 stimulated renewed attention on this interesting region of the solar system—a gap that should contain a missing planet according to the Titus-Bode law. In 1923 astronomers named the 1,000th asteroid discovered "Piazzia" to honor the Italian astronomer's important achievement.

⊠ Pickering, Edward Charles
(1846–1919)
American
Astronomer, Physicist

Starting in 1877, Edward Charles Pickering dominated American astronomy for the last quarter of the 19th century. He served as the director of the Harvard College Observatory for more than four decades, and in this capacity supervised the production of the *Henry Draper Catalogue* by ANNIE JUMP CANNON, WILLIAMINA PATON STEVENS FLEMING, HENRIETTA SWAN LEAVITT, and other astronomical "computers." This important work listed more than 225,000 stars according to their spectra, as defined by the newly introduced Harvard Classification System. He also vigorously promoted the exciting new field of astrophotography, and in 1889 discovered the first spectroscopic binary star system—two stars so visually close together that they can be distinguished only by the Doppler shift of their spectral lines.

Pickering was born on July 19, 1846, in Boston, Massachusetts. After graduating from Harvard University in 1865, he taught physics for approximately 10 years at the Massachusetts Institute of Technology (MIT). To assist in his

As director of the Harvard College Observatory, Edward Charles Pickering dominated American astronomy in the last quarter of the 19th century. He supervised the production of the *Henry Draper Catalogue* (by Anne Jump Cannon, Williamina Fleming, Henrietta Leavitt, and other astronomical "computers") and vigorously promoted the exciting new field of astrophotography. *(AIP Emilio Segrè Visual Archives)*

physics lectures, he constructed the first instructional physics laboratory in the United States. Then, in 1876, at just 30 years of age, he became a professor of astronomy and the director of the Harvard College Observatory. He remained in this position for 42 years and used it to effectively shape the course of American astronomy through the last two decades of the 19th century and the first two decades of the 20th century.

One important innovation was Pickering's decision to hire women to work as observational astronomers and "astronomical computers" at the Harvard Observatory. With the funds provided by HENRY DRAPER's widow, Ann Palmer Draper (1839–1914), Pickering employed Cannon, Fleming, Antonia Maury, Leavitt, and others to work at the observatory. He then supervised these talented and well qualified women as they produced the contents of the *Henry Draper Catalogue*. This catalogue, published in 1918, was primarily compiled by Cannon and presented the spectra classification of some 225,000 stars based on Cannon's Harvard Classification System, in which the stars were ordered in a sequence that corresponded to the strength of their hydrogen absorption lines.

In 1884 Pickering produced a new catalog called the *Harvard Photometry*. This work presented the photometric brightness of 4,260 stars using the North Star (Polaris) as a reference. Pickering revised and expanded this effort in 1908. He was an early advocate of astrophotography, and in 1903 he published the first photographic map of the entire sky. To achieve his goals in astrophotography he also sought the assistance of his younger brother, William Henry Pickering (1858–1938), who established the Harvard College Observatory's southern station in Peru. Independent of the German astronomer Hermann Vogel (1841–1907), Pickering discovered the first spectroscopic binary star system, Mizar, in 1889.

He was a dominant figure in American astronomy and a great source of encouragement to amateur astronomers. He founded the American Association of Variable Star Observers and received a large number of awards for his contributions to astronomy, including the Henry Draper Medal of the National Academy of Sciences (1888), the Rumford Prize of the American Academy of Arts and Sciences (1891), the Gold Medal of the Royal Astronomical Society (1886 and 1901), and the Bruce Gold Medal of the Astronomical Society of the Pacific (1908). He died in Cambridge, Massachusetts, on February 3, 1919.

⊠ Planck, Max Karl
(1858–1947)
German
Physicist

Modern physics, with its wonderful new approach to viewing both the most minute regions of inner space and the farthest regions of outer space, has two brilliant cofounders: Max Planck and ALBERT EINSTEIN. Many other intelligent people have built upon the great pillars of 20th-century physics—quantum theory and relativity—but these two giants of scientific achievement hold the distinction of leading the way.

Planck was born in Kiel, Germany, on April 23, 1858. His father was a professor of law at the University of Kiel and gave his son a deep sense of integrity, fairness, and the value of intellectual achievement—important traits that characterized Planck throughout his life. When Planck was a young boy, his family moved to Munich, where he received his early education. A gifted student who could easily have become a great pianist, Planck studied physics at the Universities of Munich and Berlin. In Berlin he had the opportunity to directly interact with such famous scientists as GUSTAV ROBERT KIRCHHOFF and RUDOLF JULIUS EMMANUEL CLAUSIUS. Kirchhoff introduced Planck to sophisticated, classical interpretations of blackbody radiation, while Clausius challenged him with the profound significance of the second law of thermodynamics and the elusive concept of entropy.

In 1879 Planck received his doctoral degree in physics from the University of Munich. From 1880 to 1885 he remained in Munich as a *Privatdozent* (lecturer) in physics at the university. He became an associate professor in physics at the University of Kiel in 1885 and remained in that position until 1889, when he succeeded Kirchhoff as professor of physics at the University of Berlin. His promotion to associate professor in 1885 provided Planck with the income to marry his first wife, Marie Merck—a childhood friend with whom he lived happily until her death in 1909. The couple had two sons and twin daughters. In 1910 Planck married his second wife, Marga von Hösling. He remained a professor of physics at the University of Berlin until his retirement in 1926. Good fortune and success would smile upon Planck's career as a physicist in Berlin, but tragedy would stalk his personal life.

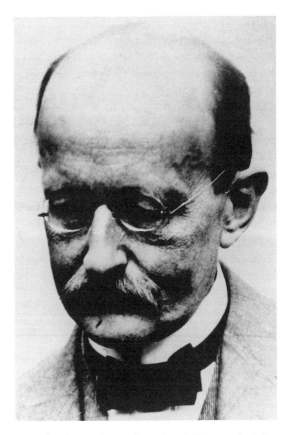

Max Planck was the gentle, cultured German physicist who introduced quantum theory in 1900, transforming physics and the world in the process. His powerful new theory suggested the transport of electromagnetic radiation in discrete energy packets or quanta. Amid great personal tragedy, Planck received the 1918 Nobel Prize in physics for his world-changing scientific accomplishment—an achievement representing one of the two great pillars of modern physics. (The other is Albert Einstein's theory of relativity.) *(AIP Emilio Segrè Visual Archives)*

While teaching at the University of Berlin at the end of the 19th century, Planck began to address the very puzzling problem involved with the emission of energy by a blackbody radiator as a function of its temperature. A decade earlier, JOSEF STEFAN had performed important heat transfer experiments concerning blackbody radiators. However, classical physics could not adequately explain his experimental observations. To make matters more puzzling, classical electromagnetic theory incorrectly predicted that a blackbody radiator should emit an infinite amount of thermal energy at very high frequencies—a paradoxical condition referred to as the "ultraviolet catastrophe."

Early in 1900, two British scientists, Lord Rayleigh (1842–1919) and SIR JAMES HOPWOOD JEANS, introduced their mathematical formula (called the Rayleigh-Jeans formula) to describe the energy emitted by a blackbody radiator as a function of wavelength. While their formula adequately predicted behavior at long wavelengths (for example, at radio frequencies), their model failed completely in trying to predict blackbody behavior at shorter wavelengths (higher frequencies)—the portion of the electromagnetic spectrum of great importance in astronomy, since stars approximate high-temperature blackbody radiators.

Planck solved the problem when he developed a bold new formula that successfully described the behavior of a blackbody radiator over all portions of the electromagnetic spectrum. To reach his successful formula, Planck assumed that the atoms of the blackbody body emitted their radiation only in discrete individual energy packets that he called quanta. In a classic paper that he published in late 1900, Planck presented his new blackbody radiation formula. He included the revolutionary idea that the energy for a blackbody resonator at a frequency (ν) is simply the product h ν, where h is a universal constant, now called Planck's constant. This paper, published in *Annalen der Physik*

(Annals of physics), contained Planck's most important work and represented a major turning point in the history of physics. The introduction of quantum mechanics had profound implications on all modern physics from the way scientists treated subatomic phenomena to the way they modeled the behavior of the universe on cosmic scales.

Yet Planck himself was a reluctant revolutionary. For years, he felt that he had only created the quantum postulate as a "convenient means" of explaining his blackbody radiation formula. However, other physicists were quick to seize Planck's quantum postulate and then to go forth and complete his revolutionary movement—displacing classical physics with modern physics. Einstein used Planck's quantum postulate to explain the photoelectric effect in 1905, and Niels Bohr (1885–1962) applied quantum mechanics in 1913 to create his world-changing model of the hydrogen atom.

Planck received the 1918 Nobel Prize in physics in recognition of his epoch-making investigations into quantum theory. He also published two important books: *Thermodynamics* (1897) and *The Theory of Heat Radiation* (1906) that summarized his major efforts in the physics of blackbody radiation. He was a well-respected physicist even after his retirement from the University of Berlin in 1926. That same year, the British Royal Society honored his contributions to physics by electing him a foreign member and awarding him its prestigious Copley Medal.

But as Planck climbed to the pinnacle of professional success, his personal life was marked with deep tragedy. At the time Planck won the Nobel Prize in physics, his oldest son, Karl, died in combat in World War I, and both his twin daughters, Margarete and Emma, died about a year apart in childbirth.

In the mid–1930s, when Adolf Hitler seized power in Germany, Planck, in his capacity as the elder statesperson for the German scientific community, bravely praised Einstein and other

German-Jewish physicists in open defiance of the ongoing Nazi persecutions. Planck even met personally with Hitler to try to stop the senseless attacks against Jewish scientists. But Hitler simply flew into a tirade and ignored the pleas from the aging Nobel laureate. As a final protest, Planck resigned in 1937 as the president of the Kaiser Wilhelm Institute—the leadership position in German science and one in which he had proudly served with great distinction since 1930. In his honor, that institution is now called the Max Planck Institute.

During the closing days of World War II, his second son, Erwin, was brutally tortured and then executed by the Nazi Gestapo for his role in the unsuccessful 1944 assassination attempt against Hitler. Just weeks before the war ended, Planck's home in Berlin was completely destroyed by Allied bombing. In the very last days of the war, U.S. troops launched a daring rescue across war-torn Germany to keep Planck and his second wife, Marga, from being captured by the advancing Russian army. The Dutch-American astronomer GERARD PETER KUIPER, was a participant in this military action that allowed Planck to live the remainder of his life in the relative safety of the Allied-occupied portion of a divided Germany. On October 3, 1947, at age 89, the brilliant physicist who ushered in a revolution in physics with his quantum theory died peacefully in Göttingen, Germany.

Plato (Aristocles, Platon)
(ca. 427–347 B.C.E.)
Greek
Philosopher

The ancient Greek philosopher born Aristocles in Athens around 427 B.C.E., was not interested in science. In fact, he is said to have despised it. Ironically, this pagan philosopher known as Plato—a nickname picked up as a child because of broad shoulders (from *Platon*, meaning "broad")—was the first to create the ideas that grew into the science of the universe—a science that would cause the death of many astronomers in the centuries to come who questioned the philosophies that became the religious "truths" of the Christian church.

Plato was a student of the philosopher Socrates (470–399 B.C.E.), who was a Sophist. This sect of philosophers considered themselves "masters of wisdom," and traveled from country to country and town to town as orators espousing their philosophies. The Sophist view was that there are two kinds of philosophy—natural and moral. Natural philosophy, that is, nature or science, was considered a complete waste of time and mental attention. The only philosophy worthy of human thought was moral philosophy because it dealt with the abstract, and abstract wisdom was the only kind of knowledge that served to strengthen the soul. Socrates took this view of moral philosophy one step further, saying the most important knowledge of all was to "know thyself." His teachings centered on this premise.

Socrates, Plato, and the others who considered themselves Sophists were changing the then-accepted views of the Ionians—a sort of competitive group of philosophers. The Ionians expanded on the views of Thales (624–546 B.C.E.), who declared that the Earth was flat and in essence afloat in an endless sea of space. This notion of a flat planet managed to resurface some centuries later, and was ultimately disproved to the "general public" once and for all—some 2,000 years after its original introduction—by Christopher Columbus. He navigated his ships by the stars, and thankfully did not fall off the "edge" of the Earth during any of his explorations.

The next evolution of the state of all things heavenly came from a student of Thales named Anaximander (610–547 B.C.E.). He introduced the idea of a sphere around the Earth in which all of the stars were held, and he expanded the shape of the Earth from a pancake to a cylinder. But since the Ionians were natural philosophers, not moral philosophers, they were attempting to explain the nature of the universe, which the Sophists were adamantly against. In fact, Socrates warned his followers that if they succumbed to the kind of thinking subscribed to by the Ionians, it would "expose [their] mental eyes to total and irreparable blindness."

In 409 Plato became a disciple of Socrates. When Socrates was executed by the government in 399, for "corrupting youth" by teaching them the "art of words," Plato left Athens, stating in his writings that the entire system would remain corrupt until "kings were philosophers or philosophers were kings."

Plato's travels took him to Italy and Africa. At some point during his journey, he became familiar with the Pythagoreans and their belief that everything could be explained with mathematics. In Plato's time, mathematics was not considered a science because it dealt with abstracts. As abstract thought, mathematics, then, was the most worthy kind of thought to ponder.

In 387 Plato found his way back to Athens and set up the first institute for higher learning. Plato's Academy, built on land once owned by a man named Academus, was dedicated to abstract thought—truly considered higher learning—and since mathematics was purely abstract, Plato became such a proponent of it that he had a sign inscribed above the Academy's entrance that read, "Let no one ignorant of mathematics enter here." This foundation of mathematics built into Plato's teachings was strong enough to hold the philosophy of the cosmos—ideas that were proposed initially by Plato then built upon by his most famous student, ARISTOTLE—for the next 1,500 years.

Plato learned in his travels that Pythagoras (ca. 582–497 B.C.E.) had gone even further with the sphere concept proposed by Anaximander.

Pythagoras claimed that the entire system was spherical—the Sun, Moon, stars, and planets—and, by the way, so was the Earth. This was truly a new concept. Plato picked up the cosmic ball and ran with it, so to speak, giving credence to the idea of an entirely spherical universe—a universe that, of course, revolved around the spherical Earth. Plato then went one step further still, describing the entire system, in whole and in every part, as a living organism.

The world according to Plato consisted of four elements—earth, air, fire, and water. The heavens consisted of one—quintessence. The perfection of the heavens was a result of the perfect one who created them. And since the heavens were perfect, it stood to reason that they had to move in the most perfect of paths—in the shape of a circle. It was reasonable, then, that they resided in perfect shapes—crystalline spheres—that held everything in place. In fact, Plato believed that the entirety of the heavens existed in perfect geometric form. With his new ideas on motion and shape, mathematics and astronomy became forever intertwined.

Plato's writings were created as dialogues—discussions between his teacher, Socrates, and one of his students—in which Plato crafted the explanations of his philosophies. Plato delved into the order of the universe in his work *Timaeus*, in which an astronomer named Timaeus explains the universe to Socrates. "What is that which always is and has no becoming; and what is that which is always becoming and never is? . . . Was the world, I say, always in existence and without beginning? Or created and had it a beginning?" It is in this writing that Plato puts forth the concept of the circular motions and spherical shape of the universe.

In another dialogue, *Republic*, Plato determines the order in which the celestial orbs lie closest to the Earth, but he changes that order in *Timaeus*, ultimately settling on the Moon, the Sun, Mercury, Venus, and Mars. He describes four concentric spheres that relate to the four elements, and then explains the sizes of the spheres as they relate to the size of Earth. In this, the Earth has a radius of one, water's radius is two, air's radius is five, and fire has a radius of 10 and is the sphere that contains the stars. He then tells the distances of the spheres of the planets from the Earth in terms of the number value he has given the four elements. For example, the distance to the moon is 8, which equals 1 + 2 + 5, or earth + water + air.

But Plato's ideas had many flaws. For starters, orbits are not circular, there are no crystalline spheres holding the universe together, and the entire cosmos does not revolve around the Earth. As his students expanded upon Plato's ideas, complications grew. One such student, Eudoxus of Cnidus, declared that there were 26 rotating spheres in the heavens. Another student, Aristotle, determined that there were in fact 54 crystal spheres holding the universe together and keeping it all moving in perfect order. As these ideas of the universe expanded, it became pretty evident that they did not always work. In order to make sure these philosophies *appeared* to work, adjustments had to be made. These adjustments were called "saving the appearances," and were performed throughout the reign of the geocentric, or Earth-centered, realm that explained the structure of the universe according to Aristotle, by way of Plato's teachings. Things did not change until the works of Nicolas Copernicus, Johannes Kepler, and Galileo Galilei became recognized as the true nature of the universe.

In *Epinomis*, another writing credited to Plato (although there has been controversy surrounding its authorship), a fifth element, aether—the quintessence—was introduced, and it is this writing that is said to bridge the transition from the four elements and Plato's universe to the expanded view of the cosmos explained by Aristotle. Here, Plato tells us that the Sun is larger than the Earth, and that the sphere of stars is immense in size.

Plato's Academy, the first university ever established, was in operation for 900 years. The Christian leader, the Roman emperor Justinian, closed the pagan institution in 529. Plato died at age 82, supposedly peacefully, in his sleep after attending a feast in honor of the wedding of a student. The crater Plato is located in the Alpine Valley region of the Moon. Nearby lie the craters Eudoxus and Aristoteles.

Ptolemy (Claudius Ptolemy, Claudius Ptolemaeus, Claudius Ptolemaios)
(ca. 100–178)
Greek
Philosopher, Astronomer, Astrologer, Cartographer, Geographer, Mathematician

As with many of the ancients, little is known about the famous Greek philosopher Claudius Ptolemy. Believed to have been a Roman citizen, and born in Alexandria, Egypt, he has been described as a man of average build who had a penchant for horsemanship. It was his mastery of natural (that is, scientific) philosophical thought and writing that made his work the prevailing doctrine of astronomy for approximately 1,500 years after his death, building on many great thinkers who came before him, including Plato, Eudoxus of Cnidus (409–356 B.C.E.), Aristotle, and of course Hipparchus (190–120 B.C.E.).

Ptolemy created ancient encyclopedic volumes on astronomy in the dawn of recorded scientific study. His goals were very clear in creating his first writing, *Megale Mathematike Syntaxis* (Great mathematical composition), which centuries later became known as *Almagest*. "We shall set out everything useful for the theory of the heavens in the proper order . . . (and) recount what has been adequately established by the ancients," he wrote.

The "ancients" were the Mesopotamians, whose work was known from about 2,000 years prior to Ptolemy's time. Their studies of the heavens included classifications of stars, determining the length of seasons, observations and theories of lunar cycles, the relationship between the Moon and planets, planetary observations (particularly of Venus), and creating the signs of the zodiac that are familiar today and often credited to Ptolemy. They also determined that the universe was made up of a system of eight concentric spheres.

Ptolemy was a student of astronomy, and it is generally agreed that Ptolemy's most influential "teacher" lived years before he was born. In the second century B.C.E., Hipparchus (ca. 160–125 B.C.E.) was a highly regarded philosopher and mathematical astronomer who worked mostly on the island of Rhodes, and was the first person to use longitude and latitude to designate the location of a city. Access to Hipparchus's work was naturally available to Ptolemy through the great libraries in Alexandria, as was the work of the philosopher Aristotle, who expanded on the work of his teacher Plato, who was a disciple of Socrates.

One of the biggest distinctions between Ptolemy and his predecessors is that most of his major works have survived. Ptolemy's writings have undergone a number of translations throughout the centuries—by Persian mathematician and astronomer Nasir ad-Din at-Tusi (1201–74); philosopher and scholar George of Trebizond (1396–1486), who wrote his own commentary on the *Almagest* that was as long as Ptolemy's work, but whose translation was criticized by the pope for its inaccuracies; and most notably German astronomer Johannes Regiomontanus (1436–76), who translated Ptolemy's work from Greek to Latin from 1460 to 1463, creating *Epitome of the Almagest*, the definitive translation of the work for the 15th and 16th centuries.

Ptolemy's astronomical work centered on his belief that the universe was geocentric in nature; in other words, the Earth was the center of the universe, around which everything else revolved. This followed Aristotle's teachings, which had become the accepted view of how the universe worked, despite the fact that, around 287 B.C.E., ARISTARCHUS OF SAMOS introduced a heliocentric model after studying at Aristotle's Lyceum in Alexandria. This concept fell out of favor and was not revived until the 16th century, when the Polish cleric NICOLAS COPERNICUS worked on simplifying Ptolemy's complicated model of the heavens.

Ptolemy's work in astronomy was his *Megale Mathematike Syntaxis* (Great mathematical composition), simply called *Megiste* (Greatest), which was translated into Arabic as *al-Majisti* (the greatest), and finally became *Almagest*. The work consists of 13 volumes—a treatise with mathematical computations, observations, and theories on all things regarding the workings of the universe. Ptolemy stated, "We shall try to note down everything which we think we have discovered up to the present time." This included mathematical computations for how the geocentric model works, using mostly trigonometry, in which he defines an approximate value of pi as 3.14166, and an introduction to a theorem (now called Ptolemy's theorem) that states that for a quadrilateral inscribed in a circle (a cyclic quadrilateral), the sum of the products of the two pairs of opposite sides are equal to the product of its two diagonals, written as $AC \times BD = (AD \times BC) + (AB \times CD)$.

In the *Almagest* Ptolemy explains that the Earth is fixed and the other heavenly spheres rotate around the Earth, and he explains how to predict the positions of the Sun, the Moon, and the planets. He includes his determination of the length of the year, a theory of the Moon, theory of eclipses, theory of the Sun, theory of planets, creates a star catalog with more than 1,000 stars, and then creates a very complicated model to fit his observations.

It is very hard to say which part of this caused the most trouble for astronomers over the centuries. The problems are, of course, based on the fact that his premise is wrong—the Earth is not at the center of the universe. Even this was obvious to Ptolemy; he tried to fix it by moving the Earth slightly off center from the spheres that rotated around it.

Other problems arose as well. The value he gave for the length of the year was 365¼—1/300 of a day, which is not accurate (the actual value is less than 1/28 of a day). He notes in *Almagest* that there are small discrepancies—that his model does not match the observed sky. He states that everything in the heavens revolves in perfect circular motion.

To give it all some sense, Ptolemy had to do things like add complicated spheres within spheres. And while moving the Earth off center was a strict violation of Aristotle's teachings, it seemed to be acceptable because it was only slightly different, and with the other adjustments, it seemed to work—for the most part. But it was this kind of fudging that would eventually arouse some heated opinions from astronomers in centuries to come. GALILEO GALILEI gave perhaps the most gentle argument against Ptolemy's model in his work *Dialogue Concerning the Two Chief World Systems, Ptolemaic and Copernican*, written in 1632. SIR ISAAC NEWTON, however, was a little more blatant in his distaste of Ptolemy's work, saying that Ptolemy was a fraud, and that his work was "a crime committed by a scientist against fellow scientists and scholars, a betrayal of the ethics and integrity of his profession" and that every observation Ptolemy wrote in *Almagest* was completely made up.

Several pieces of the *Almagest* resulted in spin-off works. *Handy Tables* gave updated calculations for planetary, solar, and lunar positions, eclipses of the Sun and Moon, and was generally used for astrology purposes. *Planetary Hypothesis* consisted of two books on Ptolemy's theories of planetary movement. *Astrology* was the culmination of *Almagest*. As Ptolemy saw it, *Almagest* was meant to explain how the heavenly bodies worked, and *Astrology* explained how they affected people's lives. Other astronomy-related titles were *Analemma*, which explained the mathematics necessary to make a sundial, and *Planesphaerium*, which delved into how to map the celestial sphere onto a plane. *Optics*, a collection of five books, explored light in terms of reflection and refraction.

Another major contribution to science was Ptolemy's *Geography*, a compilation of eight books in which he undertakes the daunting task of giving the coordinates (approximately 8,000) to map the entire known world by longitude and latitude. The three continents then known were Europe, Asia, and Africa, and the maps included the Tropics of Cancer and Capricorn. His maps did not survive, but the text did, and scholars in the 15th century re-created his maps based on his instructions in the book.

Almagest gave a historical, philosophical, and scientific perspective on the early views of astronomy, though it was fraught with mistakes. Even Ptolemy himself knew that the work had problems. The complications were huge, and in many places the text was contradictory. At least Ptolemy felt compelled to address these issues: "Let no one seeing the difficulty of our devices, find troublesome such hypotheses. For it is not proper to apply human things to divine things." *Almagest* proposed a view that, unfortunately, became dogma. But it did what Ptolemy said he was hoping to accomplish: It gave all of the details of what was known at the time. Ptolemy added original thought about planetary movements, created a star catalog that became the primary source for star maps for the next 1,400 years, and most important, defined astronomy as a science.

R

Reber, Grote
(1911–2002)
American
Radio Astronomer

From the beginning of recorded time, amateur astronomers have contributed significantly to the study of the cosmos. One such astronomer of the 20th century was Grote Reber, a pioneer in radio astronomy.

A graduate of the Illinois Institute of Technology in 1933, Reber became intrigued with the concept of radio astronomy when, during the same year, Karl Jansky announced his discovery of radio waves emanating from space. The young Reber embarked on his career as an engineer and spent the next 17 years working for radio manufacturers in Chicago, giving him the opportunity to conduct radio research on his own. Reber spent the summer of 1937 building a 31-foot (9-meter) tiltable radio telescope in his backyard in Wheaton, Illinois, becoming the first person to make such an instrument. This was during the Great Depression, and Reber reportedly spent six months' salary for the necessary parts and equipment to build the parabolic dish reflector, which he designed himself.

Over the next several years, Reber constructed three different detectors in an effort to find radio signals at the right wavelengths. It was

the third receiver that, in 1938, picked up the elusive radio emissions Reber was searching for at 160 megahertz. By 1940, and again in 1944, he had accumulated enough information to publish a series entitled "Cosmic Static" in the *Astrophysical Journal*, and the study of radio astronomy was officially under way. The year 1944 was a busy one for this important discoverer, as he also located radio emissions coming from the Sun and from the nearby Andromeda Galaxy.

Around the middle of the 20th century, Grote Reber moved his operations to an estate in Tasmania, Australia, but first he donated his original radio telescope to the National Radio Astronomy Observatory in Green Bank, Virginia, where it is still on display. His relocation to Tasmania was, as he wrote, to "build a more elaborate structure capable of being called a radio telescope." By the time he was through, Reber had constructed an array of 192 dipoles, or antennas, that spanned an area of 3,520 feet in diameter. He focused his efforts on the low-frequency radio waves that he was able to best receive in this part of the world.

Reber achieved several firsts in his field. In addition to building the first radio telescope, some of his most notable work includes classifying radio signals by density and brightness, and using his findings of radio frequency to create and publish contour maps of space. This work was significant enough to earn him the Catherine Wolfe Bruce Gold Medal, awarded by the Astronomical Society of the Pacific in 1962. Reber became only the second amateur astronomer to receive the award since the organization's beginnings in 1889.

Reber continued his work in Australia as a pioneer in long wavelength radio astronomy. The year 1962 was a big one for accolades in Reber's life. In addition to the Bruce Medal, he received an honorary doctorate from Ohio State University, the Cresson Gold Medal from the Franklin Institute of Pennsylvania, and the American Astronomical Society's Henry Norris Russell Lectureship. Additional honors include the Jansky Prize from the National Radio Astronomy Observatory (NRAO) in 1976, and the Jackson-Gwilt Medal, in 1983, from the Royal Astronomical Society.

Beginning in 1938, Reber published more than 50 papers on his work. "Reber was the first to systematically study the sky by observing something other than visible light," according to Fred Lo, director of the NRAO. Reber's scientific papers are all in the NRAO's library. In his honor, the SETI Institute in Mountain View, California, published a song in their League Songbook to the tune of *If You're Happy and You Know It*, a short version of the life and times of this beloved amateur. On December 20, 2002, Reber died in Australia, just two days before his 91st birthday.

⊠ Rees, Sir Martin John
(1942–)
British
Astrophysicist

Sir Martin John Rees was among the first astrophysicists to suggest that enormous black holes could power quasars. His investigation of the distribution of quasars helped discredit steady-state cosmology. He has also contributed to the

Sir Martin John Rees, the 15th Astronomer Royal of Great Britain, was among the first astrophysicists to suggest that enormous black holes could power quasars. His previous investigations of the distribution of quasars helped discredit steady-state cosmology. *(AIP Emilio Segrè Visual Archives, John Irwin Collection)*

theories of galaxy formation and studied the nature of the so-called dark matter of the universe. Dark matter is matter believed present throughout the universe that cannot be observed directly because it emits little or no electromagnetic radiation. Contemporary cosmological models suggest that this "missing mass" controls the ultimate fate of the universe. In 1995 Sir Rees accepted an appointment to serve Great Britain as the 15th Astronomer Royal—a position he still held as of 2004.

Rees was born on June 23, 1942, in York, England. He received his formal education at

Cambridge University. Rees graduated from Trinity College in 1963 with a bachelor of arts degree in mathematics and then continued on for his graduate work, completing a Ph.D. degree in 1967. Before becoming a professor at Sussex University in 1972, he held various postdoctoral positions in both the United Kingdom and the United States. In 1973 Rees became the Plumian Professor of Astronomy and Experimental Philosophy at Cambridge and remained in that position until 1991. During this period, he also served two separate terms (1977–82 and 1987–91) as director of the Institute of Astronomy at Cambridge. Among his many professional affiliations, Rees became a fellow of the Royal Society of London in 1979, a foreign associate of the United States National Academy of Sciences in 1982, and a member of the Pontifical Academy of Sciences in 1990. He is currently a Royal Society Professor and a fellow of King's College at Cambridge University. He is also a visiting professor at the Imperial College in London.

In 1992 he became president of the Royal Astronomical Society, and Queen Elizabeth II conferred knighthood upon him. By royal decree in 1995, Sir Martin Rees became Great Britain's Astronomer Royal, the 15th distinguished astronomer to hold this position since the English king Charles II created the post in 1675.

As an eminent astrophysicist, he has modeled quasars and studied their distribution—important work that helped discredit the steady-state universe model in cosmology. He was one of the first astrophysicists to postulate that enormous black holes might power these powerful and puzzling compact extragalactic objects that were first discovered in 1963 by MAARTEN SCHMIDT. Sir Rees maintains a strong research interest in many areas of contemporary astrophysics, including gamma ray bursts, black hole formation, the mystery of dark matter, and anthropic cosmology. One of today's most interesting cosmological issues involves the so-called anthropic principle—the somewhat controversial premise that suggests the universe evolved in just the right way after the big bang to allow for the emergence of life, especially intelligent human life.

One of Rees's greatest talents is his ability to effectively communicate the complex topics of modern astrophysics to both technical and nontechnical audiences. He has written more than 500 technical articles, mainly on cosmology and astrophysics, and several books, including *Cosmic Coincidences* (1989), and *Gravity's Fatal Attraction: Black Holes in the Universe* (1995); *Just Six Numbers* (2000), *New Perspectives in Astrophysical Cosmology* (2000), and *Our Cosmic Habitat* (2001). Rees has received numerous awards for his contributions to modern astrophysics, including the Gold Medal of the Royal Astronomical Society (1987), the Bruce Gold Medal from the Astronomical Society of the Pacific, the Science Writing Award from the American Institute of Physics (1996), the Bower Science Medal of the Franklin Institute (1998), and the Rossi Prize of the American Astronomical Society (2000).

⊠ Rossi, Bruno Benedetto
(1905–1993)
Italian/American
Physicist, Astronomer, Space Scientist

Attracted to high-energy astronomy, Bruno Rossi investigated the fundamental nature of cosmic rays in the 1930s. Using special instruments carried into outer space by sounding rockets, he collaborated with RICCARDO GIACCONI and other scientists in 1962 and discovered the X-ray sources outside the solar system. In the first decade of the Space Age, Rossi made many important contributions to the emerging fields of X-ray astronomy and space physics. To honor these important contributions, NASA renamed the X-ray astronomy satellite successfully launched in December 1995 the *Rossi X-ray Timing Explorer (RXTE)*.

As part of his investigation of cosmic rays in the 1930s, the Italian-American physicist Bruno Rossi developed pioneering electronic instrumentation that supported both nuclear particle physics and high-energy astrophysics. He collaborated with Giacconi and others in 1962 and helped discover the first X-ray sources outside the solar system. During the first decade of the Space Age, Rossi continued to make made many other important contributions to the emerging fields of X-ray astronomy and space physics. (*AIP Emilio Segrè Visual Archives*)

Bruno Rossi was born on April 13, 1905, in Venice, Italy. The son of an electrical engineer, he began his college studies at the University of Padua and received his doctorate in physics from the University of Bologna in 1927. He began his scientific career at the University of Florence and then became the chair in physics at the University of Padua, serving in that post from 1932

to 1938. However, in 1938 the Fascist regime of Benito Mussolini suddenly dismissed Rossi from his position at the university. That year, Fascist leaders went about purging the major Italian universities of many "dangerous" intellectuals who might challenge Italy's totalitarian government and question Mussolini's alliance with Nazi Germany. Like many other brilliant European physicists in the 1930s, Rossi became a political refugee from fascism. Together with his new bride, Nora Lombroso, he departed Italy in 1938.

The couple arrived in the United States in 1939, after short stays in Denmark and the United Kingdom. Rossi eventually joined the faculty at Cornell University in 1940 and remained with that university as an associate professor until 1943. In spring 1943 Rossi's official immigration status as an "enemy alien" was changed to "cleared to top secret," and he joined the many other refugee nuclear physicists at the Los Alamos National Laboratory, in New Mexico. There Rossi collaborated with other gifted scientists from Europe, as they developed the atomic bomb under the top-secret Manhattan Project. Rossi used all his skills in radiation detection instrumentation to provide his colleague, the great nuclear physicist Enrico Fermi (1901–54), an ultrafast measurement of the exponential growth of the chain reaction in the world's first plutonium bomb (called Trinity) that was tested near Alamogordo, New Mexico, on July 16, 1945. In a one-microsecond oscilloscope trace, Rossi's instrument captured the rising intensity of gamma rays from the bomb's supercritical chain reaction—marking the precise moment in world history before and after the age of nuclear weapons.

After World War II, Rossi left Los Alamos in 1946 to become a professor of physics at the Massachusetts Institute of Technology (MIT). In 1966 he became an institute professor—an academic rank at MIT reserved for scholars of great distinction. Upon retirement in 1971, the university honored his accomplishments by bestowing upon him the distinguished academic rank of institute professor emeritus.

Early in his career, Rossi's experimental investigations of cosmic rays and their interaction with matter helped establish the foundation of modern high-energy particle physics. Cosmic rays are extremely energetic nuclear particles that enter Earth's atmosphere from outer space at velocities approaching that of light. When cosmic rays collide with atoms in the upper atmosphere, they produce cascades of numerous short-lived subatomic particles like mesons. Rossi carefully measured the nuclear particles associated with such cosmic ray showers and effectively turned Earth's laboratory into one giant nuclear physics laboratory.

Rossi began his detailed study of cosmic rays in 1929, when only a few scientists had an interest in them or realized their great importance. That year, to support his cosmic ray experiments, he invented the first electronic circuit for recording the simultaneous occurrence of three or more electrical pulses. This circuit is now widely known as the Rossi coincidence circuit. It not only became one of the fundamental electronic devices used in high-energy nuclear physics research, but also was the first electronic AND circuit—a basic element in modern digital computers.

While at the University of Florence, Rossi demonstrated in 1930 that cosmic rays were extremely energetic, positively charged nuclear particles that could pass through a lead shield over a meter thick. Through years of research, Rossi helped remove much of the mystery surrounding the *Höhenstrahlung* ("radiation from above") first detected by VICTOR FRANCIS HESS in 1911–12.

By the mid-1950s large particle accelerators had replaced cosmic rays in much of contemporary nuclear particle physics research, so Rossi used the arrival of the Space Age (late 1957) to became a pioneer in two new fields within observational astrophysics: space plasma physics and X-ray astronomy. In 1958 he focused his attention on the potential value of direct measurements of ionized interplanetary gases by space probes and Earth-orbiting satellites. He and his colleagues constructed a detector (called the MIT plasma cup) that flew into space onboard NASA's *Explorer* X satellite. Launched in 1961, this instrument discovered the magnetopause—the outermost boundary of the magnetosphere, beyond which Earth's magnetic field loses its dominance.

Then, in 1962, Rossi collaborated with Riccardo Giacconi (then at American Science and Engineering, Inc.) and launched a sounding rocket from White Sands, New Mexico, with an early grazing incidence X-ray mirror as its payload. With funding from the U.S. Air Force, the scientists primarily hoped to observe any X-rays scattered from the Moon's surface as a result of interactions with energetic atomic particles from the solar wind. To their great surprise, the rocket's payload detected the first X-ray source from beyond the solar system—Scorpius X-1, the brightest and most persistent X-ray source in the sky. Rossi's fortuitous discovery of this intense cosmic X-ray source marks the beginning of extrasolar (cosmic) X-ray astronomy. Other astronomers performed optical observations of Scorpius X-1 and found a binary star system, consisting of a visually observable ordinary dwarf star and a suspected neutron star. In this type of so-called X-ray binary, matter drawn from the normal star becomes intensely heated and emits X-rays as it falls to the surface of its neutron star companion.

Rossi was a member of many important scientific societies, including the National Academy of Sciences and the American Academy of Arts and Sciences. He was a very accomplished writer. Rossi's books included *High Energy Particles* (1952), *Cosmic Rays* (1964), *Introduction to the Physics of Space* (coauthored with Stanislaw Olbert in 1970), and the insightful autobiography, *Moments in the Life of a Scientist* (1990). He received numerous awards, including the Gold Medal of the Italian Physical Society (1970), the Cresson Medal from the Franklin Institute (1974), the Rumford Award of the American Academy of Arts and Sciences (1976), and the U.S. National Medal of Science (1985). Rossi's scientific genius laid many of the foundations of high-energy physics and astrophysics. He died at home in Cambridge, Massachusetts, on November 21, 1993.

⊠ **Russell, Henry Norris**
(1877–1957)
American
Astronomer, Astrophysicist

Henry Norris Russell was one of the most influential astronomers in the first half of the 20th century. As a student, professor, and observatory director, he worked for nearly 60 years at Princeton University—a truly productive period that also included vigorous retirement activities as professor emeritus and observatory director emeritus. Primarily a theoretical astronomer, he made significant contributions in spectroscopy and to astrophysics. Independent of EJNAR HERTZSPRUNG, Russell investigated the relationship between absolute stellar magnitude and a star's spectral class. By 1913 their complementary efforts resulted in the development of the famous Hertzsprung-Russell (HR) diagram, which is of fundamental importance to all modern astronomers who wish to understand the theory of stellar evolution. Russell also performed pioneering studies of eclipsing binaries and made preliminary estimates of the relative abundance of elements in the universe. Often called "the dean of American astronomers," Russell served the astronomical community very well as a splendid teacher, writer, and research adviser.

Russell was born on October 25, 1877, in Oyster Bay, New York. The son of a Presbyterian

The American astronomer Professor Henry Norris Russell lecturing at Princeton University. Independent of the Danish astronomer Ejnar Hertzsprung, Russell developed the famous Hertzsprung-Russell (HR) diagram that helps astronomers and astrophysicists understand stellar life cycles. *(Courtesy of AIP Emilio Segrè Visual Archives, Margaret Russell Edmondson Collection)*

minister, Russell received his first introduction to astronomy at age five, when his parents showed him the 1882 transit of Venus across the Sun's disk. He completed his undergraduate education at Princeton University in 1897, graduating with the highest academic honors ever awarded by that institution—namely, insigni cum laude (with extraordinary honor). Russell remained at Princeton for his doctoral studies in astronomy, again graduating with distinction (summa cum laude) in 1900.

Following several years of postdoctoral research at Cambridge University, he returned to Princeton in 1905 to accept an appointment as an instructor in astronomy. He remained affiliated with Princeton for the remainder of his life. He became a full professor in 1911 and the following year became director of the Princeton Observatory. He remained in these positions until his retirement in 1947. Starting in 1921, Russell also held an additional appointment as a research associate of the Mount Wilson Observatory, in the San Gabriel Mountains just north of Los Angeles, California. In 1927 he received an appointment to the newly endowed C. A. Young Research Professorship at Princeton, a special honor bestowed upon him by his undergraduate classmates from the class of 1897.

Starting with his initial work at Cambridge University on the determination of stellar distances, Russell began to assemble data from different classes of stars. He noticed that these data related spectral type and absolute magnitude and soon concluded that there were actually two general types of stars: giants and dwarfs.

The Hertzsprung-Russell (H-R) diagram is actually just an innovative graph that depicts the brightness and temperature characteristics of stars. However, since its introduction by Russell at a technical meeting in 1913, it has become one of the most important tools in modern astrophysics. So-called dwarf stars, including the Sun, lie on the main sequence of the H-R diagram—the well-populated region that runs from the top left to lower right portions of this graph. Giant and supergiant stars lie in the upper right portion of the diagram above the main sequence band. Finally, huddled down in the lower left portion of the H-R diagram below the main sequence band are the extremely dense, fading cores of burned-out stars known collectively as "white dwarfs." This term is sometimes misleading because it actually applies to a variety of compact, fading stars that

have experienced gravitational collapse at the end of their life cycle.

Following his pioneering work in stellar evaluation, Russell engaged in equally significant work involving eclipsing binaries. An eclipsing binary is a binary star that has its orbital plane positioned with respect to Earth in such a way that each component star is totally or partially eclipsed by the other star every orbital period. With his graduate student HARLOW SHAPLEY, Russell analyzed the light from such stars to estimate stellar masses. Later he collaborated with another assistant, Charlotte Emma Moore Sitterly (1898–1990), in using statistical methods to determine the masses of thousands of binary stars.

Before the discovery of nuclear fusion, Russell tried to explain stellar evolution in terms of gravitational contraction and continual shrinkage and used the H-R diagram as a flowchart. When HANS ALBRECHT BETHE and other physicists began to associate nuclear fusion processes with stellar life cycles, Russell abandoned his contraction theory of stellar evolution. However, the basic information he presented in the H-R diagram still remained very useful.

In the late 1920s Russell performed a detailed analysis of the Sun's spectrum showing that hydrogen was a major constituent. He noted the presence of other elements, as well as their relative abundances. Extending this work to other stars, he postulated that most stars exhibit a similar general combination of relative elemental abundances (dominated by hydrogen and helium) that became known as the "Russell mixture." This work reached the same general conclusion that was previously suggested in 1925 by the British astronomer Cecilia Helena Payne-Gaposchkin.

Russell was an accomplished teacher, and his excellent two-volume textbook, *Astronomy*, jointly written with Raymond Dugan and John Stewart, appeared in 1926–27. It quickly became a standard text in astronomy curricula at universities around the world. Russell's *Solar System and Its Origin* (1935) served as a pioneering guide for future research in astronomy and astrophysics. Even after his retirement from Princeton in 1947, Russell remained a dominant force in American astronomy. Honored with appointments as both an emeritus professor and an emeritus observatory director, he pursued interesting areas in astrophysics for the remainder of his life.

He received recognition from many organizations, including the Gold Medal of the Royal Astronomical Society (1921), the Henry Draper Medal from the National Academy of Sciences (1922), the Rumford Prize from the American Academy of Arts and Sciences (1925), and the Bruce Gold Medal from the Astronomical Society of the Pacific (1925). He died in Princeton, New Jersey, on February 18, 1957.

Ryle, Sir Martin
(1918–1984)
British
Physicist, Radio Astronomer

Sir Martin Ryle was the British scientist who established an important center for radio astronomy at Cambridge University after World War II. By 1960 his pioneering activities in radio astronomy allowed him to introduce the technique known as aperture synthesis. He shared the 1974 Nobel Prize in physics with ANTONY HEWISH for their pioneering achievements in radio astronomy, specifically Ryle's invention of aperture synthesis and Hewish's identification of the first pulsar. Ryle also served Great Britain as the 12th Astronomer Royal from 1972 to 1982.

Ryle was born in Brighton, Sussex, England, on September 27, 1918. Son of a well-respected physician, he received education in physics at Bradfield College and Oxford University, graduating in 1939. During World War II Ryle worked on the development of radar and other

radio frequency systems for the Royal Air Force. His wartime service in the Telecommunications Research Establishment gave him valuable engineering experience that he would soon apply in innovative ways in radio astronomy. After the war Ryle received a fellowship to conduct radio wave research at the Cavendish Laboratory of the University of Cambridge. The research group he joined at the laboratory was just starting to investigate radio emissions from the Sun. He and his team developed pioneering interferometry techniques to precisely locate radio-emitting regions on the Sun, so other scientists could correlate the radio frequency emissions with visible light emissions. These were exciting research times, as Ryle began to apply wartime advances in radio frequency and electronic technologies to the embryonic field of radio astronomy. In 1947 he married Rowena Palmer and the couple eventually had three children: two daughters and a son.

It was mainly due to Ryle and his colleagues that Cambridge became one of the leading centers in the world for astronomical research. In 1948 Ryle received an appointment to lecture in physics at the university, and the following year he was elected as a fellow at Trinity College. Between 1948 and 1952 Ryle formed a winning radio astronomy research team that included Hewish. With this team, Ryle started using radio interferometry to carefully map the radio sky—an essential task that led to the establishment of radio astronomy as a very fertile field for astronomical research and discovery. He supervised the development and publication of a series of extraterrestrial radio source catalogs. By the time the *Third Cambridge Catalogue* was published in 1959, it contained the positions and strengths of more than 500 radio sources. This document, and timely updates to it, became an essential reference for all radio astronomers.

In the late 1950s Cambridge formally recognized the important work Ryle and his team were accomplishing in radio astronomy. He became the director of the Mullard Radio Astronomy Observatory in 1957, and then received an appointment to a new chair in radio astronomy. This promotion made him the university's first professor of radio astronomy.

As part of his pioneering efforts in radio astronomy in the 1950s, Ryle learned how to use several small radio telescopes to mimic the performance of a much larger instrument by selectively sampling portions of just the wave front of interest from an arriving extraterrestrial radio frequency signal. He called this invention "aperture synthesis" and constructed two major aperture synthesis radio telescopes at Cambridge to support the university's expanding astronomical effort. These two facilities were called the One Mile Telescope and the Five Kilometer Telescope.

Strange, repetitive radio signals collected by the One Mile Telescope led to the discovery and identification of the first pulsar by Hewish and his graduate student, SUSAN JOCELYN BELL BURNELL, in 1967. The Five Kilometer Telescope, consisting of four movable 13-meter dishes and four fixed 13-meter dishes, opened in 1972 and soon became the main instrument of the Mullard Radio Astronomy Observatory. It is now called the Ryle Telescope to honor the pioneering radio astronomer.

Ryle became a fellow of the Royal Society in 1952 and Queen Elizabeth II knighted him in 1966. When he shared the Nobel Prize in physics with Hewish in 1974, they became the first scientists to receive this prestigious award for achievements in astronomy. Ryle also served as the 12th Astronomer Royal of England from 1972 to 1982. His most important publication was the *Third Cambridge Catalogue* (1959)—an essential reference in radio astronomy that, among many other contributions, assisted the discovery of the first quasar. His other awards included the Gold Medal of the Royal Astronomical Society (1964), the Henry Draper Medal of the American National Academy of Sciences (1965), the Royal (Gold) Medal of the Royal Society (1973), and the Bruce Gold Medal from the Astronomical Society of the Pacific (1974).

Ryle died on October 14, 1984, in Cambridge, England. As a pioneer of radio astronomy, he developed revolutionary radio telescope systems using aperture synthesis. With these new telescopes, he and his colleagues observed some of the universe's most distant galaxies and puzzling extragalactic objects.

S

Sagan, Carl Edward
(1934–1996)
American
Astronomer, Astrophysicist, Physicist

Perhaps no other person in the 20th century was as responsible for bringing the science of astronomy and space exploration into the public eye and educating hundreds of millions of people worldwide about the nature of the cosmos as Carl Sagan.

Sagan was born in 1934 in Brooklyn, New York. In 1951 he entered the University of Chicago, where he received a bachelor's degree in 1955 and a master's degree in 1956, both in physics, and then a Ph.D. in astronomy and astrophysics in 1960. He then taught at Harvard University through the 1960s and went to Cornell University in Ithaca, New York, in 1971, where he remained for the rest of his life.

Sagan's research highlights were many. He postulated the greenhouse effect on Venus before space probes proved it. He predicted that the apparent seasonal changes on the surface of Mars were the result of windblown dust. Sagan identified organic aerosols on Titan, the major moon of Saturn. He described in great detail the long-term environmental consequences of nuclear war, and he wrote extensively on the origin of life on Earth.

A masterful storyteller and eager host of innumerable conferences, meetings, and radio and TV shows, Sagan published more than 600 scientific papers and articles and more than 20 books. His book *Dragons of Eden* won the Pulitzer Prize in 1978. The paperbound edition of his book *Pale Blue Dot: A Vision of the Human Future in Space* was a worldwide best seller. His best-selling novel *Contact* was made into a major motion picture coproduced by Sagan and his wife and frequent collaborator, Ann Druyan. All in all, books written or cowritten by Sagan were on the *New York Times* best-seller list eight times.

Sagan's translation of science into commonly understood terms and concepts is possible best remembered by his TV series on PBS, *Cosmos*, an Emmy– and Peabody Award–winning show that became the most watched series in PBS history, seen by more than 500 million people in 60 countries. The accompanying book, *Cosmos*, was on the *New York Times* best-seller list for 70 weeks and became the best-selling book about science ever published.

Needless to say, Sagan's awards and honorary degrees are almost too numerous to list. He received 22 honorary degrees from American institutions. For his leading role in NASA's Mariner, Viking, Voyager, and Galileo missions to other planets, he was awarded NASA medals for Exceptional Scientific Achievement and twice received the Distinguished Public Service and the NASA Apollo Achievement Award. Most notable of his other awards are the John F. Kennedy Astronautics Award of the American Astronautical Society; the Explorers Club 75th Anniversary Award; the Konstantin Tsiolkovsky medal of the Soviet Cosmonauts Federation; and the Mazursky Award of the American Astronomical Society. He received the highest award given by the National Academy of Sciences, the Public Welfare Medal. He was Distinguished Visiting Scientist at the Jet Propulsion Laboratory in Pasadena, California.

Sagan was editor of *Icarus*—a professional journal devoted to planetary research—for 12 years, and was a cofounder of the Planetary Society, the largest space-interest organization in the world, with 100,000 members.

At the time of his death on December 20, 1996, of pneumonia resulting from myelodisplasia in a hospital in Seattle, Washington, Sagan was the David Duncan Professor of Astronomy and Space Sciences, and the director of the Laboratory for Planetary Studies at Cornell University. The asteroid 2709 Sagan is named after him.

⊠ **Scheiner, Christoph**
(1573–1650)
German
Astronomer, Mathematician

Born in Wald, near Mindelheim, in what is now southwest Germany, Christoph Scheiner is best known for his protracted controversy with GALILEO GALILEI over the nature of sunspots— Scheiner thought they were shadows cast by the Sun's satellites, while Galileo was convinced they were irregular phenomena on the Sun's surface. In his later years, Scheiner gave in and admitted that Galileo was right.

As a boy, Scheiner entered the Jesuit Latin school in Augsburg, then continued his studies at the Jesuit college in Landsberg before he entered the then-young Jesuit order in 1595. Scheiner studied mathematics, Hebrew, and metaphysics at the university in Ingolstadt, and in 1610 he joined the faculty at the Jesuit College in Ingolstadt as a professor of mathematics and Hebrew.

At first, Scheiner became an expert in the mathematics of sundials, but when the telescope came into wide use he obtained one and turned his studies toward the Sun. Using colored glass and observing for the most part on somewhat foggy days, Scheiner first observed sunspots in 1611. As the chief proponents of traditional Aristotelian cosmology, and hailed by many as the intellectual champions of the Catholic Church, the Jesuits and the Catholic Church in general adhered to the "purity" of the universe and PTOLEMY's geocentric view of the Earth's position in the heavens. However, the Copernican heliocentric view was becoming popular, and scientists such as Galileo were beginning to prove him right.

Largely because of his religion's insistence that the heavens were unchangeable and the Sun was "virginal," Scheiner concluded that sunspots were shadows cast on the Sun's surface by small planets closely circling it inside Mercury's orbit. Using a pseudonym, he sent three letters to his friend Marcus Wexler, an Augsburg banker and Jesuit patron, describing his conclusions, and Wexler showed the letters, called "Three Letters on Solar Spots," to Galileo, inviting him to comment on them.

Thus began a correspondence in which the two scientists argued voluminously with one another. Galileo, to Scheiner's consternation, insisted that sunspots were actually on the Sun's surface, changing their shape and appearing and reappearing at irregular intervals. That the Sun was not perfect was a heretical position, and contributed seriously to Galileo's falling out with Rome and the Jesuit order. Nevertheless, and although he seems to have taken pains not to insult Scheiner, Galileo was ultimately proven correct, and in his publication *Rosa Ursina,* which was to remain the standard work on sunspots for more than a century, Scheiner eventually capitulated to Galileo.

After his first observations, Scheiner continued to study sunspots for 15 years, and in *Rosa Ursina* he also described new methods of representing the motions of sunspots across the Sun's surface. Scheiner also published several works on atmospheric refraction and performed detailed studies on the optics of the human eye, describing the functions of the retina, pupil, and iris. He elaborated on the previous optical work of JOHANNES KEPLER, and built what he called a "helioscope" for projecting images of the Sun, improving on the projection methods described by Galileo and Castelli. This instrument is the earliest known equatorially mounted projector.

While conducting his sunspot research, Scheiner was a mathematics instructor to Archduke Maximilian, brother of Emperor Rudolph II. This endeared him to royalty and in 1621 he became confessor to Archduke Karl, brother of the new emperor, Ferdinand II. At this time he founded a new Jesuit college at Neisse, in Silesia. When the archduke died in 1624, Scheiner journeyed to Rome, remaining there for eight years and continuing to observe the Sun at the observatory of the Gregoriana. After charting the movements of sunspots, he calculated the rotation of the Sun at 25 days near the equator and 31 days at the poles. Using this information, he then calculated the equator and the rotation axis, discovered solar flares, and observed the granulation of the photosphere. While in Rome, he published the famous *Rosa Ursina.*

In 1633 Scheiner went back to Bavaria, where he spent the rest of his life in the area of Vienna, both supervising the new college in Neisse and working on a refutation of Copernican theory. The latter work went virtually unknown and is reported to have little effect on the Copernicus versus Ptolemy argument. Scheiner died on June 18, 1650, in Neisse.

⊠ **Schiaparelli, Giovanni Virginio**
(1835–1910)
Italian
Astronomer

Giovanni Schiaparelli was the 19th-century Italian astronomer whose misinterpreted words stimulated undue excitement about civilization on Mars. An excellent astronomer, he had carefully observed Mars in the 1870s and made a detailed map of its surface, including some straight markings that he dutifully described as *canali*— meaning "channels" in Italian. Unfortunately, when his description of Martian surface features was translated into English, the word *canali* became "canals." Some astronomers, most notably the wealthy American astronomer PERCIVAL LOWELL, completely misunderstood the true meaning of Schiaparelli's observations and launched a zealous telescopic search for the supposed canals that represented the handiwork of a neighboring intelligent alien civilization on the Red Planet.

Schiaparelli was born on March 14, 1835, in Savigliano, Italy. He graduated from the University of Turin in 1854 with a degree in civil engineering. After graduation, he engaged in the private study of astronomy, foreign languages, and mathematics. In 1856 he received an appointment to teach mathematics in an elementary school in Turin—an experience that he did not particularly enjoy. By 1857, government officials of the Piedmont region of Italy recognized his talents and interest in astronomy and so they provided Schiaparelli with a modest fellowship to pursue additional efforts in observational astronomy. Much to his parents' dismay, he abandoned his teaching appointment, accepted the small stipend, and studied astronomy under Johann Encke (1791–1865) in Berlin and Otto

Struve (1819–1905) at the Pulkova Observatory in St. Petersburg, Russia.

When Schiaparelli returned to Italy in 1860, he became the second astronomer at the Brera Observatory in Milan. Two years later, he became the observatory's director and retained that position until he retired in 1900. In 1861 Schiaparelli, an exceptionally skilled observer, used the antiquated equipment he found at the Brera Observatory as best he could and managed to discover the asteroid Hesperia. Over the years he struggled to upgrade the observatory facilities, while still making significant contributions to astronomy. He linked the periodic meteor shower called the Perseids to the remnants of a comet, and he conducted observational and theoretical studies concerning the shapes of comet tails. Through careful and patient observation, he also showed that Venus and Mercury rotated very slowly.

He also studied the planet Mars and made detailed observations during its 1877 opposition, when the Red Planet approached within about 56 million kilometers of Earth. From these careful observations, Schiaparelli made a detailed map of Mars, even using his knowledge of classical literature and the Bible to name various surface features. Schiaparelli's keen eye for details caused him to note certain straight features as *canali*. The mistaken translation of *canali* appearing in Schiaparelli's 1877 Mars report spurred another astronomer, the wealthy Lowell, to build a private observatory at Flagstaff, Arizona, and diligently study Mars in an attempt to prove through a set of carefully selected (but improperly interpreted) observations that the "canals" on Mars were manifestations of the presence of an ancient Martian civilization. Lowell even suggested the unfounded, but very popular, hypothesis that this race of Martians was struggling to transport water from the poles to water-starved regions of their dying planet.

Schiaparelli never endorsed Lowell's extensive extrapolation of some of his best work in solar system astronomy. Yet, this erroneous interpretation of his *canali* in an otherwise excellent observational report about Mars is how most people remember the Italian astronomer. If Schiaparelli had never published his now infamous 1877 Mars report, he might still have won many international awards. He received the Gold Medal of the Royal Astronomical Society in 1872 and the Bruce Gold Medal from the Astronomical Society of the Pacific in 1902.

In 1900 he decided to retire as directory of the Brera Observatory because of his failing eyesight. However, even though he could no longer practice astronomical observing, he still contributed to the field of astronomy by researching the history of ancient astronomy and writing the book *Astronomy in the Old Testament* in 1903. He died on July 4, 1910, in Milan, Italy.

⊠ Schmidt, Maarten
(1929–)
Dutch/American
Astronomer, Astrophysicist

Modern astronomers are constantly discovering that the universe is filled with interesting and strange objects. In 1963 the joy of discovering the unusual, touched astronomer Maarten Schmidt as he analyzed the enormous redshift in radio source 3C 273 and found the first quasar. Schmidt's analysis of the unexpectedly large redshift in the hydrogen lines of this strange object's spectrum indicated that it was very young and very distant and traveling away from Earth at more than 15 percent of the speed of light.

Schmidt was born in Groningen, the Netherlands, on December 28, 1929. He completed his undergraduate education at the University of Groningen in 1949 and then earned his doctoral degree in 1956 at the University of Leiden under the research supervision of JAN HENDRIK OORT. After performing about three years of postdoctoral research at the University of Leiden, Schmidt came to the United States to accept an appointment at the California Institute of Technology (Caltech) as a staff member at the Hale Observatories (Palomar Observatory) northeast of San Diego. Schmidt initially accepted this position primarily to continue investigating the structure and dynamics of the Milky Way Galaxy. But when the German-American astrophysicist Rudolph Minkowski (1895–1976) retired, Schmidt assumed responsibility for his project of taking optical spectra of distant objects that had been found to emit radio wave signals.

At the time, astrophysicists were puzzling over several compact radio sources, originally identified in SIR MARTIN RYLE's *Third Cambridge Catalogue*, published in 1959. In the early 1960s the American astronomer Alan Sandage (b. 1926), discovered a mysterious class of radio-loud objects (that is, celestial objects that emit detectable radio wave signals) with an ultraviolet excess in color. At the time, no one really knew what these strange objects might be, because they were starlike in size and had variable brightness, yet were radio sources. It was not until radio source 3C 273 was identified as one of these puzzling and bizarre radio sources that interesting things began to happen in the astronomical community. Astronomers originally called the objects "quasi-stellar radio sources"—quasars—because of their starlike appearance (hence "quasi-stellar") and the fact they were strong radio sources. However, subsequent discoveries forced astrophysicists to change the term to "quasi-stellar object" (QSO)—expanding the meaning to include a compact extragalactic object that appears like a point source of light but emits more energy than 100 or so galaxies.

In 1963 Schmidt examined the corresponding optical spectrum of the brightest quasar, called 3C 273. His insight and calculations helped unravel this cosmic mystery, and for this reason he is generally credited with identifying the first quasar. He recognized that certain broad emission lines in the corresponding optical spectrum of radio source 3C 273 were actually hydrogen lines in the Balmer series that had been redshifted an enormous amount, approximately that of an object receding from Earth at about 16 percent of the velocity of light. Assuming that this enormous redshift was the result of an expanding universe, Schmidt estimated 3C 273 to be about a billion light-years away. This strange, compact object was not very far away. It also had the energy output of perhaps 100 galaxies.

In 1964 Schmidt became a professor of astronomy at Caltech and remained affiliated with that institution as of 2004. Following his breakthrough work on identifying 3C 273 as the first quasar, Schmidt studied the evolution and distribution of quasars throughout the observable universe. He soon discovered that quasars were more abundant when the universe was younger, a fact that challenged the steady-state universe hypothesis and supported Big Bang cosmology. His current research interests center around the spatial distribution and luminosity function of quasars in the radio frequency, optical, and X-ray portions of the electromagnetic spectrum. He also maintains an interest in the nature of the extragalactic X-ray background, the statistics associated with gamma-ray bursts, and the distribution of mass within the Galaxy.

Since identifying the first quasar in 1963, Schmidt's astronomical endeavors have earned him numerous awards and international recognition. These awards include the Rumford Prize of the American Academy of Arts and Sciences (1968), the Jansky Prize from the National Radio Astronomy Observatory of the United States (1979), the Gold Medal of the Royal Astronomical Society (1980), and the Bruce Gold Medal from the Astronomical Society of the Pacific (1992). Professor Schmidt continues to perform creative research at the frontiers of astrophysics at Caltech.

⊠ **Schwabe, Samuel Heinrich**
(1789–1875)
German
Astronomer (Amateur)

Professional astronomers are not the only people who have made interesting astronomical discoveries. Samuel Heinrich Schwabe was a German pharmacist who spent most of his free time as an amateur astronomer. Starting in 1825, he began making systematic observations of the Sun and continued this activity for many years. He was primarily searching for the hypothetical planet of Vulcan believed to be inside the orbit of Mercury. However, instead of finding this fictitious planet, he discovered that sunspots have a cycle of about 10 years or so. He announced his findings in 1843, but German scientists generally ignored his discovery. It was only after the sunspot cycle was rediscovered and confirmed by professional astronomers in the 1850s that Schwabe received some recognition for his excellent work.

Schwabe was born on October 25, 1789, in the town of Dessau, near Berlin, Germany. He studied pharmacology in Berlin and then returned to Dessau in 1812 to assume management of his family's apothecary shop. While studying to become a pharmacist, Schwabe also became interested in astronomy and soon immersed himself in various astronomical studies as a hobby. In 1825, according to one biographical anecdote, Schwabe won his first telescope in a lottery. He soon ordered a more powerful viewing instrument from the famous German optician JOSEPH VON FRAUNHOFER.

In October 1825 Schwabe began his systematic observations of the Sun—a dedicated effort that extended for more than 17 years. His primary objective was to discover the planet Vulcan that some professional astronomers postulated resided close to the Sun inside the orbit of Mercury. While trying to catch a glimpse of this nonexistent planet as it traveled across the face of the Sun, Schwabe also noticed and sketched sunspots—thereby establishing a very careful record of their population and patterns as a function of time. He made very detailed sunspot records, but the sunspots themselves were of only secondary interest. Weather conditions permitting, he observed the Sun almost daily.

By 1843, however, his quest for Vulcan proved fruitless. At this point, inspiration struck Schwabe. As he reviewed almost 17 years of carefully recorded sunspot data, he suddenly noticed that there was an approximate 10-year (now established more precisely as 11 years) periodicity in the number of sunspots on the solar disk. Later that year, he announced this important astronomical observation in an article titled "Solar Observations during 1843" that appeared in the German journal *Astronomische Nachrichten*. At first, the German scientific establishment completely ignored Schwabe's discovery, possibly because he was an amateur astronomer with no formal academic status in physics or astronomy.

But Schwabe's careful record of sunspots and his conjecture about an apparent 10-year periodicity emerged out of obscurity and gained long-overdue scientific recognition in 1851 when the famous German naturalist Baron Alexander von Humboldt (1769–1859) included these data in the third volume of his monumental work, *Kosmos*, a vast encyclopedia of natural knowledge. Once Humboldt recognized the value of Schwabe's work, other scientists immediately began to examine the sunspot cycle and to relate the periodicity of solar activity to geomagnetic activity of Earth. In a very real sense, this rediscovery of Schwabe's sunspot cycle hypothesis represents the beginning of modern solar-terrestrial physics. Schwabe, the pharmacist-astronomer, personally likened his serendipitous discovery to the biblical account of Saul, who went out in search of his father's donkey and found a kingdom.

In 1831 Schwabe became the first astronomer, amateur or professional, to provide a detailed description and drawing of Jupiter's Great Red Spot. The British Royal Astronomical Society awarded Schwabe its prestigious Gold Medal in 1857 and elected him as a foreign member in 1868—an honor rarely bestowed on an amateur astronomer. He died in Dessau, Germany, on April 11, 1875. As part of his legacy to astronomy, the Swiss astronomer Johann Rudolf Wolf (1816–93) immediately built upon Schwabe's many years of detailed solar disk observations and was able to announce in 1857 a refined value for sunspot periodicity at slightly more than 11 years.

⊠ Schwarzschild, Karl
(1873–1916)
German
Astronomer, Astrophysicist

Karl Schwarzschild was the talented German astronomer who started black hole astrophysics. He did this by applying ALBERT EINSTEIN's relativity theory to a very high-density object: a point mass. In 1916 while voluntarily serving in the German army in Russia, he developed the concept of the Schwarzschild radius—the zone or "event horizon" around a superdense, gravitationally collapsing star from which nothing, not even light itself, can escape.

He was born in Frankfurt, Germany, on October 9, 1873. The son of a prosperous Jewish businessman, Schwarzschild showed his mathematical aptitude and interest in astronomy early in life. At age 16 he published a significant celestial mechanics paper that addressed the problem of binary orbits. Encouraged to pursue a scientific career, he began his studies at the

The German astronomer Karl Schwarzschild poses here in full academic regalia early in the 20th century. Decades ahead of his time, Schwarzschild brilliantly applied Einstein's relativity theory to very high-density objects and singularities (point masses)—thereby starting black hole astrophysics. In 1916 he introduced the idea of an event horizon (now called the Schwarzschild radius) around a super-dense, gravitationally collapsing star, from which nothing, not even light, can escape. *(Photo by Robert Bein, courtesy AIP Emilio Segrè Visual Archives)*

University of Strasbourg and then transferred to the University of Munich in 1893, completing his doctor of philosophy degree there in 1896.

After graduation Schwarzschild worked in Vienna from 1896 to 1899 at the Kuffner Observatory. He then spent some time lecturing and writing before he joined the faculty at the University of Göttingen in 1901 as an associate professor. He remained at the university until 1909, quickly reaching rank of full professor and becoming the director of the observatory. During this period, he explored the possibility that space might be non-Euclidean (that is, have curvature), he investigated radiative equilibrium processes in the Sun's outer atmosphere, and he promoted the use of astrophotography—particularly to support the study of the relationship between a star's spectral type and its color.

In 1909 he left the University of Göttingen to accept the position of director of the Astrophysical Observatory at Potsdam. Although already in his early 40s when World War I broke out, he volunteered for military service in the Kaiser's imperial army. Starting in 1914, he served with the German army in Belgium, France, and finally Russia. While on duty in Russia in 1916, he stayed in touch with scientific developments by writing two excellent papers that dealt with Einstein's newly introduced (1915) general theory of relativity. One of Schwarzschild's papers presented the first exact solution of Einstein's general gravitational equations, providing a description of how gravity curves space and time around a single, very compact object. In his second paper, Schwarzschild introduced the concept of the event horizon around a black hole, beyond which nothing, not even light, can escape. Schwarzschild demonstrated that according to general relativity, if an object's gravity is sufficiently strong, the space-time continuum around that body becomes so highly curved and warped that light itself cannot escape from the object. He further explained that this condition occurs when the radius of a massive body is less than a certain critical value, now called the Schwarzschild radius. Today, astrophysicists call the Schwarzschild radius the "event horizon" or "boundary of no return" for a black hole. Anything that falls inside this radius can never escape. The event horizon represents the start of a special region that is totally disconnected from normal space and time. Schwarzschild's papers, written weeks before his

death, represent the theoretical start of black hole astrophysics.

Unfortunately, the brilliant German astronomer contracted a fatal skin disease while serving in Russia during World War I. The German army discharged the stricken scientist and shipped him home, where he died shortly after his return to Potsdam, on May 11, 1916. However, his son, the German-born American astrophysicist Martin Schwarzschild (1912–97), would continue making important contributions to astronomy. Following in his father's scientific footsteps, Martin left Germany after World War II and became a professor at Princeton University, where he pioneered the use of digital computers to perform theoretical studies of stellar structure and evolution.

⊠ **Secchi, Pietro Angelo**
(1818–1878)
Italian
Astronomer, Spectroscopist

The 19th-century astronomer and Jesuit priest Pietro Angelo Secchi was the first person to systematically apply spectroscopy to astronomy. By the mid-1860s he completed the first major astronomical spectroscopic survey and then published a catalog that contained the spectra of more than 4,000 stars. After examining these data, in about 1867 he placed stellar spectra into four basic classes. This represented what was later referred to as the Secchi classification system. He also supported advances in astrophotography, primarily by photographing solar eclipses—pioneering work that assisted in the study of solar phenomena such as prominences.

Secchi was born on June 29, 1818, in the Lombardian town of Reggio, Italy. In 1833 he entered a religious order known as the Society of Jesus (Jesuits), and began lecturing on physics and mathematics at the Jesuit Roman College in 1839. He started his formal theological studies in 1844 and was ordained as a priest in the Roman Catholic Church in 1847. Due to extreme antireligious political unrest that occurred in Rome in 1848, he left Italy and traveled to Great Britain and then on to the United States, where he taught natural sciences at Georgetown University in Washington, D.C. By 1849 order was restored in Rome and Secchi returned to become a professor of astronomy and the director of the observatory at the Roman College. Finding the old observatory in great disrepair and poorly equipped, he started construction of a new observatory dome on top of the firm vault of the Church of Saint Ignatius at the college. From this refurbished observatory he became one of the first astronomers to carry out research that concentrated on the physical properties of the stars rather than simply their positions. He also conducted pioneering work in meteorology and terrestrial magnetism.

Soon after ROBERT WILHELM BUNSEN and GUSTAV ROBERT KIRCHHOFF introduced the concept of astronomical spectroscopy in the early 1860s, Secchi became the first astronomer to adapt spectroscopy to astronomy in a rigorous, systematic manner. In the mid-1860s he conducted the first spectroscopic survey of the heavens by using visual observation to their spectra to classify more than 4,000 stars. He divided these stars into four general groups, introducing a system that later became known as the Secchi classification. His pioneering work anticipated the more precise photographic work of EDWARD CHARLES PICKERING, ANNIE JUMP CANNON, and others who developed the Harvard Classification System, which appeared in the 1890s.

Secchi was also extremely active in solar physics. He investigated prominences during solar eclipses, using both visual and spectroscopic observations. He proved that prominences were features on the Sun itself and not an ancillary phenomenon associated with the Moon or the eclipse period. His influential monograph *Le Soleil* (The Sun) first appeared in Paris in 1870. Secchi enjoyed a wide international reputation as a skilled observer and excellent astronomer. He was consequently allowed to operate his observatory in Rome despite the political turmoil and anti-Jesuit politics that swept Italy in the early 1870s. He enjoyed membership in the British Royal Society, the Royal Astronomical Society, the French Academy of Sciences, and the Imperial Russian Academy of St. Petersburg. Pietro Secchi, a true pioneer of astronomical spectroscopy, died in Rome on February 26, 1878.

⊠ Shapley, Harlow
(1885–1972)
American
Astronomer

Early in his career, astronomer Harlow Shapley single-handedly more than doubled the size of the known universe in 1914, when he used a detailed study of variable stars (especially Cepheid variables) to establish more accurate dimensions for the Milky Way Galaxy. He discovered that the Sun was actually two-thirds of the way out in the rim of our spiral galaxy. Up until then, astronomers assumed the Sun enjoyed a favored location near the center of the Galaxy. In 1920 Shapley engaged in a well-publicized, though inconclusive, debate with fellow astronomer HEBER DOUST CURTIS concerning the nature of distant spiral nebula and the true size of the Milky Way Galaxy. As director of the Harvard College Observatory, Shapley made many useful contributions to astronomy, including studies of the Magellanic Clouds and clusters of galaxies.

Shapley was born on November 2, 1885, in Nashville, Missouri. He started out in professional life as a journalist and then decided to switch to astronomy. He changed careers during that exciting and intellectually turbulent period in the

The American astronomer Harlow Shapley at his desk at the Harvard College Observatory. In 1914 Shapley started a detailed study of variable stars in an effort to establish more accurate dimensions for the Milky Way Galaxy. This work encouraged him to engage in a great public debate with fellow astronomer Heber Curtis. They disagreed about the actual size of the Milky Way and the true nature of distant spiral nebulas. In 1929, however, Edwin Hubble presented his hypothesis of an expanding universe filled with spiral and other types of distant galaxies, collapsing many of Shapley's key debate points and resolving the highly contested discussions. *(AIP Emilio Segrè Visual Archives, Shapley Collection)*

vast expanding universe. Prior to entering the University of Missouri, Shapley had worked for two years as a newspaper reporter. However, once exposed to astronomy during his undergraduate education, he embraced the field and quickly became one of its rising young stars. He completed his undergraduate degree in astronomy at the University of Missouri in 1910, followed a year later by his master of arts degree from the same institution. Shapley then went to Princeton University where he studied for his Ph.D. under the great American astronomer HENRY NORRIS RUSSELL.

Shapley's doctoral research involved a careful study of the orbits of 90 eclipsing binaries. It was a major effort, supervised by Russell, through which Shapley essentially created a special branch of binary star astronomy. In the process of completing his doctoral research, Shapley distinguished Cepheid variables from eclipsing binaries and then correctly suggested Cepheid variability was due to pulsations.

Following completion of his doctoral degree in 1914, Shapley worked at the Mount Wilson Observatory, near Los Angeles, California, until 1921. There he investigated globular clusters and was able to calibrate the period-luminosity relation for Cepheid variables discovered by HENRIETTA SWAN LEAVITT in 1912. He boldly and correctly postulated that the distribution of these globular clusters served as an outline for the Milky Way Galaxy. Using Cepheid variables as an astronomical "yardstick," he speculated that the Milky Way Galaxy was actually much larger than previously believed. He further noted that because of the observed asymmetric distribution of these globular clusters, the Sun was not near the center of the Galaxy, as had been assumed, but rather resided many light-years away from the center. Shapley, initially unaware of the dimming influence of interstellar gas and dust, estimated that the Sun was about 50,000 light-years from the galactic center. He later corrected this overestimate and placed the Sun at

early part of the 20th century when great astronomical discoveries, including his important contributions, displaced the Sun from its long-assumed position near the center of the Galaxy. Astronomers also recognized that the Milky Way Galaxy was but one of millions of others in a about 30,000 light-years from the galactic center—a value independently confirmed by JAN HENDRIK OORT and now generally accepted. Shapley's pioneering work at Mount Wilson also prepared him to participate in debates about the size of the Galaxy and the true nature of distant spiral galaxies.

Many astronomers, including Shapley, opposed Curtis's hypothesis about the extragalactic nature of spiral nebulas. On April 26, 1920, Curtis and Shapley engaged in their famous "Great Debate" on the scale of the universe and the nature of the Milky Way Galaxy at the National Academy of Sciences in Washington, D.C. At the time, the vast majority of astronomers considered the extent of the Milky Way Galaxy synonymous with the size of the universe—that is, most thought the universe was just one big galaxy. However, because of incomplete knowledge, neither astronomer involved in this highly publicized debate was completely correct. Curtis argued that spiral nebulas were other galaxies similar to the Milky Way—a bold hypothesis later proven to be correct—but he also suggested that the Galaxy was small and that the Sun was near its center, both of which ideas were subsequently proven incorrect. Shapley, on the other hand, incorrectly opposed the hypothesis that spiral nebulas were other galaxies. He argued that the Galaxy was very large (in fact, much larger than it actually is) and that the Sun was far from the galactic center.

It was not until the mid-1920s that the great American astronomer EDWIN POWELL HUBBLE, helped resolve one of the main points of controversy. He did this by using the behavior of Cepheid variable stars to estimate the distance to the Andromeda Galaxy. Hubble showed that this distance was much greater than the size of the Milky Way Galaxy proposed by Shapley. So, as Curtis had suggested, the great spiral nebula in Andromeda, like other spiral nebulas, could not be part of the Milky Way and therefore must be separate. In the 1930s astronomers such as Oort proved that Shapley's comments were more accurate concerning the actual size of the Milky Way and the Sun's relative location within it. Therefore, when viewed from the perspective of science history, both eminent astronomers had inconclusively argued their positions using partially faulty and fragmentary data. The most important consequence of the Curtis-Shapley debate was that it triggered a new wave of astronomical inquiry in the 1920s. This burst of inquiry encouraged astronomers, Hubble among them, to determine the true size of the Milky Way and to recognize that it is but one of many other galaxies in an incredibly vast, expanding universe.

Upon the death of EDWARD CHARLES PICKERING, Shapley was offered the directorship of the Harvard College Observatory. He accepted, left Mount Wilson in 1921, and remained as the director of the Harvard Observatory until his retirement in 1952. During his directorship, Shapley created an outstanding graduate school in astronomy, performed his own studies of the Magellanic Clouds, and wrote many interesting articles and books about astronomy, including the popular autobiographical work *Through Rugged Ways to the Stars* (1969). In the 1930s Shapley discovered the first two dwarf galaxies within the Local Group and also discovered the grouping of clusters of galaxies, or superclusters—the most prominent of which is now called the Shapley concentration.

Shapley's contributions to astronomy earned numerous awards, including the Henry Draper Medal of the National Academy of Sciences (1926), the Rumford Prize of the American Academy of Arts and Sciences (1933), the Gold Medal of the Royal Astronomical Society (1934), and the Bruce Gold Medal of the Astronomical Society of the Pacific (1939). The astronomer who helped displace the Sun from its historically assumed position near the center of the universe died in Boulder, Colorado, on October 20, 1972.

⊠ Sitter, Willem de
(1872–1934)
Dutch
Astronomer, Mathematician

Willem de Sitter was born May 6, 1872, at Sneek, in Friesland, a province in the northern part of the Netherlands. His early education was at the Gymnasium at Arnhem, where his father was a judge. In 1891 he entered the University of Groningen, where he intended to study mathematics. However, while there he became interested in physics and astronomy and worked in the Groningen Astronomical Laboratory, then under the direction of the famous Professor JACOBUS CORNELIUS KAPTEYN.

In 1896 SIR DAVID GILL, who was the Royal (British) Astronomer at the observatory at the Cape of Good Hope, South Africa, paid a visit to Kapteyn at the Groningen laboratory and was introduced to de Sitter as one of Kapteyn's brilliant mathematics students. At the time, de Sitter was at a measuring instrument, making computations on a photographic plate. The following morning, de Sitter was having breakfast in his room when he received a summons to meet with Gill in the laboratory. There, with Mrs. Kapteyn acting as interpreter, Gill invited de Sitter to visit the South African observatory and offered him a job as assistant. After explaining that he was a mathematician and not an astronomer, and requesting (and receiving) another year's time in order to complete his doctoral degree, de Sitter agreed. Thus was born the astronomy career of de Sitter.

Arriving at the Cape observatory in August 1897, de Sitter began his initial work in parallax and triangulation computations with a heliometer. He observed the colors of stars and was first to determine that stars near the plane of the Milky Way are always bluer than those away from the plane. But his main interest soon became observations of the moons of Jupiter. Combining his mathematical talents and his newly acquired astronomy techniques, de Sitter studied the motions of the major moons of Jupiter and calculated their masses and the mass of Jupiter itself.

After working with Gill for more than two years, de Sitter returned to the Groningen Astronomical Laboratory. In 1908 he was appointed professor of astronomy at the University of Leiden, the oldest university in the Netherlands, established in 1575, and whose observatory, established in 1633, is the oldest operating university observatory in the world.

In this position, de Sitter made observations in 1913 of double-star systems and became the first astronomer to prove that the velocity of light had nothing to do with the velocity of its source. Also while at Leiden, de Sitter became familiar with the newly promulgated theory of general relativity of ALBERT EINSTEIN, then director of the Kaiser Wilhelm Institute of Physics in Berlin. Einstein visited de Sitter at Leiden, but the political situation in Europe, as war was fast approaching, prevented the two scientists from working together. Einstein began sending his correspondence to de Sitter, who in turn sent them to SIR ARTHUR STANLEY EDDINGTON at the Royal Observatory in Greenwich, England.

De Sitter then published a series of three papers entitled "On Einstein's Theory of Gravitation and Its Astronomical Consequences." The first two papers explained Einstein's theories, and the third proposed his own interpretation of the astronomical consequences of Einstein's theory. One biographer states that while Einstein, who knew little of astronomy, wondered how astronomy could be applied to his theory, de Sitter became the first scientist to apply the theory of general relativity to astronomy. Indeed, de Sitter's solution in 1917 to Einstein's field equations showed that a practically empty universe was expanding, instead of contracting, as was the common theory at the time.

Later, with the war over and the political climate eased, de Sitter and Einstein published a joint paper in which they proposed there may be in the universe large amounts of matter that does not emit light and consequently has not been detected. This became known as "dark matter" and has since been shown to exist; however, its exact nature is still under investigation.

In 1919 de Sitter was named director of the Leiden Observatory and, under the guidance of Kapteyn, expanded and modernized it adding an astrophysical division dedicated to the spectroscopy and photometry of stars, and a theoretical division, which he headed himself. De Sitter also had the wisdom to hire EJNAR HERTZSPRUNG, who would succeed him and co-invent the famous Hertzsprung-Russell diagram of stars, still in use today. While director at Leiden, de Sitter also hired the man who would become the most famous of all Leiden astronomers, JAN HENDRIK OORT.

De Sitter was awarded the Watson Medal of the National Academy of Science in 1929, the Bruce Gold Medal of the Astronomical Society of the Pacific and the Gold Medal of the Royal Astronomical Society in 1931. During the period 1925–28 he was president of the International Astronomical Union.

The chance meeting in 1896 of Gill and de Sitter changed the course of the latter's life from mathematics to theoretical astronomy and gave the world new knowledge of Jupiter and its moons and the first interpreter and analyst of Einstein's revolutionary theories. De Sitter died of pneumonia on November 19, 1934, at Leiden.

⊠ Somerville, Mary Fairfax
(1780–1872)
British
*Astronomer, Geographer,
Mathematician*

Self-taught, with only one year of formal education at a boarding school when she was 10 years old, the prolific Mary Somerville became one of the most scientifically influential persons of the 19th century and, as one contemporary observer put it, "one of the most remarkable women of her generation."

She was born in 1780 at Jedburgh, Roxburghshire, Scotland, to Margaret Charters Fairfax and naval officer Sir William George Fairfax, who later became vice-admiral of the Scottish navy. However, Jedburgh was just a stopover; the family home was in Burntisland, in the county of Fife.

The social mores of the time dictated that boys got the education and girls learned needlepoint. Somerville's mother did teach her to read, using the family Bible, but saw no reason for her daughter to learn how to write. When she was 10, Somerville was sent to Miss Primrose's Boarding School, where she spent a dismal and unhappy year, but did learn to write. When she returned home, she had such a craving for knowledge that she began to read every book in the house, to the point where she was criticized by family members for her "unladylike" preoccupation. Consequently, she was sent to a special school to learn needlework. She also took piano lessons and had lessons in painting from the artist Alexander Naysmyth.

Visiting an uncle in Jedburgh, Mary casually mentioned to him that she was teaching herself Latin. The uncle, perhaps the only family member to appreciate her quest for knowledge, encouraged her, and together they studied Latin whenever she visited.

In her early teens she became interested in mathematics when her art teacher, Naysmyth, taught her the basics of perspective in painting by using Euclid's *Elements*. She immediately read the whole book, which led to her interest in astronomy. However, it was not proper for young ladies to buy science books in those days, so Somerville prevailed upon her brother to smuggle her his science books. She also began visiting her brother's tutor to learn more.

Her next passion was algebra. While reading puzzles in a friend's woman's magazine, she found her curiosity piqued by the strange symbols in an algebra problem, and she borrowed algebra texts from the tutor. She became so immersed in algebra and other mathematical branches that her parents worried about her health.

At age 24 she married a Russian naval officer, Samuel Grieg, who had a low opinion of her interest in education. They settled in London, and when Grieg died, only three years later, he left her financially independent and she seized her newfound free time to pursue her interest on her own. Now the mother of two children, she had a circle of friends in London who encouraged her pursuit of science, including a professor of natural philosophy at Edinburgh University named John Playfair, who introduced her by mail to William Wallace, a professor of mathematics at the Royal Military College at Great Marlow. In their letters he taught her how to solve mathematical problems to the point where she eventually won a silver medal for her solution to a problem published in *Mathematical Repository*. In accepting the medal, she sought the advice of the editor of the magazine for the proper course of mathematical study, which led to her reading of SIR ISAAC NEWTON's *Principia* and PIERRE SIMON MARQUIS DE LAPLACE's *Mécanique Celeste*, as well as several other prominent astronomical and mathematical textbooks.

In a big turning point in her life, she married William Somerville in 1812. He was a surgeon in the Royal Navy who was supportive of her interest in science. By this time Somerville also was fluent in French and Greek, and in Edinburgh both she and her husband studied geology and traveled in a close circle of physicists and mathematicians. William Somerville was transferred to London and accepted into the Royal Society, which led them into another tight circle of the leading scientists of the day associated with the University of Cambridge and several prominent scientific societies, both in London and on the Continent.

Somerville published her first paper in 1826, at age 46: "The Magnetic Properties of the Violet Rays of the Solar Spectrum." It attracted interest at the time, but its premise was eventually refuted. However, it gained her her first attention as a serious scientific investigator, and the next year Lord Brougham, of the Society for the Diffusion of Useful Knowledge, asked Somerville to translate *Mécanique Celeste* for popular consumption. Instead, she explained in minute detail the mathematics used by the Frenchman Laplace, which was unfamiliar to most English mathematicians of the day. Her "translation" eventually became a ponderous tome, too big for the society to publish, and it was published in 1831 as *The Mechanism of the Heavens*.

The book was a smashing success both critically and financially, and boosted her reputation immensely. She spent a year in Paris, meeting even more prominent scientists, and wrote her second book, *The Connection of the Physical Sciences*. Her discussion in the book of a theoretical planet perturbing Uranus led JOHN COUCH ADAMS to his discovery of Neptune. This produced a shower of honors. She and CAROLINE LUCRETIA HERSCHEL were the first women elected to the Royal Astronomical Society in 1835. The Société de Physique et d'Histoire Naturelle de Genève gave her an honorary membership, and the Royal Irish Academy did the same.

In 1838 William's health deteriorated and the family moved to Italy, where William survived for 22 more years while Mary wrote books at a prolific rate. Her most important later publication was *Physical Geography*, which was published when she was 68, in 1848, and it became her most successful textbook, used in secondary schools and universities until the turn of the 20th century. This book got her elected to the

American Geographical and Statistical Society and the Italian Geographical Society. In 1870, at 90 years of age, she received the Victoria Gold Medal of the Royal Geographical Society.

Somerville did more to publicize mathematics and astronomy than any other woman of her time, and, for obvious reasons, was a devout supporter of women's suffrage and women's education. When John Stuart Mill, the great British philosopher and economist, encouraged the British parliament to give women the vote, he amassed a huge petition to submit to them. The first name on his list was Mary Fairfax Somerville.

⊠ Slipher, Vesto Melvin
(1875–1969)
American
Astronomer

Behind many great scientific breakthroughs, there is usually some generally unrecognized pathfinder who first marked the trail for others to travel and achieve technical glory. The astronomer Vesto Slipher was the pathfinder for the famous American astronomer EDWIN POWELL HUBBLE. In 1912 Slipher began his important series of spectroscopic studies involving the light from objects called "spiral nebula" at the time, which are now recognized as distant galaxies. He noticed that the spectroscopic data he collected at the Lowell Observatory exhibited interesting Doppler shift phenomena. The predominant number of spectral redshifts he observed in the spiral nebulas under study suggested that these objects were receding from Earth at very high speed. Although his data were generally ignored when he presented them in the late 1910s, his pioneering spectroscopic studies provided the framework around which Hubble and other astronomers eventually developed the modern concept of an expanding universe.

The American astronomer Vesto Slipher worked at the Lowell Observatory and in about 1912 performed the initial spectroscopic studies of spiral nebulas (galaxies) that revealed Doppler redshifts. His work became the foundation upon which Edwin Hubble and others pursued the concept of an expanding universe. Later, as director of the Lowell Observatory, he hired a young Kansas farm boy, Clyde Tombaugh, and assigned him the arduous task of searching for Percival Lowell's Planet X. Tombaugh delivered on February 18, 1930, when he discovered the planet Pluto. *(AIP Emilio Segrè Visual Archives)*

Slipher was born on November 11, 1875, in Mulberry, Indiana. He received all his education at the University of Indiana at Bloomington. There he received his bachelor of arts degree in mechanics and astronomy in 1901, his M.A. degree in 1903, and his Ph.D. degree in 1909.

In August 1901 Slipher started working at the Lowell Observatory, in Flagstaff, Arizona. He became the assistant of the observatory in 1915, and when PERCIVAL LOWELL died in 1916, Slipher became the acting director of the observatory. In 1926, he became its director and served in that position until his retirement in 1954.

Slipher was not only an excellent spectroscopist but he also proved to be a competent administrator. One of his early decisions as director was to hire a young farm boy named CLYDE WILLIAM TOMBAUGH in the late 1920s and assign him the task of searching the night skies for Lowell's Planet X. On March 13, 1930, in his capacity as director of the Lowell Observatory, Slipher had the pleasure of announcing to the astronomical world that Tombaugh had discovered the planet Pluto.

But about three decades earlier, Slipher had made his own great contribution to astronomy at the Lowell Observatory—a generally overlooked discovery that changed the trajectory of modern cosmology. Starting in about 1912, he used spectroscopic measurements and exposure times as long as 80 hours to measure the enormous radial velocities of spiral nebulas. Slipher determined the radial velocities of these so-called spiral nebulas (now known to be galaxies beyond the Milky Way) by measuring the displacement of their spectral lines. The Doppler effect, for example, will shift spectral lines toward the red (longer wavelength) portion of the visible spectrum if the object is going away (receding) from Earth and toward the blue or violet portion of the spectrum if the object is approaching at high velocity.

In 1913 Slipher examined the spectral lines from the great spiral nebula in Andromeda (also known as Messier catalogue number M31 or the Andromeda Galaxy) and, to his surprise, discovered this object was approaching Earth (indicated by blueshifted spectral lines) at approximately 300 kilometers per second. He extended this work to many other spiral nebulas and soon discovered that out of the 40 he had carefully observed, 36 displayed redshifted spectral lines and only four blueshifted spectral lines. Slipher also measured radial velocities in excess of many thousands of kilometers per second.

The mystery presented by these data puzzled him and other astronomers including HARLOW SHAPLEY. However, unable to resolve the mystery astronomers preferred to simply ignore these important data, often suggesting that Slipher's experimental technique was faulty or his interpretation of the data in error. But one great astronomer did not ignore the incredibly important implications of Slipher's work. Hubble used Slipher's work as a starting point to develop his cosmology of an expanding universe filled with millions of galaxies receding from Earth. Hubble presented these ideas in the late 1920s and changed forever our view of the vast size and dynamic nature of the universe.

As a very skilled astronomer, Slipher applied spectroscopic techniques at the Lowell Observatory to prove the existence of vast quantities of dust and gas in interstellar space and to investigate the atmospheric composition of the gaseous giant planets Jupiter, Saturn, Uranus, and Neptune. In the early 1930s he also investigated the phenomenon of zodiacal light—the faint cone of light extending upward from Earth's horizon in the direction of the ecliptic caused by the reflection of sunlight from tiny pieces of interplanetary dust in orbit around the Sun.

Numerous awards marked Slipher's contributions to astronomy, including the Lalande Prize and Gold Medal from the French Academy of Sciences (1919), the Henry Draper Medal of the National Academy of Sciences (1932), the Gold Medal of the Royal Astronomical Society (1933), and the Bruce Gold Medal from the Astronomical Society of the Pacific (1935). The astronomer whose pioneering spectroscopic studies of spiral nebulas led to our modern understanding of the expanding universe died in Flagstaff, Arizona, on November 8, 1969, just four days short of his 94th birthday.

Stefan, Josef
(1835–1893)
Austrian
Physicist

In about 1879 Josef Stefan experimentally demonstrated that the energy radiated per unit time by a blackbody was proportional to the fourth power of that body's absolute temperature. When LUDWIG BOLTZMANN provided the theoretical foundations for this relationship in 1884, the collaboration of the two Austrian physicists resulted in the formulation of the famous Stefan-Boltzmann law—a physical principle of great importance to astronomers and astrophysicists.

Stefan was born on March 24, 1835, in St. Peter, near Klagenfurt, Austria. He received his childhood education in Klagenfurt and then entered the University of Vienna to study physics and mathematics. Upon graduation in 1858, he received an appointment as lecturer in mathematical physics at the university. By 1866 he earned the rank of professor of mathematics and physics and also became the director of the Physical Institute in Vienna. He stayed affiliated with the university for the remainder of his life. He was an excellent experimental physicist and a very well-liked instructor. One of his most famous students was Boltzmann.

As an experimental physicist, Stefan had a wide range of research interests, including optical interference, the kinetic theory of gases, electromagnetic induction, diffusion, thermomagnetic phenomena, and the cooling rate of hot bodies. He is best remembered for his pioneering work with respect to radiation heat transfer. Stefan carefully investigated thermal energy losses from hot bodies, such as heated platinum wires, and he became the first scientist to effectively quantify the phenomenon of radiation heat transfer. In 1879 he noted that the rate of thermal energy loss was proportional to the fourth power of an object's absolute temperature. For example, if the temperature of an object doubled, say from 1,000 to 2,000 kelvins, the amount of thermal energy it radiated increased 16-fold.

In 1884 one of his former students, Boltzmann, provided the theoretical basis for Stefan's empirical observations. Boltzmann used thermodynamic principles to develop the famous physical law that is now called the Stefan-Boltzmann law. This law states that the luminosity, or radiant power, of a blackbody is proportional to the fourth power of the blackbody's absolute temperature. Astronomers and astrophysicists often use this law to describe and compare the radiant properties and temperatures of stars, because to a good approximation stars closely approximate the behavior of blackbody radiators. For convenience, astronomers will also frequently use the luminosity, radius, and absolute surface temperature of the Sun as their reference and then apply the Stefan-Boltzmann law to form a ratio that compares the corresponding values for another star to this solar baseline.

During his lifetime, Stefan was honored with special appointments, including those of Rector Magnificus at the University of Vienna in 1876 and vice president of the Viennese Academy of Sciences in 1885. He died on January 7, 1893, in Vienna, Austria. His radiation heat transfer research provided a major breakthrough in the understanding of blackbody radiation and set the stage for MAX KARL PLANCK and his development of the quantum theory of thermal radiation.

T

Tereshkova, Valentina
(1937–   )
Russian
Cosmonaut

The early race into space between the Soviets and the Americans produced a win for the Soviets and for women in space exploration when Valentina Tereshkova became the first woman to fly into space on June 16, 1963.

From her humble beginnings in the Russian village of Masslenikovo, where she was born to farmworkers on March 6, 1937, young Tereshkova could easily have been considered least likely to become a world-renowned figure. At age 18, the young woman left school to work in a textile mill. In hindsight, one of the most important decisions to her future career came when she joined a parachutists' club. Her first jump, at age 22, would do far more than propel her from a plane. It would ultimately help launch her into the Russian space program.

In 1961 Tereshkova decided to become a cosmonaut. Her timing was good: Since Gherman Titov had recently made a flight into space, the Russian government was contemplating something new—sending a woman into space. Fortunately for Tereshkova, they decided that this woman must be a parachutist.

One of only five women to be considered for the role, Tereshkova was selected for the *Vostok* 6 flight in May 1963 by Premier Khrushchev himself, who was pushing the space race to beat out the United States with as many Soviet "firsts" as possible. The following month, she had her historic moment in space. The event lasted 70 hours, 50 minutes, allowing Tereshkova to orbit the Earth once every 88 minutes. Parachuting came into play when, upon reentering the atmosphere, she jumped from the spacecraft and landed in central Asia.

While she never ventured into space again, Tereshkova played a pivotal role for women in astronomy as the first woman to physically enter this new frontier. She has since received two Order of Lenin awards, as well as the United Nations Gold Meal of Peace, the Simba International Women's Movement Award, and the Joliot-Curie Gold Medal.

Tombaugh, Clyde William
(1906–1997)
American
Astronomer

Clyde Tombaugh was the "farm boy astronomer" who discovered the planet Pluto on February 18,

1930. He did this while working as a junior astronomer at the Lowell Observatory near Flagstaff, Arizona. His success came through very hard work, perseverance, and Tombaugh's skilled use of the "blinking comparator"—an innovative approach to astrophotography based on the difference in photographic images taken a few days apart. His discovery of the elusive trans-Neptunian planet also helped fulfill the dreams of the observatory's founder, PERCIVAL LOWELL, who had predicted the existence of a "Planet X" decades earlier.

Clyde Tombaugh was born on February 4, 1906, on a farm near Streator, Illinois. When he was in secondary school, his family moved to the small farming community of Burdett in the western portion of Kansas. When Tombaugh graduated from Burdett High School in 1925, he had to abandon any immediate hope of attending college, because crop failures had recently impoverished his family. While growing up on the farm, he acquired a very strong interest in amateur astronomy from his father. To complement this interest, he taught himself mathematics and physics. In 1927 he constructed several homemade telescopes, including a 9-inch reflector built from discarded farm machinery, auto parts, and handmade lenses and mirrors. Using this reflector telescope under the dark skies of western Kansas, he spent evenings observing and made detailed drawings of Mars and Jupiter. Searching for advice and comments about his observations, Tombaugh submitted his drawings in 1928 to the professional planetary astronomers at the Lowell Observatory in Flagstaff, Arizona. To his great surprise, the astronomers were very impressed with his detailed drawings, which revealed his great talent as a careful and skilled astronomical observer.

So despite the fact that Tombaugh lacked a formal education in astronomy, VESTO MELVIN SLIPHER, the director of the Lowell Observatory, hired him in 1929 as a junior astronomer. Slipher also placed Tombaugh on a rendezvous trajectory with astronomical history when he assigned the young "farm boy astronomer" to the monumental task of searching the night sky for Planet X. As early as 1905, Lowell had noticed subtle perturbations in the orbits of Neptune and Uranus, which he attributed to the gravitational tug of some undetected planet farther out in the solar system. After several more years of investigation, Lowell boldly predicted in 1915 the existence of this "Planet X" and began to conduct a detailed photographic search in candidate sections of the sky from his newly built observatory in Flagstaff. Unfortunately, Lowell died in 1916 without finding his mysterious Planet X.

Using a 13-inch photographic telescope at the Lowell Observatory, Tombaugh embarked on a systematic search for the planet. He worked through the nights in a cold, unheated dome, making pairs of exposures of portions of the sky with time intervals of two to six days. He then carefully examined these astrographs (star photographs) under a special device called the blink-comparator in the hope of detecting the small shift in position of one faint point of light among hundreds of thousands of points of light—the sign of Planet X among a field of stars. On the nights of January 23 and 29, 1930, Tombaugh made two such photographs of the region of the sky near the star Delta Geminorum. On the afternoon of February 18, 1930, he triumphed. While viewing these photographic plates under the blink-comparator, Tombaugh detected the telltale shift of a faint starlike object. Slipher and other astronomers at the observatory verified the results and rejoiced in Tombaugh's discovery. However, caution was urged and time was taken for Tombaugh to confirm this important discovery. The discovery of Pluto was confirmed with subsequent photographic measurements and announced to the world on March 13, 1930.

With the announcement of this discovery, Tombaugh joined an exclusive group of astronomers who observed and then named major planets in the solar system. Of the many thousands of astronomers in history who have searched the heavens with telescopes, only the following accompany Tombaugh in this very elite group—SIR WILLIAM HERSCHEL, who discovered Uranus in 1781, and JOHANN GOTTFRIED GALLE, who discovered Neptune in 1846 (credit for this discovery is shared by JOHN COUCH ADAMS and URBAIN-JEAN-JOSEPH LEVERRIER). In keeping with astronomical tradition, Tombaugh had the right as the planet's discoverer to give the distant celestial body a name. He chose Pluto, god of darkness and the underworld in Roman mythology.

Pluto is a tiny, frigid world, very different from the gaseous giant planets that occupy the outer solar system. For that reason there is some debate within the astronomical community as to whether Pluto is really a "major planet" or perhaps should be regarded as an icy moon that escaped from Neptune or perhaps a large Kuiper belt object. As more and more trans-Neptunian objects are discovered beyond the orbit of Pluto, this debate will most likely intensify in the 21st century. But for now, Pluto is regarded as the ninth major planetary body in the solar system and Tombaugh's discovery represents a marvelous accomplishment in 20th-century observational astronomy.

After discovering Pluto, Tombaugh remained with the Lowell Observatory for the next 13 years. During this period, he also entered the University of Kansas on scholarship in 1932 to pursue the undergraduate education he had been forced to delay because of financial constraints. In 1934, while attending the university, he met and married Patricia Edson of Kansas City. The couple remained married for more than 60 years and raised two children. While observing during the summers at the Lowell Observatory, Tombaugh earned his bachelor of science degree in astronomy in 1936 and then his master of arts degree in astronomy in 1939. He frequently told the story of his perplexed astronomy professor

who did not want a "planet discoverer" in his basic astronomy course.

Upon graduation, he returned to the Lowell Observatory and continued a rigorous program of sky watching that resulted in his meticulous cataloging of more than 30,000 celestial objects, including hundreds of variable stars, thousands of new asteroids, and two comets. He also engaged in a search for possible small natural satellites encircling Earth. However, save for the Moon, he could not find any natural satellites of Earth that were large enough or bright enough to be detected by means of photography.

During World War II, Tombaugh taught navigation to military personnel from 1943 to 1945 at Arizona State College (now called Northern Arizona University). Following World War II, the Lowell Observatory did not rehire him as an astronomer, because of funding reductions. Instead Tombaugh went to work for the military at the White Sands Missile Range, in Las Cruces, New Mexico. There he supervised the development and installation of the optical instrumentation used during the testing of ballistic missiles, including WERNHER VON BRAUN's V-2 rockets that had been captured in Germany by the U.S. Army at the close of the war.

Tombaugh left his position at the White Sands Missile Range in 1955 and joined the faculty at New Mexico State University in Las Cruces. He helped this university establish a planetary astronomy program and remained an active observational astronomer. Upon retirement from the university in 1973, he went on extensive lecture tours throughout the United States and Canada, accompanied by his wife, to raise money for scholarships in astronomy at New Mexico State University. Several weeks before his 91st birthday, Tombaugh died on January 17, 1997, in Las Cruces.

Tombaugh was recognized around the world as the discoverer of Pluto, and the only American to discover a planet. He became a professor emeritus in 1973 at New Mexico State University and published several books, including *The Search for Small Natural Earth Satellites* (1959) and *Out of the Darkness: The Planet Pluto* (1980), coauthored with Patrick Moore.

⊠ Tsiolkovsky, Konstantin Eduardovich
(1857–1935)
Russian
Physicist, (Theoretical) Rocket Engineer

The nearly deaf Russian schoolteacher Konstantin Eduardovich Tsiolkovsky was a theoretical rocket expert and space travel pioneer light-years ahead of his time. At the beginning of the 20th century, Tsiolkovsky worked independent of ROBERT HUTCHINGS GODDARD and HERMANN JULIUS OBERTH, toward their common vision: the use of rockets for interplanetary travel. This brilliant schoolteacher lived a simple life in isolated, rural towns in czarist Russia. Yet he wrote with such uncanny accuracy about modern rockets and space that he cofounded astronautics. Primarily a theorist, Tsiolkovsky never constructed any of the rockets he proposed in his remarkably prophetic books but they inspired many future Russian cosmonauts, space scientists, and rocket engineers, including SERGEI PAVLOVICH KOROLEV, whose powerful rockets helped fulfill Tsiolkovsky's predictions.

Tsiolkovsky was born on September 17, 1857, in the village of Izhevskoye, in the Ryazan Province of Russia. His father, Eduard Ignatyevich Tsiolkovsky, was a Polish noble by birth, but now in exile working as a provincial forestry official. His mother, Mariya Yumasheva Tsiolkovskaya was Russian and Tartar. At age nine a near-fatal attack of scarlet fever left him almost totally deaf. With his loss of hearing, he adjusted to a lonely, isolated childhood in which books became his friends. He also learned to educate himself and in the process acquired a high degree of self-reliance.

At age 16 Tsiolkovsky ventured to Moscow, where he studied mathematics, astronomy, mechanics, and physics. He used an ear trumpet to listen to lectures and struggled with a meager allowance of just a few kopecks (pennies) each week for food. Three years later Tsiolkovsky returned home. He soon passed the schoolteacher's examination and began his teaching career at a rural school in Borovsk, located about 100 kilometers from Moscow. In Borovsk he met and married his wife, Varvara Sokolovaya. He remained a provincial schoolteacher in Borovsk for more than a decade. Then, in 1892, Tsiolkovsky moved to another teaching post in Kaluga, where he remained until he retired in 1920.

As he began his teaching career in rural Russia, Tsiolkovsky also turned his fertile mind to science, especially concepts about rockets and space travel. Despite his severe hearing impairment, Tsiolkovsky's tenacity and self-reliance allowed him to become an effective teacher and also to make significant contributions to the fields of aeronautics and astronautics.

But teaching in rural Russian villages in the late 19th century physically isolated Tsiolkovsky from the mainstream of scientific activities, both in his native country and elsewhere in the world. He independently worked out the kinetic theory of gases in 1881, then proudly submitted a manuscript concerning this original effort to the Russian Physico-Chemical Society. Unfortunately for Tsiolkovsky, the famous chemist Dmitri Mendeleyev (1834–1907) informed him that the theory had been developed a decade earlier. However, the originality and quality of Tsiolkovsky's paper impressed Mendeleyev and the other reviewers, and they invited him to become a member of the society.

While teaching in Borovsk, Tsiolkovsky used his own meager funds to construct the first wind tunnel in Russia. He did this so he could experiment with airflow over various streamlined bodies. He also began making models of gas-filled, metal-skinned dirigibles. His interest in aeronautics served as a stimulus for his more visionary work involving the theory of rockets and their role in space travel. As early as 1883, he accurately described the weightlessness condition of space in an article entitled "Free Space." In his 1895 book, *Dreams of Earth and Sky*, Tsiolkovsky discussed the concept of an artificial satellite orbiting Earth. By 1898 he correctly linked the rocket to space travel and concluded that it would have to be a liquid-fueled rocket to achieve the necessary escape velocity. The escape velocity is the minimum velocity an object must acquire in order to overcome the gravitational attraction of a large celestial body, such as planet Earth. To completely escape from Earth's gravity, for example, a spacecraft would need to reach a minimum velocity of approximately 11 kilometers per second.

Many of the fundamental principles of astronautics were described in his seminal work, *Exploration of Space by Reactive Devices*. This important theoretical treatise showed that space travel was possible using the rocket. Another pioneering concept found in the book is a design for a liquid-propellant rocket that used liquid hydrogen and liquid oxygen. Tsiolkovsky delayed publishing the important document until 1903. One possible reason for the delay is the fact that Tsiolkovsky's son, Iganty, had committed suicide in 1902.

Because of Tsiolkovsky isolation his important work in aeronautics and astronautics went essentially unnoticed by the world. Few in Russia cared about space travel in those days and he never received significant government funding to pursue any type of practical demonstration of his innovative concepts. His suggestions included the space suit, space stations, multistage rockets, large habitats in space, the use of solar energy, and closed life support systems.

In the first two decades of the 20th century, things seemed to go from bad to worse in Tsiolkovsky's life. In 1908 an overflowing river flooded his home and destroyed many of his notes and scientific materials. Undaunted, he salvaged what he could, rebuilt, and pressed on with teaching and writing about space.

Following the Russian Revolution of 1917, the new Soviet government grew interested in rocketry and rediscovered Tsiolkovsky's amazing work, honoring him for his previous achievements in aeronautics and astronautics and encouraging him to continue his pioneering research. He received membership in the Soviet Academy of Sciences in 1919, and the government granted him a pension for life in 1921 in recognition of his teaching and scientific contributions. Tsiolkovsky used the free time of retirement to continue to make significant contributions to astronautics. In 1924 he published his book *Cosmic Rocket Trains*, in which he recognized that a single-stage rocket would not be powerful enough to escape Earth's gravity on its own, so he developed the concept of a staged rocket, which he called a rocket train. Tsiolkovsky's visionary writings inspired many future cosmonauts and aerospace engineers, including Korolev.

Tsiolkovsky died in Kaluga on September 19, 1935. His epitaph conveys the important message "Mankind will not remain tied to Earth forever." As part of the Soviet Union's centennial celebration of Tsiolkovsky's birth, one of Korolev's envisioned powerful rockets launched *Sputnik 1*, the world's first artificial satellite—starting the Space Age on October 4, 1957.

U

Urey, Harold Clayton
(1893–1981)
American
Chemist, Exobiologist

Harold Urey was the American physical chemist who won the 1934 Nobel Prize in chemistry for his discovery of the hydrogen isotope deuterium. In World War II he contributed to uranium isotopic enrichment efforts as part of the U.S. atomic bomb program known as the Manhattan Project. Then, starting in the early 1950s, he made a very exciting contribution to the emerging field of planetary sciences by conducting one of the first experiments in exobiology—a classic experiment often known as the Urey-Miller experiment.

He was born on April 29, 1893, in Walkerton, Indiana, the son of a minister and the grandson of one of the pioneers who originally settled the area. Following his early education in rural schools and high school graduation in 1911, Urey taught for three years in country schools. Then, in 1914, he enrolled in the University of Montana and earned his bachelor of science degree in zoology in 1917. He spent the next two years as an industrial research chemist and then returned to Montana to teach chemistry. In 1921 he entered the University of California and graduated in 1923 with a Ph.D. in chemistry.

With funding from an American-Scandinavian Foundation fellowship, Urey traveled to Denmark, where he spent a year in postdoctoral research at Niels Bohr's Institute of Theoretical Physics. Upon his return to the United States, he became an associate professor in chemistry at Johns Hopkins University.

In 1929 Urey accepted an appointment as an associate professor in chemistry at Columbia University. There, in 1931, while engaging in research on diatomic gases, he devised a method to concentrate any possible heavy hydrogen isotopes by the fractional distillation of liquid hydrogen. His efforts led to the discovery of deuterium (D), the nonradioactive heavy isotope of hydrogen that forms heavy water (D_2O). He became a full professor at Columbia in 1934. From 1940 to 1945 Urey also served as the director of War Research, Atomic Bomb Project, at Columbia University. After World War II he moved to the University of Chicago and served there as a distinguished professor of chemistry until 1955.

While at the University of Chicago in the early 1950s, he made a dramatic break away from his previous Nobel Prize–winning efforts in terrestrial chemistry and began to investigate the possible origins of life on Earth and elsewhere in the universe from the perspective of extraterrestrial chemistry. As one of the first exobiologists, he introduced some of the basic ideas in this field

The American chemist and Nobel laureate Harold Urey at work in his laboratory. Early in his career, Urey discovered the nonradioactive isotope of ordinary hydrogen called deuterium. He received the 1934 Nobel Prize in chemistry for this accomplishment. Later in his career, he became one of the earliest exobiologists as he investigated the origins of life on Earth and the possibility of life in other worlds. In 1953, together with his student Stanley Miller, Urey performed the classic exobiology experiment in which gaseous mixtures that simulated Earth's primitive atmosphere were subjected to various energy sources. To their pleasant surprise, life-forming organic compounds (amino acids) appeared in the solutions. *(Argonne National Laboratory, courtesy AIP Emilio Segrè Visual Archives)*

with his 1952 book *The Planets: Their Origin and Development*. In 1953, together with his graduate student Stanley Miller (b. 1930), Urey performed the famous exobiology experiment that is now widely known as the "Urey-Miller experiment." They created gaseous mixtures, simulating Earth's primitive atmosphere, and then subjected these mixtures to various energy sources, such as ultraviolet radiation and lightning discharges. Within days, life-precursor organic compounds known as amino acids began to form in some of the test beakers.

The intriguing question that other scientists began to seriously ask after the Urey-Miller experiment was: "If life started in a primordial chemical soup here on Earth, does (or can) life start on other worlds that have similar primitive environments?" The question of whether life is unique to Earth or common throughout the Galaxy remains unanswered in the 21st century.

In 1958 Urey accepted a position at the University of California in La Jolla, and he remained with that institution until he retired in 1970. He died on January 5, 1981, in La Jolla.

The 1934 Nobel Prize in chemistry highlighted his achievements as a terrestrial chemist, but his innovative work in extraterrestrial chemistry may have far greater technical consequences. Thanks to Urey's supervision of pioneering experiments at the University of Chicago, exobiology has become a credible part of contemporary space science programs—most notably reflected today by NASA's continued search for microscopic life on Mars and the space organization's planned search for possible alien life in the suspected subsurface liquid water ocean on the Jovian moon Europa.

Van Allen, James Alfred
(1914–　)
American
Physicist, Space Scientist

The pioneering space scientist James Van Allen placed the Iowa cosmic ray experiment on the first U.S satellite, *Explorer 1*, and with this instrument discovered the inner portion of Earth's trapped radiation belts in early 1958. Today, space scientists call this distinctive zone of magnetically trapped atomic particles around Earth the Van Allen radiation belts in his honor.

Van Allen was born in Mount Pleasant, Iowa, on September 7, 1914. During his sophomore year at Iowa Wesleyan College, he made his first measurements of cosmic ray intensities. After graduation in 1935, he attended the University of Iowa, where he earned his master's degree in 1936 and completed his doctoral degree in physics in 1939. From 1939 to 1942 Van Allen worked at the Carnegie Institution of Washington as a physics research fellow in the department of terrestrial magnetism. As a Carnegie Fellow, he received valuable cross-training in geomagnetism, cosmic rays, nuclear physics, solar-terrestrial physics, and ionospheric physics. All of this scientific cross-training prepared him well for his leading role as

the premier U.S. space scientist at the beginning of the Space Age.

In 1942 Van Allen transferred to the Applied Physics Laboratory at Johns Hopkins University to work on rugged vacuum tubes. He received a wartime commission that autumn in the U.S. Navy, and served as ordnance specialist for the remainder of World War II. One of his primary contributions was the development of an effective radio proximity fuse—one that detonated an explosive shell when the ordnance came near its target. Following combat duty in the Pacific, he returned to the laboratory at Johns Hopkins University in Maryland. One afternoon in March 1945, he quite literally ran into his future wife, Abigail, during a minor traffic accident. Six months later both drivers were married and their fortuitous traffic encounter eventually yielded five children and a house full of grandchildren.

In the postwar research environment, Van Allen began applying his wartime engineering experience to miniaturize the rugged new electronic equipment. He used this small, but tough, equipment in conjunction with his pioneering rocket and satellite scientific instrument payloads. By spring 1946 the navy transferred Lieutenant Commander Van Allen to its inactive reserve and he resumed his war-interrupted research work at Johns Hopkins. He remained at the Applied Physics Laboratory until 1950, when he returned to the University of Iowa as head of the physics department.

While at Johns Hopkins, Van Allen performed a series of preliminary space science experiments that anticipated his great discovery at the dawn of the U.S. space program. He designed and constructed rugged, miniaturized instruments to collect geophysical data at the edge of space, using rides on captured German V-2 rockets, Aerobee sounding rockets, and even rockets launched from high altitude balloons (rockoons). One of his prime interests was the measurement of the intensity of cosmic rays and any other energetic particles arriving at the top of Earth's atmosphere from outer space. He carried these research interests back to the University of Iowa and over the years established an internationally recognized space physics program.

On October 4, 1957, the Soviet Union shocked the world by sending the first artificial satellite, *Sputnik 1*, into orbit around Earth. In addition to starting the Space Age, this satellite forced the United States into a hotly contested space technology race. In a desperate attempt to win the race, both the Soviets and the Americans started pumping large quantities of money into the construction of military and scientific (civilian) space systems.

Fortune often favors the prepared. As a gifted scientist, Van Allen was well prepared for the great opportunity that suddenly came his way. With the dramatic failure of the first *Vanguard* rocket vehicle at Cape Canaveral, Florida, in early December 1957, senior U.S. government officials made an emergency decision to launch the country's first satellite with a military rocket. A scientific payload that was rugged enough to fly on a rocket was needed at once. Van Allen's Iowa cosmic-ray experiment was available and quickly selected to become the principal component of the payload on *Explorer 1*. He responded to this great opportunity by providing a rugged Geiger-Muller tube to the Jet Propulsion Laboratory, which was on contract with the U.S. Army to construct the upper stage-spacecraft portion of *Explorer 1*. Van Allen's scientific instrument was sturdy enough to survive the ride into space on a rocket and then start collecting interesting geophysical data that was transmitted back to Earth by the host spacecraft.

All was ready on January 31, 1958. The U.S. Army hastily pressed a Jupiter C rocket into service as a launch vehicle, with a cleverly improvised configuration designed by WERNHER MAGNUS VON BRAUN. Flawlessly, the rocket rumbled into orbit from Cape Canaveral, Florida. Soon the first U.S. satellite, *Explorer 1*, traveled in orbit around Earth. As the satellite glided though space, Van Allen's ionizing radiation detector, primarily designed to measure cosmic ray intensity, jumped off scale. As a great surprise to Van Allen and other scientists, this instrument had unexpectedly detected the inner portion of Earth's trapped radiation belts.

The *Explorer II* spacecraft, launched into orbit on March 26, 1958, carried an augmented version of Van Allen's Iowa cosmic-ray instrument. That spacecraft harvested an enormous quantity of data about the radiation environment in space and confirmed the presence of trapped energetic charged-particle belts (mainly electrons and protons) within Earth's magnetosphere. These belts are now called the Van Allen radiation belts. Soon an armada of spacecraft poked and probed the region of outer space near Earth, defining the extent, shape, and composition of Earth's trapped radiation belts. Van Allen became an instant scientific celebrity. Because of his widely recognized space science accomplishments, he has since met eight U.S. presidents.

In the 1960s and 1970s he assisted the National Aeronautics and Space Administration in planning, designing, and operating

Pioneering space scientist James Van Allen was part of the jubilant team that launched the first American satellite, *Explorer I*, on January 31, 1958. Appearing in this photograph are: (left) William H. Pickering, former director of the Jet Propulsion Laboratory, which built and operated the satellite; (center) James Van Allen, who designed and built the satellite's space science instruments; and (right) Wernher von Braun, who built the launch vehicle that successfully sent the satellite into orbit around Earth. They are holding a model of *Explorer I*, whose instruments made scientific history by detecting Earth's trapped radiation belts—now called the Van Allen belts. *(NASA)*

energetic particle instruments on planetary exploration spacecraft to Mars, Venus, Jupiter, and Saturn. Since *Explorer 1*, Van Allen and his team of researchers and graduate students at the University of Iowa have actively participated in many other pioneering space exploration missions. Members of this group have published major papers dealing with Jupiter's intensely powerful magnetosphere, the discovery and preliminary survey of Saturn's magnetosphere, and the energetic particles population in interplanetary space.

Van Allen retired from the department of physics and astronomy at the University of Iowa in 1985. However, as the Carver Professor of Physics Emeritus, he still pores over interesting space physics data in his office in the campus building appropriately named Van Allen Hall. His space physics efforts not only thrust the University of Iowa into international prominence but he also served his nation as a truly inspirational scientific hero.

W

⊠ **Wilson, Robert Woodrow**
(1936–)
American
Physicist, Radio Astronomer

Robert Wilson collaborated with ARNO ALLEN PENZIAS at Bell Laboratories in the mid-1960s and made the most important discovery in 20th-century cosmology. By detecting cosmic microwave background radiation, they provided the first empirical evidence in support of the Big Bang theory presented as a doctoral thesis in 1948 by RALPH ASHER ALPHER.

Robert Woodrow Wilson was born in Houston, Texas, on January 10, 1936. His childhood visits to the company shop with his father, a chemist working for the oil industry, provided Wilson with his lifelong interest in electronics and machinery. In 1957 he received his bachelor of arts degree with honors in physics from Rice University in Houston and then enrolled for graduate work in physics at the California Institute of Technology (Caltech). At the time Wilson was not decided about what area of physics he wanted to pursue for his doctoral research. However, fortuitous meetings and discussions on campus led him to become involved in interesting work at Caltech's Owens Valley Radio Observatory—then undergoing development in the Sierra Nevada near Bishop,

California, and today the largest university-operated radio observatory in the world. For Wilson, graduate research in radio astronomy provided a nice mixture of electronics and physics. His decision to pursue this area of research placed him on a highly productive career path in radio astronomy. However, in 1958 he briefly delayed his entry into radio astronomy to return to Houston so he could court and marry his wife, Elizabeth Rhoads Sawin.

In 1959 Wilson took his first astronomy courses at Caltech and began working at the Owens Valley Radio Observatory during breaks in the academic calendar. He completed the last part of his thesis research under the supervision of MAARTEN SCHMIDT, who was exploring quasars at the time. Following graduation with his doctoral degree, Wilson stayed on at Caltech as a postdoctoral researcher and completed several other projects. In 1963 he joined Bell Laboratories and soon met Penzias, the only other radio astronomer at the laboratory.

Between 1963 and 1965 the two collaborated in applying a special antenna six meters in diameter to investigate the problem of radio noise from the sky. At this time, Bell Laboratories had a general interest in detecting and resolving any cosmic radio noise problems that might adversely affect the operation of the early communication satellites. Expecting radio noise to be less severe at shorter wavelengths, Wilson and Penzias were surprised to discover that the sky was uniformly "bright" at 7.3 centimeters wavelength in the microwave region. Discussions with Robert Dicke (1916–97) in spring 1965 indicated that Wilson and Penzias had stumbled upon the first direct evidence of the cosmic microwave background—the "Holy Grail" of Big Bang cosmology. Wilson and Penzias estimated the detected cosmic "radio noise" to be about three kelvins. Measurements performed by NASA's *Cosmic Background Explorer (COBE)* spacecraft between 1989 and 1990 indicated that the cosmic microwave background closely approximates a blackbody radiator at a temperature of 2.735 kelvins. For this great (though serendipitous) discovery, Wilson shared one half of the 1978 Nobel Prize in physics with Penzias. The other half of the award that year went to the Russian physicist Pyotr Kapitsa for unrelated work in low-temperature physics.

In the late 1960s and early 1970s, Wilson and Penzias used techniques in radio astronomy at millimeter wavelengths to investigate molecular species in interstellar space. They were excited to unexpectedly find large quantities of carbon monoxide (CO) behind the Orion Nebula, and soon found that this molecule was rather widely distributed throughout the Galaxy.

From 1976 to 1990 Wilson served as the head of the Radio Physics Research Department at Bell Laboratories. He was elected to the U. S. National Academy of Sciences in 1979. In addition to the 1978 Nobel Prize, Wilson also received the Herschel Medal from the Royal Astronomical Society (1977) and the Henry Draper Medal from the American National Academy of Sciences (1977). Wilson's major contribution to astronomy was his codiscovery of the cosmic microwave background—the lingering remnant of the Big Bang explosion that started the universe.

⊠   **Wolf, Maximilian (Max Wolf)**
(1863–1932)
German
Astronomer

Max Wolf was an asteroid hunter who pioneered the use of astrophotography to help him in his search for these elusive solar system bodies. He discovered more than 200 minor planets, including asteroid Achilles in 1906—the first of the Trojan group of asteroids. This group of minor planets moves around the Sun in Jupiter's orbit, but at the Lagrangian libration points of 60 degrees ahead and 60 degrees behind the giant planet.

Wolf was born on June 21, 1863, in Heidelberg, Germany, the city where he spent almost his entire lifetime. His father was a respected physician and the family was comfortably wealthy. When Wolf was a young boy, his father encouraged him to develop an interest in science and astronomy by building him a private observatory adjacent to the family's residence. His father's investment was a good one. Wolf completed his doctoral degree in mathematical studies at the University of Heidelberg in 1888 and then went to the University of Stockholm for two years of postdoctoral work. He returned to Heidelberg in 1890 and began lecturing in astronomy at the university as a *Privatdozent* (unpaid instructor). In 1893 Wolf received a dual appointment from the University of Heidelberg to serve as a special professor of astrophysics and to direct the new observatory being constructed at nearby Königsstuhl. In 1902 he became the chair of astronomy at the university and, along with his directorship of the Königsstuhl Observatory, remained in these positions until his death.

Wolf performed important spectroscopic work concerning nebulas and published 16 lists of nebulae containing a total of about 6,000 celestial objects. But he is best remembered as being the first astronomer to use photography to find asteroids. In 1891 he demonstrated the ability to photograph a large region of the sky with a telescope that followed the "fixed" stars precisely as the Earth rotated. The stars therefore appeared as points, while the minor planets he hunted appeared as short streaks on the photographic plate. On September 11, 1898, he discovered his first asteroid with this photographic technique and named the celestial object Brucia (also called asteroid 323) to honor the American philanthropist Catherine Wolfe Bruce. She had donated money for Wolf's new telescope at the Königsstuhl Observatory. She also established the Bruce Gold Medal as the award presented annually to an astronomer in commemoration of his or her lifetime achievements.

By the end of 1898, Wolf had discovered nine other minor planets using astrophotography. Over the next three decades, this technique allowed him to personally discover more than 200 new asteroids, including asteroid 588, Achilles. Wolf found and named Achilles in 1906, noting that it orbited the Sun at the L_4 Lagrangian point 60 degrees ahead of Jupiter. The Trojan group is the collection of minor planets found at the two Lagrangian libration points lying on Jupiter's orbital path around the Sun. The name comes from the fact that many of these asteroids were named after the mythical heroes of the Trojan War.

Wolf was an avid asteroid hunter and a pioneer in astrophotography. In recognition of his contributions to astronomy, he received the Gold Medal of the Royal Astronomical Society in 1914 and the Bruce Gold Medal from the Astronomical Society of the Pacific in 1930. He held membership in both the American Astronomical Society and the Royal Astronomical Society of Great Britain. He died in Heidelberg on October 3, 1932.

Z

⊗ **Zwicky, Fritz**
(1898–1974)
Swiss
Astrophysicist

Shortly after EDWIN POWELL HUBBLE introduced the concept of an expanding universe in the late 1920s, Fritz Zwicky pioneered the early search for the "dark matter" (that is, the missing mass) of the universe. While collaborating with WILHELM HEINRICH WALTER BAADE in 1934, he coined the term *supernova* to describe certain very cataclysmic nova phenomena. Zwicky, a visionary with an extremely eccentric and often caustic personality, also postulated that the by-product of this supernova explosion would be a neutron star.

Fritz Zwicky was born on February 14, 1898, in Varna, Bulgaria. A lifelong Swiss citizen, he spent his childhood in Mollis, a village in the Canton (state) of Glarus, Switzerland. He received his education in physics at the Federal Institute of Technology in Zürich, graduating with a Ph.D. in 1922. While a graduate student, Zwicky interacted with the Swiss physicist Auguste Piccard (1884–1962) and performed his dissertation research under a Nobel laureate, the Dutch-American physical chemist Peter Deybe (1884–1966).

Since ALBERT EINSTEIN was a physics instructor at the Federal Institute in Zurich when Zwicky was also enrolled there as a student, science historians often suggest that he was a student of Einstein. Without question, Zwicky was a very creative and intelligent physicist, so he could certainly have engaged in interesting scientific discussions with Einstein, as both a student in Switzerland and then after graduation, when he became a professor at the California Institute of Technology (Caltech). However, the truth is not known. What is clear is that Zwicky had a dark side to his technical genius. He was very abrasive and fond of using rough language and intimidating colleagues and students alike. Only a few extremely tolerant individuals, such as Baade, could or would overlook his eccentric and difficult personality in order to encounter the bursts of genius that occasionally could be found within Zwicky's characteristically caustic comments.

At the invitation of the Nobel laureate physicist Robert Millikan (1868–1953), Zwicky came to the United States in 1925 to work as a postdoctoral fellow at Caltech. Claiming he enjoyed the mountains near Pasadena, Zwicky remained affiliated with that institution and stayed in California for the rest of his life. However, despite living and working for almost five decades in the United States, Zwicky retained his Swiss citizenship and never chose to apply for U.S. citizenship.

The Swiss astrophysicist Fritz Zwicky lecturing at the California Institute of Technology (Caltech). In the late 1920s he pioneered the early search for the "dark matter" (that is, the missing mass) of the universe. While collaborating with Walter Baade in 1934, he coined the term *supernova* to describe certain very cataclysmic nova phenomena. Zwicky, a visionary with an extremely eccentric and often caustic personality, also postulated that the by-product of this supernova explosion would be a neutron star. *(AIP Emilio Segrè Visual Archives)*

In 1927 he became a professor of theoretical physics and retained that academic position at Caltech until he became a professor of astrophysics there in 1942. He retired from Caltech in 1972 given the status of professor emeritus in recognition of his lifelong service to the university and to astrophysics. While at Caltech Zwicky interacted with many of the world's best astronomers, including Baade, who came to Southern California to use the Mount Wilson Observatory, located in the nearby San Gabriel Mountains. On campus at Caltech, Zwicky's

antics are legendary. When encountering an unfamiliar student on campus, Zwicky would often suddenly stop, stare in his or her face, and loudly inquire: "Who the hell are you?" Similarly, during a lecture on campus, he might suddenly stand up and leave the room, pausing at the door to turn around and inform the startled speaker that he (Zwicky) had already solved the particular problem under discussion. Without question, many on the campus breathed a sigh of relief when Zwicky committed much of his time between 1943 and 1946 to direct research at the Aerojet Corporation in Azusa, California, in support of jet engine development and other defense-related activities.

Despite his extensive involvement with astronomers, many of them also fell victim to his aggressive social behavior. While working at the Mount Wilson Observatory or at the Palomar Observatory, Zwicky would often inject one of his favorite exclamations, that all astronomers were "spherical bastards." Why spherical? Well, according to Zwicky, astronomers "looked the same when viewed from any direction." But behind this difficult personality was a gifted astrophysicist who helped establish the framework of modern cosmology.

In 1933, while investigating the Coma Cluster—a rich cluster of galaxies in the Constellation Coma Berenices (also known as Berenice's Hair)—Zwicky observed that the velocities of the individual galaxies in this cluster were so high that they should definitely have escaped from each other's gravitational attraction long ago. Zwicky, never afraid of controversy, quickly came to the pioneering conclusion that the amount of matter actually in that cluster had to be much greater than what could be accounted for by the visible galaxies alone—an amount estimated to be only about 10 percent of the mass needed to gravitationally bind the galaxies in the cluster. Zwicky then focused a great deal of his attention on the problem of this "missing mass" or, as it is now more commonly called, the problem of "dark matter."

To support his research in dark matter, Zwicky and his collaborators started to use the 18-inch Schmidt telescope at the Palomar Observatory near San Diego in 1936 to investigate and then catalog numerous clusters of galaxies. Between 1961 and 1968 Zwicky published the results of this intense effort as a six-volume work called *Catalog of Galaxies and of Clusters of Galaxies*. Sometimes simply referred to as the Zwicky catalog, this important work contained about 10,000 clusters (classified as either compact, medium compact, or open) and approximately 31,000 galaxies.

While investigating bright novas with Baade in 1934, Zwicky coined the term *supernova* to describe the family or new class of novas that appeared to be far more energetic and cataclysmic than any of the more frequently encountered novas. The classical nova phenomenon involves a sudden and unpredictable increase in the brightness (perhaps up to 10 orders of magnitude) of a binary star system. In a typical spiral galaxy like the Milky Way, astrophysicists expect about 25 such eruptions to occur each year. As Zwicky noted, the supernova was a much more violent and rare stellar event. A star undergoing a supernova explosion temporarily shines with a brightness that is more than 100 times more luminous than an ordinary nova. The supernova occurs in a galaxy like the Milky Way only about two or three times per century. From 1937 to 1941 Zwicky engaged in an extensive search for such events beyond the Milky Way and personally discovered 18 extragalactic supernovas. Until then, only about 12 such events had been recorded in the entire history of astronomy.

Since this early work by Zwicky and Baade, other astrophysicists have discovered more than 800 extragalactic supernovas. Despite their brilliance, not all supernovas are visible to naked-eye observers on Earth. The most famous naked-eye supernovas include the very bright new star reported by Chinese and Korean astronomers in 1054; Tyco's star, appearing in 1572; Kepler's star, appearing in 1604; and supernova 1987A, the most recent event, which took place in the Large Magellan Cloud on February 23, 1987. Zwicky recognized that supernovas were incredibly interesting, yet violent celestial events. During a supernova explosion, the dying star might become more than 1 billion times brighter than the Sun.

In 1932 the British physicist Sir James Chadwick (1891–1974) experimentally discovered the neutron. Zwicky recognized the astronomical significance of Chadwick's discovery and boldly presented an extremely visionary "neutron star" hypothesis in 1934. Specifically, Zwicky proposed that a neutron star might be formed as a by-product of a supernova explosion. He reasoned that the violent explosion and subsequent gravitational collapse of the dying star's core should create a compact object of such incredible density that all the electrons and protons become so closely packed together that they become neutrons. However, his daring hypothesis was essentially ignored for about three decades. Then, in 1968, astronomers confirmed the presence of the first suspected neutron star, a young optical pulsar, in the cosmic debris of the great supernova event of 1054. The discovery proved that the eccentric Zwicky was years ahead of his contemporaries in suggesting a linkage between supernovas and neutron stars.

Zwicky also published the creative book *Morphological Cosmology* in 1957. He received the Gold Medal from the Royal Astronomical Society in 1973 for his contributions to modern cosmology. After spending almost five decades in the United States, Zwicky died on February 8, 1974, in Pasadena, California.

Entries by Field

ASTRONAUTICS
Braun, Wernher Magnus von
Ehricke, Krafft Arnold
Goddard, Robert Hutchings
Hohmann, Walter
Korolev, Sergei Pavlovich
Lagrange, Comte Joseph-Louis
Laplace, Pierre-Simon, Marquis de
Leverrier, Urbain-Jean-Joseph
Oberth, Hermann Julius
Tereshkova, Valentina
Tsiolkovsky, Konstantin Eduardovich

ASTRONOMY
Adams, John Couch
Adams, Walter Sydney
Alfonso X
Ångström, Anders Jonas
Arago, Dominique-François-Jean
Argelander, Friedrich Wilhelm August
Aristarchus of Samos
Aryabhata
Baade, Wilhelm Heinrich Walter
Barnard, Edward Emerson
Battani, al-
Bayer, Johannes

Beg, Muhammed Taragai Ulugh
Bell Burnell, Susan Jocelyn
Bessel, Friedrich Wilhelm
Bode, Johann Elert
Bok, Bartholomeus Jan
Bradley, James
Brahe, Tycho
Burbidge, Eleanor Margaret Peachey
Campbell, William Wallace
Cannon, Annie Jump
Cassini, Giovanni Domenico
Charlier, Carl Vilhelm Ludvig
Clark, Alvan Graham
Copernicus, Nicolas
Curtis, Heber Doust
Douglass, Andrew Ellicott
Drake, Frank Donald
Draper, Henry
Dyson, Sir Frank Watson
Fleming, Williamina Paton Stevens
Galilei, Galileo
Galle, Johann Gottfried
Gill, Sir David
Hall, Asaph
Halley, Sir Edmund
Hawking, Sir Stephen William
Herschel, Caroline Lucretia

Herschel, Sir John Frederick William
Herschel, Sir William
Hertzsprung, Ejnar
Hewish, Antony
Hubble, Edwin Powell
Huggins, Sir William
Jeans, Sir James Hopwood
Jones, Sir Harold Spencer
Kant, Immanuel
Kapteyn, Jacobus Cornelius
Kepler, Johannes
al-Khwarizmi, Abu
Kuiper, Gerard Peter
Laplace, Pierre-Simon, Marquis de
Leverrier, Urbain-Jean-Joseph
Lockyer, Sir Joseph Norman
Lowell, Percival
Messier, Charles
Newcomb, Simon
Olbers, Heinrich Wilhelm Matthäus
Oort, Jan Hendrik
Penzias, Arno Allen
Piazzi, Giuseppe
Pickering, Edward Charles
Ptolemy
Reber, Grote
Rossi, Bruno Benedetto
Russell, Henry Norris

Ryle, Sir Martin
Sagan, Carl Edward
Scheiner, Christoph
Schiaparelli, Giovanni
    Virginio
Schmidt, Maarten
Schwabe, Samuel Heinrich
Schwarzschild, Karl
Secchi, Pietro Angelo
Shapley, Harlow
Sitter, Willem de
Slipher, Vesto Melvin
Somerville, Mary Fairfax
Tombaugh, Clyde William
Wilson, Robert Woodrow
Wolf, Maximilian

ASTROPHYSICS

Alpher, Ralph Asher
Ambartsumian, Viktor
    Amazaspovich
Bell Burnell, Susan Jocelyn
Bethe, Hans Albrecht
Burbidge, Eleanor Margaret
    Peachey
Chandrasekhar,
    Subrahmanyan
Eddington, Sir Arthur Stanley
Giacconi, Riccardo
Hale, George Ellery
Lemaître, Abbé Georges-
    Édouard
Rees, Sir Martin John
Russell, Henry Norris
Sagan, Carl Edward
Schmidt, Maarten
Schwarzschild, Karl
Zwicky, Fritz

CHEMISTRY

Arrhenius, Svante August
Bunsen, Robert Wilhelm
Cavendish, Henry
Urey, Harold Clayton

COSMOLOGY

Alfvèn, Hannes Olof Gösta
Alpher, Ralph Asher
Gamow, George
Hawking, Sir Stephen William
Hubble, Edwin Powell
Lemaître, Abbé Georges-
    Édouard

ENGINEERING

Braun, Wernher Magnus von
Ehricke, Krafft Arnold
Hohmann, Walter
Korolev, Sergei Pavlovich

ENVIRONMENTAL SCIENCE

Douglass, Andrew Ellicott

EXOBIOLOGY

Arrhenius, Svante August
Urey, Harold Clayton

HISTORY

Alfonso X

LITERATURE

Aratus of Soli

MATHEMATICS

Adams, John Couch
Arago, Dominique-François-
    Jean
Aristarchus of Samos
Aryabhata
Battani, al-
Beg, Muhammed Taragai
    Ulugh
Brahe, Tycho
Cassini, Giovanni Domenico
Charlier, Carl Vilhelm Ludvig
Dirac, Paul-Adrien-Maurice
Doppler, Christian Andreas
Euler, Leonhard
Galilei, Galileo

Halley, Sir Edmund
Hawking, Sir Stephen
    William
Herschel, Sir John Frederick
    William
Jeans, Sir James Hopwood
Kepler, Johannes
al- Khwarizmi, Abu
Lagrange, Comte Joseph-Louis
Laplace, Pierre-Simon,
    Marquis de
Leverrier, Urbain-Jean-Joseph
Newcomb, Simon
Newton, Sir Isaac
Oberth, Hermann Julius
Ptolemy
Scheiner, Christoph
Sitter, Willem de
Somerville, Mary Fairfax

OPTICAL SCIENCE AND INSTRUMENTATION

Clark, Alvan Graham
Draper, Henry
Fraunhofer, Joseph von
Hale, George Ellery
Lippershey, Hans
Newton, Sir Isaac
Scheiner, Christoph

PHILOSOPHY

Aristotle
Bruno, Giordano
Herschel, Sir John Frederick
    William
Kant, Immanuel
Plato
Ptolemy

PHYSICS

Alfvén, Hannes Olof Gösta
Alpher, Ralph Asher
Alvarez, Luis Walter
Anderson, Carl David

Ångström, Anders Jonas
Arago, Dominique-François-
    Jean
Boltzmann, Ludwig
Cavendish, Henry
Clausius, Rudolf Julius
Compton, Arthur Holly
Dirac, Paul Adrien Maurice
Einstein, Albert
Euler, Leonhard
Fowler, William Alfred
Fraunhofer, Joseph von
Galilei, Galileo
Gamow, George
Goddard, Robert Hutchings
Halley, Sir Edmund
Hawking, Sir Stephen William

Hertz, Heinrich Rudolf
Hess, Victor Francis
Hubble, Edwin Powell
Jeans, Sir James Hopwood
Kirchhoff, Gustav Robert
Lockyer, Sir Joseph
    Norman
Michelson, Albert Abraham
Newton, Sir Isaac
Oberth, Hermann Julius
Penzias, Arno Allen
Pickering, Edward Charles
Planck, Max Karl
Rossi, Bruno Benedetto
Ryle, Sir Martin
Sagan, Carl Edward
Stefan, Josef

Tsiolkovsky, Konstantin
    Eduardovich
Van Allen, James Alfred
Wilson, Robert Woodrow

SPACE SCIENCES
Alfvén, Hannes Olof Gösta
Van Allen, James Alfred
Rossi, Bruno Benedetto

SPECTROSCOPY
Bunsen, Robert Wilhelm
Fraunhofer, Joseph von
Huggins, Sir William
Kirchhoff, Gustav Robert
Lockyer, Sir Joseph Norman
Secchi, Pietro Angelo

Entries by Country of Birth

ARMENIA
Ambartsumian, Viktor
   Amazaspovich

AUSTRIA
Boltzmann, Ludwig
Doppler, Christian
   Andreas
Hess, Victor Francis
Stefan, Josef

BELGIUM
Lemaître, Abbé Georges-
   Édouard

BULGARIA
Zwicky, Fritz

CANADA
Newcomb, Simon

DENMARK
Brahe, Tycho
Hertzsprung, Ejnar

EGYPT
Ptolemy

FRANCE
Arago, Dominique-François-
   Jean

Laplace, Pierre-Simon,
   Marquis de
Leverrier, Urbain-Jean-Joseph
Messier, Charles

GERMANY (INCLUDES PRUSSIA)
Baade, Wilhelm Heinrich
   Walter
Bayer, Johannes
Bessel, Friedrich Wilhelm
Bethe, Hans Albrecht
Bode, Johann Elert
Braun, Wernher Magnus von
Bunsen, Robert Wilhelm
Ehricke, Krafft Arnold
Einstein, Albert
Fraunhofer, Joseph von
Galle, Johann Gottfried
Herschel, Caroline Lucretia
Herschel, Sir William
Hertz, Heinrich Rudolf
Hohmann, Walter
Kant, Immanuel
Kepler, Johannes
Kirchhoff, Gustav Robert
Lippershey, Hans
Michelson, Albert Abraham
Olbers, Heinrich Wilhelm
   Matthäus
Penzias, Arno Allen
Planck Max Karl

Scheiner, Christoph
Schwabe, Samuel Heinrich
Schwarzschild, Karl
Wolf, Maximilian

GREAT BRITAIN
Adams, John Couch
Bell Burnell, Susan Jocelyn
Bradley, James
Burbidge, Eleanor Margaret
   Peachey
Cavendish, Henry
Dirac, Paul Adrien Maurice
Dyson, Sir Frank Watson
Eddington, Sir Arthur Stanley
Fleming, Williamina Paton
   Stevens
Gill, Sir David
Halley, Sir Edmund
Hawking, Sir Stephen William
Herschel, Sir John Frederick
   William
Hewish, Antony
Huggins, Sir William
Jeans, Sir James Hopwood
Jones, Sir Harold Spencer
Lockyer, Sir Joseph Norman
Newton, Sir Isaac
Rees, Sir Martin John
Ryle, Sir Martin
Somerville, Mary Fairfax

GREECE
Aratus of Soli
Aristarchus of Samos
Aristotle
Plato

INDIA
Aryabhata
Chandrasekhar, Subrahmanyan

ITALY
Bruno, Giordano
Cassini, Giovanni Domenico
Galilei, Galileo
Giacconi, Riccardo
Lagrange, Comte Joseph-Louis
Piazzi, Giuseppe
Rossi, Bruno Benedetto
Schiaparelli, Giovanni
    Virginio
Secchi, Pietro Angelo

LITHUANIA
Argelander, Friedrich
    Wilhelm

NETHERLANDS
Bok, Bartholomeus Jan
Kapteyn, Jacobus
    Cornelius
Kuiper, Gerard Peter
Oort, Jan Hendrik
Schmidt, Maarten
Sitter, Willem de

PERSIA
Beg, Muhammed Taragai Ulugh
al-Khwarizmi, Abu

POLAND
Clausius, Rudolf Julius
    Emmanuel
Copernicus, Nicolas

ROMANIA
Oberth, Hermann Julius

RUSSIA
Tereshkova, Valentina
Tsiolkovsky, Konstantin
    Eduardovich

SPAIN
Alfonso X

SWEDEN
Alfvèn, Hannes Olof Gösta
Ångström, Anders Jonas
Arrhenius, Svante August
Charlier, Carl Vilhelm Ludvig

SWITZERLAND
Euler, Leonhard

SYRIA
Adams, Walter Sydney

TURKEY
Battani, al-

UKRAINE
Gamow, George
Korolev, Sergei Pavlovich

UNITED STATES
Alpher, Ralph Asher
Alvarez, Luis Walter
Anderson, Carl David
Barnard, Edward Emerson
Campbell, William
    Wallace
Cannon, Annie Jump
Clark, Alvan Graham
Compton, Arthur Holly
Curtis, Heber Doust
Douglass, Andrew Ellicott
Drake, Frank Donald
Draper, Henry
Fowler, William Alfred
Goddard, Robert Hutchings
Hale, George Ellery
Hall, Asaph
Hubble, Edwin Powell
Lowell, Percival
Pickering, Edward Charles
Reber, Grote
Russell, Henry Norris
Sagan, Carl Edward
Shapley, Harlow
Slipher, Vesto Melvin
Tombaugh, Clyde William
Urey, Harold Clayton
Van Allen, James Alfred
Wilson, Robert Woodrow

Entries by Country of Major Scientific Activity

AUSTRALIA
Reber, Grote

AUSTRIA
Boltzmann, Ludwig
Doppler, Christian Andreas
Hess, Victor Francis
Stefan, Josef

BELGIUM
Lemaître, Abbé Georges-
    Édouard

DENMARK
Brahe, Tycho
Hertzsprung, Ejnar

EGYPT
Ptolemy

FRANCE
Arago, Dominique-François-
    Jean
Cassini, Giovanni Domenico
Lagrange, Comte Joseph-
    Louis
Laplace, Pierre-Simon,
    Marquis de
Leverrier, Urbain-Jean-
    Joseph
Messier, Charles

GERMANY
Argelander, Friedrich
    Wilhelm August
Bayer, Johannes
Bessel, Friedrich Wilhelm
Bode, Johann Elert
Bunsen, Robert Wilhelm
Clausius, Rudolf Julius
    Emmanuel
Fraunhofer, Joseph von
Galle, Johann Gottfried
Hertz, Heinrich Rudolf
Hohmann, Walter
Kant, Immanuel
Kepler, Johannes
Kirchhoff, Gustav Robert
Oberth, Hermann Julius
Olbers, Heinrich Wilhelm
    Matthäus
Planck, Max Karl
Scheiner, Christoph
Schwabe, Samuel Heinrich
Schwarzschild, Karl
Wolf, Maximilian

GREAT BRITAIN
Adams, John Couch
Bell Burnell, Susan Jocelyn
Bradley, James
Cavendish, Henry
Chandrasekhar, Subrahmanyan

Dirac, Paul Adrien Maurice
Dyson, Sir Frank Watson
Eddington, Sir Arthur
    Stanley
Halley, Sir Edmund
Herschel, Caroline Lucretia
Herschel, Sir William
Herschel, Sir John Frederick
    William
Hewish, Antony
Huggins, Sir William
Jeans, Sir James Hopwood
Jones, Sir Harold Spencer
Lockyer, Sir Joseph Norman
Newton, Sir Isaac
Rees, Sir Martin John
Ryle, Sir Martin
Somerville, Mary Fairfax

GREECE
Aratus of Soli
Aristarchus of Samos
Aristotle
Plato

INDIA
Aryabhata

ITALY
Bruno, Giordano
Galilei, Galileo

Piazzi, Giuseppe
Schiaparelli, Giovanni
    Virginio
Secchi, Pietro Angelo

NETHERLANDS
Kapteyn, Jacobus Cornelius
Lippershey, Hans
Oort, Jan Hendrik
Sitter, Willem de

PERSIA
Beg, Muhammed Taragai
    Ulugh
al-Khwarizmi, Abu

POLAND
Copernicus, Nicolas

RUSSIA
Ambartsumian, Viktor
    Amazaspovich
Euler, Leonhard
Korolev, Sergei Pavlovich
Tereshkova, Valentina
Tsiolkovsky, Konstantin
    Eduardovich

SPAIN
Alfonso X

SOUTH AFRICA
Gill, Sir David
Herschel, Sir John Frederick
    William
Sitter, Williem de

SWEDEN
Alfvén, Hannes Olof Gösta
Ångström, Anders Jonas
Arrhenius, Svante August
Charlier, Carl Vilhelm
    Ludvig
Dirac, Paul Adrien Maurice

SWITZERLAND
Einstein, Albert

SYRIA
Battani, al-

UNITED STATES
Adams, Walter Sydney
Alpher, Ralph Asher
Alvarez, Luis Walter
Anderson, Carl David
Baade, Wilhelm Heinrich
    Walter
Barnard, Edward Emerson
Bethe, Hans Albrecht
Bok, Bartholomeus Jan
Braun, Wernher
    Magnus von
Burbidge, Eleanor Margaret
    Peachey
Campbell, William Wallace
Cannon, Annie Jump
Chandrasekhar,
    Subrahmanyan
Clark, Alvan Graham
Compton, Arthur Holly
Curtis, Heber Doust
Douglass, Andrew Ellicott

Drake, Frank Donald
Draper, Henry
Ehricke, Krafft Arnold
Einstein, Albert
Fleming, Williamina Paton
    Stevens
Fowler, William Alfred
Gamow, George
Giacconi, Riccardo
Goddard, Robert
    Hutchings
Hale, George Ellery
Hall, Asaph
Hubble, Edwin Powell
Kuiper, Gerard Peter
Lowell, Percival
Michelson, Albert
    Abraham
Newcomb, Simon
Penzias, Arno Allen
Pickering, Edward Charles
Reber, Grote
Rossi, Bruno
    Benedetto
Russell, Henry Norris
Sagan, Carl Edward
Schmidt, Maarten
Shapley, Harlow
Slipher, Vesto Melvin
Tombaugh, Clyde
    William
Urey, Harold Clayton
Van Allen, James Alfred
Wilson, Robert
    Woodrow
Zwicky, Fritz

Entries by Year of Birth

499–400 B.C.E.
Plato

399–300 B.C.E.
Aratus of Soli
Aristarchus of Samos
Aristotle

100–0 B.C.E.
Ptolemy

400–499
Aryabhata

700–799
al-Khwarizmi, Abu

800–900
Battani, al-

1200–1299
Alfonso X

1300–1399
Beg, Muhammed Taragai
    Ulugh

1400–1499
Copernicus, Nicolas

1500–1599
Bayer, Johannes
Brahe, Tycho
Bruno, Giordano
Galilei, Galileo
Kepler, Johannes
Lippershey, Hans
Scheiner, Christoph

1600–1699
Bradley, James
Cassini, Giovanni Domenico
Halley, Sir Edmund
Newton, Sir Isaac

1700–1799
Arago, Dominique-François-
    Jean
Argelander, Friedrich Wilhelm
    August
Bessel, Friedrich Wilhelm
Bode, Johann Elert
Cavendish, Henry
Euler, Leonhard
Fraunhofer, Joseph von
Herschel, Caroline Lucretia
Herschel, Sir William
Herschel, Sir John Frederick
    William
Kant, Immanuel
Lagrange, Comte Joseph-Louis

Laplace, Pierre-Simon,
    Marquis de
Messier, Charles
Olbers, Heinrich Wilhelm
    Matthäus
Piazzi, Giuseppe
Schwabe, Samuel Heinrich
Somerville, Mary Fairfax

1800–1809
Doppler, Christian Andreas

1810–1819
Adams, John Couch
Ångström, Anders Jonas
Bunsen, Robert Wilhelm
Galle, Johann Gottfried
Leverrier, Urbain-Jean-Joseph
Secchi, Pietro Angelo

1820–1829
Clausius, Rudolf Julius
    Emmanuel
Hall, Asaph
Huggins, Sir William
Kirchhoff, Gustav Robert

1830–1839
Clark, Alvan Graham
Draper, Henry
Lockyer, Sir Joseph Norman

Newcomb, Simon
Schiaparelli, Giovanni
    Virginio
Stefan, Josef

1840–1849
Boltzmann, Ludwig
Gill, Sir David
Pickering, Edward Charles

1850–1859
Arrhenius, Svante August
Barnard, Edward Emerson
Fleming, Williamina Paton
    Stevens
Hertz, Heinrich Rudolf
Kapteyn, Jacobus Cornelius
Lowell, Percival
Michelson, Albert Abraham
Planck, Max Karl
Tsiolkovsky, Konstantin
    Eduardovich

1860–1869
Campbell, William Wallace
Cannon, Annie Jump
Charlier, Carl Vilhelm Ludvig
Douglass, Andrew Ellicott
Dyson, Sir Frank Watson
Hale, George Ellery
Wolf, Maximilian

1870–1879
Adams, Walter Sydney
Curtis, Heber Doust
Einstein, Albert

Hertzsprung, Ejnar
Jeans, Sir James Hopwood
Russell, Henry Norris
Schwarzschild, Karl
Sitter, Willem de
Slipher, Vesto Melvin

1880–1889
Eddington, Sir Arthur
    Stanley
Goddard, Robert Hutchings
Hess, Victor Francis
Hohmann, Walter
Hubble, Edwin Powell
Shapley, Harlow

1890–1899
Baade, Wilhelm Heinrich
    Walter
Compton, Arthur Holly
Jones, Sir Harold Spencer
Lemaître, Abbé Georges-
    Édouard
Oberth, Hermann Julius
Urey, Harold Clayton
Zwicky, Fritz

1900–1909
Alfvén, Hannes Olof Gösta
Ambartsumian, Viktor
    Amazaspovich
Anderson, Carl David
Bethe, Hans Albrecht
Bok, Bartholomeus Jan
Dirac, Paul Adrien Maurice
Gamow, George

Korolev, Sergei Pavlovich
Kuiper, Gerard Peter
Oort, Jan Hendrik
Rossi, Bruno Benedetto
Tombaugh, Clyde William

1910–1919
Alvarez, Luis Walter
Braun, Wernher Magnus von
Burbidge, Eleanor Margaret
    Peachey
Chandrasekhar,
    Subrahmanyan
Ehricke, Krafft Arnold
Fowler, William Alfred
Reber, Grote
Ryle, Sir Martin
Van Allen, James Alfred

1920–1929
Alpher, Ralph Asher
Hewish, Antony
Schmidt, Maarten

1930–1939
Drake, Frank Donald
Giacconi, Riccardo
Penzias, Arno Allen
Sagan, Carl Edward
Tereshkova, Valentina
Wilson, Robert Woodrow

1940–1949
Bell Burnell, Susan Jocelyn
Hawking, Sir Stephen William
Rees, Sir Martin John

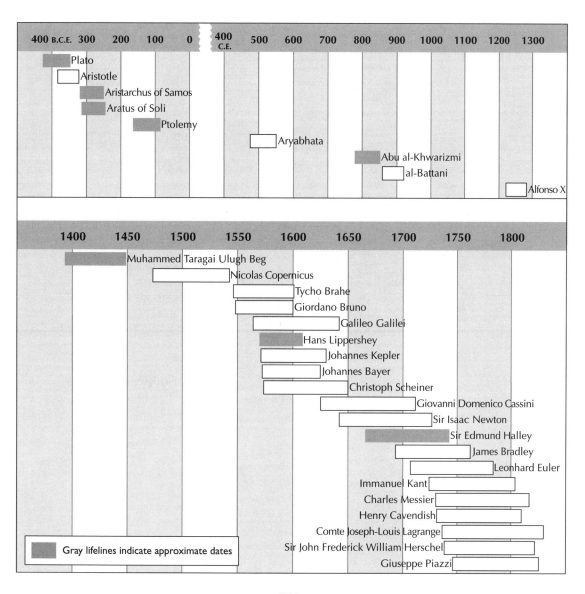

| 400 B.C.E. | 300 | 200 | 100 | 0 | 400 C.E. | 500 | 600 | 700 | 800 | 900 | 1000 | 1100 | 1200 | 1300 |

Plato
Aristotle
Aristarchus of Samos
Aratus of Soli
Ptolemy
Aryabhata
Abu al-Khwarizmi
al-Battani
Alfonso X

| 1400 | 1450 | 1500 | 1550 | 1600 | 1650 | 1700 | 1750 | 1800 |

Muhammed Taragai Ulugh Beg
Nicolas Copernicus
Tycho Brahe
Giordano Bruno
Galileo Galilei
Hans Lippershey
Johannes Kepler
Johannes Bayer
Christoph Scheiner
Giovanni Domenico Cassini
Sir Isaac Newton
Sir Edmund Halley
James Bradley
Leonhard Euler
Immanuel Kant
Charles Messier
Henry Cavendish
Comte Joseph-Louis Lagrange
Sir John Frederick William Herschel
Giuseppe Piazzi

Gray lifelines indicate approximate dates

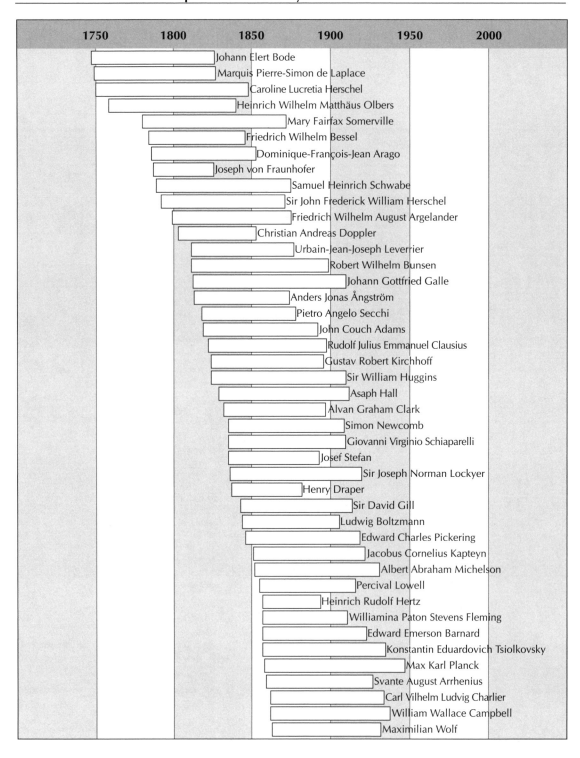

1750	1800	1850	1900	1950	2000

Johann Elert Bode

Marquis Pierre-Simon de Laplace

Caroline Lucretia Herschel

Heinrich Wilhelm Matthäus Olbers

Mary Fairfax Somerville

Friedrich Wilhelm Bessel

Dominique-François-Jean Arago

Joseph von Fraunhofer

Samuel Heinrich Schwabe

Sir John Frederick William Herschel

Friedrich Wilhelm August Argelander

Christian Andreas Doppler

Urbain-Jean-Joseph Leverrier

Robert Wilhelm Bunsen

Johann Gottfried Galle

Anders Jonas Ångström

Pietro Angelo Secchi

John Couch Adams

Rudolf Julius Emmanuel Clausius

Gustav Robert Kirchhoff

Sir William Huggins

Asaph Hall

Alvan Graham Clark

Simon Newcomb

Giovanni Virginio Schiaparelli

Josef Stefan

Sir Joseph Norman Lockyer

Henry Draper

Sir David Gill

Ludwig Boltzmann

Edward Charles Pickering

Jacobus Cornelius Kapteyn

Albert Abraham Michelson

Percival Lowell

Heinrich Rudolf Hertz

Williamina Paton Stevens Fleming

Edward Emerson Barnard

Konstantin Eduardovich Tsiolkovsky

Max Karl Planck

Svante August Arrhenius

Carl Vilhelm Ludvig Charlier

William Wallace Campbell

Maximilian Wolf

	1750	1800	1850	1900	1950	2000

Annie Jump Cannon
Andrew Ellicott Douglass
George Ellery Hale
Sir Frank Watson Dyson
Willem de Sitter
Heber Doust Curtis
Ejnar Hertzsprung
Karl Schwarzschild
Vesto Melvin Slipher
Walter Sydney Adams
Sir James Hopwood Jeans
Albert Einstein
Walter Hohmann
Sir Arthur Stanley Eddington
Robert Hutchings Goddard
Victor Francis Hess
Harlow Shapley
Henry Norris Russell
Edwin Powell Hubble
Sir Harold Spencer Jones
Arthur Holly Compton
Wilhelm Heinrich Walter Baade
Harold Clayton Urey
Abbé-Georges-Édouard Lemaître
Hermann Julius Oberth
Fritz Zwicky
Jan Hendrik Oort
Paul Adrien Maurice Dirac
George Gamow
Gerard Peter Kuiper
Carl David Anderson
Bruno Benedetto Rossi
Bartholomeus Jan Bok
Clyde William Tombaugh
Hans Albrecht Bethe
Sergei Pavlovich Korolev
Hannes Olof Gösta Alfvén
Viktor Amazaspovich Ambartsumian
Subrahmanyan Chandrasekhar
Luis Walter Alvarez
William Alfred Fowler
Grote Reber
Wernher Magnus von Braun

	1750	1800	1850	1900	1950	2000

James Alfred Van Allen

Krafft Arnold Ehricke

Sir Martin Ryle

Eleanor Margaret Peachey Burbidge

Ralph Asher Alpher

Antony Hewish

Maarten Schmidt

Frank Drake

Riccardo Giacconi

Arno Allen Penzias

Carl Edward Sagan

Robert Woodrow Wilson

Valentina Tereshkova

Sir Martin John Rees

Sir Stephen William Hawking

Susan Jocelyn Bell Burnell

Abbott, David, ed. *Biographical Dictionary of Scientists: Astronomy*. New York: Peter Bedrick Books, 1984.

Abetti, Giorgio. *The History of Astronomy*. New York: Abelard-Schuman, 1952.

The American Scientific Association Astronomy Cosmology Page. Available on-line. URL: http://www.asa3.org/ASA/topics/Astronomy-Cosmology/PSCF9-95philippidis.html. Accessed March 10, 2004.

American Women's History: A Research Guide. Middle Tennessee State University. Available on-line. URL: http://www.mtsu.edu/~kmiddlet/history/women.html. Accessed March 10, 2004.

Angelo, Joseph A., Jr. *The Dictionary of Space Technology*. Rev. Ed. New York: Facts On File, 2003.

———. *The Facts On File Space and Astronomy Handbook*. New York: Facts On File, 2002.

———. *Encyclopedia of Space Exploration*. New York: Facts On File, 2000.

Asimov, Isaac. *Asimov's Biographical Encyclopedia of Science and Technology*. 2d ed. New York: Doubleday & Co., 1982.

———. *Asimov's Chronology of Science and Discovery*. New York: Harper & Row, 1989.

Australian Academy of Science, "Biographical Memoirs." Available on-line. URL: http://www.asap.unimelb.edu.au/bsparc/aasmemoirs. Accessed March 10, 2004.

Barnett, Lincoln. *The Universe and Dr. Einstein*. New York: Bantam Books, William Sloane Associates, 1948.

Bartky, Walter. *Highlights of Astronomy*. Chicago: University of Chicago Press, 1935.

Beatty, J. Kelly, Carolyn Peterson, and Andrew L. Chaikin, eds. *The New Solar System*. 4th ed. Cambridge, U.K.: Cambridge University Press, 1999.

Bertotti, B., et al. *Modern Cosmology in Retrospect*. Cambridge, U.K.: Cambridge University Press, 1990.

Boas, Marie. *The Scientific Renaissance 1450–1630*. New York: Harper & Brothers, 1962.

Boslough, John. *Masters of Time: Cosmology at the End of Innocence*. Reading, Mass.: Addison-Wesley, 1992.

Bova, Ben. *In Quest of Quasars: An Introduction to Stars and Starlike Objects*. Toronto, Ont.: Grolier-Macmillan Canada Ltd., 1969.

Brahmin World. Available on-line. URL: http://www.brahminworld.com. Accessed March 10, 2004.

Calder, Nigel. *Einstein's Universe*. New York: Random House, 1990.

NASA. "Cassini Mission to Saturn." Available on-line. URL: http://saturn.jpl.nasa.gov/index.cfm. Accessed March 10, 2004.

Chaikin, Andrew. *Space*. London: Carlton Publishing Group, 2002.

Chaisson, Eric, and Steve McMillan. *Astronomy Today*. 4th ed. Upper Saddle River, N.J.: Prentice Hall, 2001.

Harvard University. "Chandra X-Ray Observatory Center." Available on-line. URL: http://chandra.harvard.edu. Accessed March 10, 2004.

Chapman, Allan. *Astronomical Instruments and Their Users: Tycho Brahe to William Lassell*. London, U.K.: Ashgate Publishing, 1996.

Chown, Marcus. *The Universe Next Door: The Making of Tomorrow's Science*. Oxford, U.K.: Oxford University Press, 2002.

Collins, Michael. *Carrying the Fire*. New York: Cooper Square Publishers, 2001.

Consolmagno, Guy J., et al. *Turn Left at Orion: A Hundred Night Objects to See in a Small Telescope—And How to Find Them*. Cambridge, U.K.: Cambridge University Press, 2000.

Cornford, Francis MacDonald. *Plato's Cosmology*. Cambridge, Mass.: Hackett Publishing, 1997.

Damon, Thomas D. *Introduction to Space: The Science of Spaceflight*. 3d ed. Malabar, Fla.: Krieger Publishing Co., 2000.

Danielson, Dennis. *The Book of the Cosmos: Imagining the Universe from Heraclitus to Hawking*. Cambridge, Mass.: Perseus Publishing, 2001.

Delsemme, Armand H. *Our Cosmic Origins: From the Big Bang to the Emergence of Life and Intelligence*. Cambridge, U.K.: Cambridge University Press, 1998.

De Pater, Imke, and Jack Jonathan Lissauer. *Planetary Sciences*. Cambridge, U.K.: Cambridge University Press, 2001.

Dickinson, Terence. *The Universe and Beyond*. 3d ed. Toronto, Ont.: Firefly Books Ltd., 1999.

Dickinson, Terence, Timothy Ferris, and Victor Costanza. *NightWatch: A Practical Guide to Viewing the Universe*. 3d ed. Toronto, Ont.: Firefly Books Ltd., 1998.

Duhem, Pierre, and Roger Ariew, eds. *Medieval Cosmology*. Chicago: University of Chicago Press, 1985.

Dyson, Freeman J. *The Sun, the Genome, and the Internet: Tools of Scientific Revolutions*. Oxford, U.K.: Oxford University Press, 2000.

Ebbighausen, E. G. *Astronomy*. Merrill Physical Science Series. Columbus, Ohio: Charles E. Merrill Books, 1966.

Education Resource Directory (NASA) for Space Science. Available on-line. URL: http://teachspacescience.stsci.edu/cgi-bin/ssrtop.plex. Accessed March 10, 2004.

Englebert, Phillis, and Diane L. Dupuis. *The Handy Space Answer Book*. Detroit, Mich.: Visible Ink Press, 1998.

European Space Agency. Available on-line. URL: http://www.esrin.esa.it. Accessed March 10, 2004.

Evans, James. *History and Practice of Ancient Astronomy*. Oxford, U.K.: Oxford University Press, 1998.

Ezell, Edward Clinton, and Linda Neuman Ezell. *On Mars: Exploration of the Red Planet 1958–1978*. NASA SP-4212. Washington, D.C.: National Aeronautics and Space Administration, 1984.

Ferris, Timothy. *Coming of Age in the Milky Way*. New York: William Morrow & Company, 1988.

———. *Seeing in the Dark*. New York: Simon & Schuster, 2002.

———. *The Whole Shebang: A State-of-the-Universe(s) Report*. New York: Simon & Schuster, 1998.

Fimmel, Richard O., James Van Allen, and Eric Burgess. *Pioneer: First to Jupiter, Saturn, and Beyond*. NASA SP-446. Washington, D.C.: National Aeronautics and Space Administration, 1980.

Fowler, Michael. *How the Greeks Used Geometry to Understand the Stars*. University of Virginia. Available on-line. URL: http://www.phys.virginia.edu/classes/109N/lectures/greek_astro.htm. Accessed March 10, 2004.

French, Bevan M., and Stephan P. Maran, eds. *A Meeting with the Universe: Science Discoveries from the Space Program*. NASA EP-177. Washington, D.C.: National Aeronautics and Space Administration, 1981.

Friedman, Herbert. *The Astronomer's Universe: Stars, Galaxies, and Cosmos*. New York: W. W. Norton & Company, 1990.

Galilei, Galileo. *Discoveries and Opinions of Galileo*. New York: Doubleday & Company, 1989.

Galileo Project, Rice University. Available on-line. URL: http://es.rice.edu/ES/humsoc/Galileo. Accessed March 10, 2004.

Gatland, K. *Space Technology: A Comprehensive History of Space Exploration*. London: Salamander, 1980.

Girvetz, Harry, and George Geiger, Horld Hanitz, and Bertram Morris. *Science, Folklore, and Philosophy*. New York: Harper & Row, 1966.

Goldsmith, Donald. *The Astronomers: Companion Book to the PBS Television Series*. New York: St. Martin's Press, 1991.

Goldstein, Thomas, *Dawn of Modern Science: From the Arabs to Leonardo da Vinci*. Boston: Houghton Mifflin Company, 1980.

Greene, Brian. *The Elegant Universe*. New York: Knopf Publishing Group, 2000.

Gribbin, John. *In Search of the Edge of Time: Black Holes, White Holes, Wormholes*. New York: Penguin, 1999.

———. *The Case of the Missing Neutrinos: and Other Curious Phenomena of the Universe*. New York: Fromm International, 1998.

Hathaway, Nancy. *The Friendly Guide to the Universe*. New York: Penguin Books, 1995.

Hawking, Stephen W. *The Universe in a Nutshell*. New York: Bantam Books, 2001.

———. *A Brief History of Time: From the Big Bang to Black Holes*. New York: Bantam Books, 1998.

———. *The Illustrated A Brief History of Time: Updated and Expanded Edition*. New York: Bantam Books, 1988.

———. *On the Shoulders of Giants: The Great Works of Physics and Astronomy*. Philadelphia, Pa.: Running Press, 2002.

Hawking, Stephen, and Roger Penrose. *The Nature of Space and Time*. Princeton, N.J.: Princeton University Press, 2000.

Heath, Sir Thomas L. *Greek Astronomy*. Volume 10 of *The Library of Greek Thought*. London: J. M. Dent & Sons, Ltd., 1932. Reprint, New York: Dover Publications, 1991.

High Altitude Observatory. Available on-line. URL: http://www.hao.ucar.edu/public/education/sp/images/aristarchus.html. Accessed March 10, 2004.

History of Mathematics, Simon Frasier University. Available on-line. URL: http://www.math.sfu.ca/histmath. Accessed March 10, 2004.

Hoskin, Michael, ed. *The Cambridge Illustrated History of Astronomy*. Cambridge, U.K.: Cambridge University Press, 1997.

———. *Concise History of Astronomy*. Cambridge, U.K.: Cambridge University Press, 1999.

Hoyle, Fred. *Astronomy: A History of Man's Investigation of the Universe*. London: Rathbone Books Limited, 1962.

Huffner, Charles M., Frederick Trinklein, and Mark Bunge. *An Introduction to Astronomy*. New York: Holt, Reinhart and Winston, 1967.

Illingworth, Valerie, and John O. E. Clark. *The Dictionary of Astronomy*. New York: Facts On File, 2000.

The Internet Encyclopedia of Philosophy. Available on-line. URL: http://www.iep.utm.edu. Accessed March 10, 2004.

Jet Propulsion Laboratory, California Institute of Technology. Available on-line. URL: http://www.jpl.nasa.gov/engineers_scientists. Accessed March 10, 2004.

Jones, Barrie W. *Discovering the Solar System*. New York: John Wiley & Sons, 1999.

Kahn, Charles H. *Anaximander and the Origins of Greek Cosmology*. Cambridge, Mass.: Hackett Publishing Co., 1994.

Kauffman, Stuart. *At Home in the Universe: The Search for the Laws of Self-Organization and Complexity*. Oxford, U.K.: Oxford University Press, 1996.

Kaufmann, William J., III. *Universe*. New York: W. H. Freeman and Company, 1985.

Lightman, Alan, and Roberta Brawer. *Origins: The Lives and Worlds of Modern Cosmologists*. Cambridge, Mass.: Harvard University Press, 1990.

Kepler, Johannes. *Kepler's Dream*. Berkeley: University of California Press, 1965.

Al-Khalili, Jim S. *Black Holes, Wormholes and Time Machines*. London: Institute of Physics Publishing, 1990.

Kirshner, Robert P. *The Extravagant Universe: Exploding Stars, Dark Energy, and the Accelerating Cosmos*. Princeton, N.J.: Princeton University Press, 2002.

Leslie, John, ed. *Modern Cosmology and Philosophy*. New York: Prometheus Books, 1998.

Levin, Janna. *How the Universe Got Its Spots: Diary of a Finite Time in a Finite Space*. Princeton, N. J.: Princeton University Press, 2002.

Ley, Willy. *Watchers of the Skies: An Informal History of Astronomy from Babylon to the Space Age*. New York: Viking Press, 1966.

Liddle, Andrew. *An Introduction to Modern Cosmology*. New York: John Wiley & Sons, 1999.

Lunar and Planetary Institute. Available on-line. URL: http://www.lpi.usra.edu. Accessed March 10, 2004.

Maran, Stephen P. *Astronomy for Dummies*. New York: John Wiley & Sons, 1999.

Mather, John C., and John Boslough. *The Very First Light*. New York: Basic Books, 1998.

Matloff, Gregory L. *The Urban Astronomer: A Practical Guide for Observers in Cities and Suburbs*. New York: John Wiley & Sons, 1991.

McCluskey, Stephen C. *Astronomies and Cultures in Early Medieval Europe*. Cambridge, U.K.: Cambridge University Press, 2000.

Miller, Ron. *The Dream Machines: An Illustrated History of the Spaceship in Art, Science and Literature*. Malabar, Fla.: Krieger Publishing Co., 1993.

Moche, Dinah L. *Astronomy: A Self Teaching Guide*. 5th ed. New York: John Wiley & Sons, 2000.

Moore, Patrick. *Astronomers' Stars*. New York: W. W. Norton & Company, 1989.

Moore, Scott. *Aristotle*. Baylor University. Available on-line. URL: http://www.baylor.edu/~Scott_Moore/aristotle.html. Accessed March 10, 2004.

Morrison, Philip, et al., eds. *The Search for Extraterrestrial Intelligence*. NASA SP-419. Washington, D.C.: National Aeronautics and Space Administration, 1977.

Murdin, Paul. *Encyclopedia of Astronomy and Astrophysics*. New York: Nature Publishing Group, 2000.

Murray, Carl D., et al. *Solar System Dynamics*. Cambridge, U.K.: Cambridge University Press, 1999.

Narlikar, Jayant Vishnu. *Introduction to Cosmology*. Cambridge, U.K.: Cambridge University Press, 2000.

National Academy of Sciences Biographical Memoirs. Available on-line. URL: http://bob.nap.edu/html/biomems. Accessed March 10, 2004.

NASA Astrophysics Data System. Available on-line. URL: http://adswww.harvard.edu. Accessed March 10, 2004.

Newell, Homer E. *Beyond the Atmosphere: Early Years of Space Science*. NASA SP-4211. Washington, D.C.: National Aeronautics and Space Administration, 1980.

North, John. *Norton History of Astronomy and Cosmology*. New York: W. W. Norton, 1995.

———. *The Measure of the Universe: A History of Modern Cosmology*. New York: Dover Publications, 1990.

Pannekoek, A. *A History of Astronomy*. New York: Interscience Publishers, 1961.

Parton, James. *Eminent Women: A Series of Sketches of Women Who Have Won Distinction by Their Genius and Achievements as Authors, Artists, Actors, Rulers, or within the Precincts of Home*. Philadelphia: Edgewood Publishing, ca. 1850.

Pasachoff, Jay M. *Contemporary Astronomy*. 2d ed. Philadelphia: Sanders College Publishing, 1981.

———. *Peterson First Guide to Astronomy*. New York: Houghton Mifflin Co., 1998.

Plait, Philip C. *Bad Astronomy: Misconceptions and Misuses Revealed, from Astrology to the Moon Landing "Hoax."* New York: John Wiley & Sons, 2002.

Pomerans, A. J., and René Taton, ed. *History of Science: Ancient and Medieval Science*. New York: Basic Books, 1963.

———. *History of Science: Science in the Nineteenth Century*. New York: Basic Books, 1965.

———. *History of Science: Science in the Twentieth Century*. New York: Basic Books, 1966.

Rees, Martin. *Before the Beginning: Our Universe and Others*. Reading, Mass.: Addison-Wesley, 1997.

———. *Just Six Numbers*. New York: Basic Books, 2001.

———. *Our Cosmic Habitat*. Princeton, N.J.: Princeton University Press, 2001.

Ridpath, Ian, ed. *The Illustrated Encyclopedia of the Universe*. New York: Watson-Guptill Publications, 2001.

Ruth, A. B. *John Couch Adams and the Discovery of Neptune*. St. John's College, University of Cambridge. Available on-line. URL: http://www.joh.cam.ac.uk/publications/eagle97/Eagle97-John.html. Accessed March 10, 2004.

Sagan, Carl. *Carl Sagan's Cosmic Connection: An Extraterrestrial Perspective*. 2d ed. Cambridge, U.K.: Cambridge University Press, 2000.

Sagan, Carl, and Carol Sagan. *Pale Blue Dot*. New York: Random House, 1997.

Schechner, Sara J. *Comets, Popular Culture and the Birth of Modern Cosmology*. Princeton, N.J.: Princeton University Press, 1999.

Schilpp, Paul Arthur, ed. *Albert Einstein: Philosopher-Scientist*. New York: MJF Books, 1970.

Seeds, Michael A. *Horizons: Exploring the Universe*. 6th ed. Pacific Grove, Calif.: Brooks/Cole Publishing, 1999.

Silk, Joseph. *The Big Bang*. 3d ed. New York: W. H. Freeman, 2001.

Skyserver. Sloan Digital Sky Survey. Available on-line. URL: http://skyserver.fnal.gov/en/astro/universe/universe.asp. Accessed March 10, 2004.

Sobel, Dava. *Galileo's Daughter: A Historical Memoir of Science, Faith, and Love*. New York: Walker & Co., 1999.

Solar System Exploration. Available on-line. URL: http://sseforum.jpl.nasa.gov/index.cfm. Accessed March 10, 2004.

Space Science at the National Aeronautics and Space Administration. Available on-line. URL: http://science.nasa.gov. Accessed on March 10, 2004.

Space Telescope Science Institute. Available on-line. URL: http://www.stsci.edu/resources. Accessed March 10, 2004.

Spangenburg, Ray, and Diane K. Moser. *Science Frontiers: 1946 to the Present*. New York: Facts On File, 2004.

———. *Wernher von Braun: Space Visionary and Rocket Engineer*. New York: Facts On File, 1995.

Spires, David N., et al. *Beyond Horizons: A Half Century of Air Force Space Leadership*. Rev. ed. Washington, D.C.: U.S. Government Printing Office, 1998.

Spudis, P. *The Once and Future Moon*. Washington, D.C.: Smithsonian Institution Press, 1996.

St. Andrews University. The Mactutor History of Mathematicians Archive. Available on-line. URL: http://turnbull.mcs.st-and.ac.uk/history. Accessed March 10, 2004.

Starry Messenger University of Cambridge. Available on-line. URL: http://www.hps.cam.ac.uk/starry/tables.html. Accessed March 10, 2004.

Stephenson, Bruce. *Kepler's Physical Astronomy*. New York: Springer-Verlag, 1997.

Stephenson, Bruce, et al., eds. *The Universe Unveiled: Instruments and Images through History*. Cambridge, U.K.: Cambridge University Press, 2000.

Stewart, John. *Moons of the Solar System: An Illustrated Encyclopedia*. Jefferson, N.C.: McFarland & Company, 1991.

Sutton, George Paul. *Rocket Propulsion Elements*. 7th ed. New York: John Wiley & Sons, 2000.

Tenn, Joseph. *The Bruce Medalists*. Sonoma State University, Department of Astronomy, Santa Rosa, Calif. Available on-line. URL: http://www.phys-astro.sonoma.edu/BruceMedalists. Accessed March 10, 2004.

Todd, Deborah. *The Facts On File Algebra Handbook*. New York: Facts On File, 2003.

Union College Department of Physics. Available on-line. URL: http://idol.union.edu/~alpherr. Accessed March 10, 2004.

Universe: An Educational Forum. Available on-line. URL: http://cfa-www.harvard.edu/seuforum. Accessed March 10, 2004.

University of Missouri St. Louis Math and Computer Science Department. Available on-line. URL: http://www.cs.umsl.edu/~sanjiv/aryabhat.html. Accessed March 10, 2004.

University of Washington. Available on-line. URL: http://faculty.washington.edu/petersen/alfonso/esctra13.htm. Accessed March 10, 2004.

Vanderbilt University, Department of Physics and Astronomy, *Astronomy as a Tool of Islam*. Available on-line. URL: http://www.physics.vanderbilt.edu/astrocourses/ast203/islam_astronomy.html. Accessed March 10, 2004.

Walker, Christopher, ed. *Astronomy before the Telescope*. London: British Museum Press, 1996.

Weedman, Daniel W. *Quasar Astronomy*. Cambridge, U.K.: Cambridge University Press, 1986.

Weinberg, Steven. *The First Three Minutes: A Modern View of the Origin of the Universe*. 2d ed. Cambridge, Mass.: Perseus Publishing, 1993.

Westfall, Richard. *The Life of Isaac Newton*. Cambridge, U.K.: Cambridge University Press, 1993.

Wheeler, J. Craig. *Cosmic Catastrophes: Supernovae, Gamma-Ray Bursts, and Adventure in Hyperspace*. Cambridge, U.K.: Cambridge University Press, 2000.

Wilson, Fred L. *Science and Human Values: Aristotle*. Rochester Institute of Technology. Available online. URL: http://www.rit.edu/~flwstv/aristotle1.html. Accessed March 10, 2004.

Wolfram Research. Available on-line. URL: http://scienceworld.wolfram.com. Accessed March 10, 2004.

Yerkes Observatory University of Chicago. Available on-line. URL: http://astro.uchicago.edu/yerkes. Accessed March 10, 2004.

Yount, Lisa, *Twentieth-Century Women Scientists*. New York: Facts On File, 1996.

Zeilik, Michael. *Astronomy: The Evolving Universe*. 8th ed. New York: John Wiley & Sons, 1999.

INDEX

Note: Page numbers in **boldface** indicate main topics. Page numbers in *italic* refer to illustrations.